The Environmental Communication Yearbook

Volume 3

THE ENVIRONMENTAL COMMUNICATION YEARBOOK
VOLUME 3

EDITOR

Stephen P. Depoe
University of Cincinnati

EDITORIAL BOARD MEMBERS

The Environmental Communication Yearbook

Volume 3

Edited by

STEPHEN P. DEPOE
University of Cincinnati

Routledge
Taylor & Francis Group

NEW YORK AND LONDON

This edition published 2011 by Routledge

Routledge
Taylor & Francis Group
711 Third Avenue
New York, NY 10017

Routledge
Taylor & Francis Group
2 Park Square, Milton Park
Abingdon, Oxon OX14 4RN

First issued in paperback 2012

Routledge is an imprint of the Taylor & Francis Group, an informa business

ISBN13: 978-0-415-65239-1 (PBK)
ISBN13: 978-0-805-85914-0 (HBK)

Cover design by Tomai Maridou

**CIP information for this book can be obtained by contacting
the Library of Congress**

Contents

v

Preface

The field of environmental communication has grown significantly since its inception over 20 years ago. Eight biennial Conferences on Communication and Environment have been held since 1991, attracting scholars and practitioners from around the world to share current scholarship. An online Environmental Communication Network (http://www.esf.edu/ecn/) has flourished for a number of years. A robust number of books and essays related to environmental communication studies are published annually in high-quality, peer-reviewed outlets. Colleges and universities across the United States, Europe, and elsewhere now offer undergraduate and graduate courses in environmental communication, supported by the recent publication of high-quality textbooks, and doctoral programs in communication are producing a number of scholars each year who pursue research in the area.

The popularity of environmental communication as an area of study is due to a number of factors, including the continuing salience of environmental issues in contemporary politics, as well as the variety of theoretical and methodological frameworks that can be employed to examine communication about the environment. These factors were illustrated in a recent attempt by the National Communication Association's Environmental Communication Division to describe the contours of environmental communication studies:

> We believe all communication involves an environmental dimension, because symbolic and natural systems are mutually constituted. Humans are one part of the broader ecosystems and cultures we inhabit, both shaping and shaped by our corpo-

real, intellectual, spiritual, emotional, and physical alienation from and proximity to those spaces and communities. To explore these rich and significant connections, we encourage both qualitative and quantitative scholarship and pedagogy that showcases and advances our understanding of the production, reception, contexts, or processes of human communication regarding environmental issues. Areas of interest include, but are not limited to: environmental participatory processes, environmental representations and discourses circulated through media, rhetorical analyses of environmental controversies and advocacy in public culture, cultural studies approaches to popular "green" or "eco-" practices, historical case studies of environmental events, organizational analyses of environmental and anti-environmental institutions, interpersonal/relational dimensions shaping human and non-human relations, risk communication about environmental decision-making, and psychological/cognitive research regarding environmental attitudes and behaviors.[1]

Since the publication of the first volume of *The Environmental Communication Yearbook* in 2004, we have been committed to disseminating environmental communication scholarship that reflects the scope and importance of issues and concerns outlined in the description just given. Volume 3 continues in this tradition. We are hopeful that the *Yearbook* is making a contribution to the many important conversations about environmental communication theory, criticism, and practice that are occurring in classrooms, laboratories, board rooms, meeting halls, and public gatherings around the world.

We also look forward to Volume 4 of *The Environmental Communication Yearbook*, which will include essays submitted in response to special Calls for Papers that address the following subject areas:

• *Rhetoric of Science and the Discourse of Environmental Advocacy: Theoretical and Critical Connections.* Scholars in the areas of risk communication, rhetoric of science, and science and technology studies have contributed much to the understanding of how scientific arguments function in human societies. Many of the positions taken by various sides of contemporary environmental debates (e.g., global warming, energy policy, protection of endangered species) are grounded in competing claims concerning the substance and validity of various scientific findings, and in divergent assumptions about the nature and role of scientific argument itself. Essays accepted for publication will explore the theoretical and critical dimensions of scientific argument as it unfolds in environmental discourse. Literature reviews, theoretical expositions, and case studies are all welcome.

• *Forum: The Future of Environmental Communication Scholarship.* 2006 marks the 10th anniversary of the formation of the Environmental Communication

---

[1]This description, which can be found on the Web site of the National Communication Association (NCA) under the "About NCA—units and divisions" link (www.natcom.org), was prepared in 2005 by members of the NCA Environmental Communication Division. It is cited here not as an authoritative definition of environmental communication, but as an illustration of the breadth and possibilities offered by the field.

Commission with the National Communication Association; the 15th anniversary of the first Conference on Communication and Environment (originally called the Conference on the Discourse of Environmental Advocacy); and the 25th anniversary of Christine Oravec's classic essay on John Muir and the rhetoric of the sublime published in the *Quarterly Journal of Speech*, a work that marks for many the beginning of the environmental communication field. Essays accepted for publication, including both literature reviews and shorter reflections, will assess the current health and future prospects of environmental communication within the broader communication discipline, as well as its significance across other academic disciplines and contexts.

For more information about this special call for papers, consult our Web site at www.erlbaum.com/ecy.htm.

ACKNOWLEDGMENTS

Finally, I want to acknowledge the efforts of our Editorial Board members and guest reviewers. We are very pleased that the Editorial Board for *The Environmental Communication Yearbook* has grown from 30 to 50 outstanding scholars in environmental communication studies. The growth in the number and diversity of reviewers is a sign of growing interest in the *Yearbook* project. The thoughtful work of our reviewers contributed to a high-quality set of essays accepted for this volume. In addition, I would like to thank the following individuals who served as guest reviewers for Volume 3: Jennifer Good, Brock University; Dennis Jaehne, San Jose State University; M. Jimmie Killingsworth, Texas A&M University; and Anne Marie Todd, San Jose State University.

Thanks also to the many scholars who submitted manuscripts to the *Yearbook*, as well as the many institutions and individuals who have supported the project through purchasing the *Yearbook* for their own libraries and personal collections. And, finally, I give special thanks to Ms. Autumn Garrison, who performed outstanding service as an Editorial Assistant for Volume 3.

Contributing to *The Environmental Communication Yearbook*

The Environmental Communication Yearbook is a multidisciplinary forum through which a broad audience of academics, professionals, and practitioners can share and build theoretical, critical, and applied scholarship addressing environmental communication in a variety of contexts. This peer-reviewed annual publication invites submissions that showcase and/or advance our understanding of the production, reception, contexts, or processes of human communication regarding environmental issues. Theoretical expositions, literature reviews, case studies, cultural and mass media studies, best practices, and essays on emerging issues are welcome, as are both qualitative and quantitative methodologies. Areas of topical coverage will include:

- *Participatory processes:* public participation, collaborative decision making, dispute resolution, consensus-building processes, regulatory negotiations, community dialogue, building civic capacity.
- *Journalism and mass communication:* newspaper, magazine, book and other forms of printed mass media; advertising and public relations; media studies; and radio, television and Internet broadcasting.
- *Communication studies:* rhetorical/historical case studies, organizational analyses, public relations/issues management; interpersonal/relational dimensions, risk communication, and psychological/cognitive research, all of which examine the origins, content, structure, and outcomes of discourse about environmental issues.

Submissions are accepted on an ongoing basis for inclusion in subsequent volumes of the *Yearbook* to be published annually.

The Environmental Communication Yearbook is intended for use by researchers, scholars, students, and practitioners in environmental communication, journalism, rhetoric, public relations, mass communication, risk analysis, political science, environmental education, environmental studies and public administration; and by policymakers and others interested in environmental issues and the communication channels used for discourse and information dissemination on the topic.

INFORMATION FOR CONTRIBUTORS

Manuscripts should conform to current guidelines established by the American Psychological Association. Authors are encouraged to submit their work electronically to the e-mail address listed. Essays submitted electronically should be sent in WORD or PDF format. If submitting via regular mail, authors need to send four copies of the manuscript, plus cover page with contact information, to Stephen P. Depoe, Editor, *Environmental Communication Yearbook*, Department of Communication, P.O. Box 210184, University of Cincinnati, Cincinnati OH 45221-0184 (e-mail: depoesp@email.uc.edu). For more information, including information about Volume 4's Special Calls for Papers, consult our Web site at www.erlbaum .com/ecy.htm.

CHAPTER ONE

I, Me, Mine: On the Rhetoric of Water Wars in the Pacific Northwest

Mark P. Moore
Oregon State University

During a period of near drought in the spring and summer of 2001, angry farmers in the Klamath Basin of southern Oregon protested against the withholding of essential irrigation water, provided previously by the United States Bureau of Reclamation since 1907, for growing crops in the otherwise dry uplands of what became known as the Klamath Project. In accordance with the Endangered Species Act, the federal government withheld irrigation water to protect threatened coho salmon and endangered Kaptu (known also as "the sucker") fish in the Upper Klamath Lake, which serves as the source of irrigation water in the nearby basin. While farmers protested in the end of June, one headgate to a canal that provided irrigation water for 180,000 acres of farmland was opened illegally. On July 4, during a "Klamath Tea Party," protesters used a cutting torch to open gates that allowed water to flow from the Upper Klamath Lake, and in mid-July, more than 10 farmers opened the headgates of the project's main canal. In response to such desperate acts, ones that included the violation of federal law and destruction of federal property, no responsible parties were identified and no arrests were made. In fact, local authorities chose to look the other way after a total of four illegal headgate openings at the project.

In light of constant protests and illegal headgate openings, Interior Secretary Gale Norton announced in late July that 75,000 acre-feet of water would be released for irrigation, despite the previously invoked Endangered Species Act that protected the endangered and threatened fish in the region. With much of the growing season over, the water released by Norton provided little more than a symbolic victory for farmers and only a temporary respite for the federal government in general and Bureau of Reclamation in particular, a bureau now controlling a water system that is overly stressed with demands that it cannot meet. Demand for water forced the Bureau into an unusual, if not paradoxical, position. Although created to supply Klamath Basin with irrigation water, court rulings in 2001 forced the Bureau to hold that water for Native American tribes and endangered fish. The problem is even more complex than water for irrigation. In addition to the Native American treaty agreements for water rights, farmers in Klamath Basin must also compete with growing urban centers that also receive subsidized water, not to mention power companies with hydroelectric dams and other beneficiaries, such as manufacturing plants, transportation and shipping industries, fisheries, and conservation wetlands. As noted by Bud Ullman, a water rights attorney for the Klamath tribes, "There is a lot more to water management today than delivering water to farms" (cited in Milstein, 2002b, p. A4).

As demands for water increase in quantity, they also increase for quality. In other words, issues surrounding water not only include those of abundance turning to scarcity, but also purity giving way to pollution. For example, as farmers in the Klamath Basin protested in July of 2001, the Environmental Protection Agency (EPA) activated the Total Maximum Daily Load (TMDL) provision of the Clean Water Act in Oregon to control nonpoint source pollution in residential urban areas, beginning with Portland's Tualatin Watershed. Urbanization has contributed to the water quality problems in the Tualatin Watershed (and its river basin) through domestic sewage, canneries, slaughterhouses, meat packing, tannery works, paperboard plants, dog-food processing, and milk product manufacturing (Cass & Miner, 1993). Also, construction of impervious surfaces such as roads, roofs, and parking lots cause poor groundwater recharge and generate increased pollution in the water from runoff that contains fertilizers and other pollutants (Shively, 1993). Residential lifestyle needs for maintaining gardens and lawns (from a growing population in the region) also raise phosphorus and nitrogen levels in the water from fertilizer and pesticide use. Finally, with growing needs from urban development and other demands, water supply has become a growing concern not only for the Tulatin River Basin (as just one example of a region with urban demands that reduce quality and are becoming harder to meet), but for the world in general. As Weiss (2003) reported from a 600-page study, "The world's reserves of clean, fresh water are shrinking quickly and posing serious threats to public health, political stability and the environment, according to a massive analysis released by the United Nations" (p. A7).

This chapter examines the public discourse by and about stakeholders that constitutes the "water wars" in Oregon surrounding Interior Secretary Norton's decision to release water for the Klamath Basin farms in July of 2001, despite enforcement of the Endangered Species Act that withheld water at the time to protect threatened and endangered fish. The chapter focuses on the outcry for water rights in the war that also follows from what Kenneth Burke (1984) describes in

Attitudes Toward History as a strategic advantage provided by law, politics, and governmental action upon which our democracy is based. With this overemphasis on rights, there is a lack of consideration for the "corrective feature" in the conflict that Burke (1984) identified as "duties" or "obligations" (p. 55). Burke views the partisanship and incompleteness that follows an overemphasis on rights within the context of a burlesque frame of reference that is not "well-rounded," but is necessary to consider from a rhetorical perspective in order to rise above it, that is, to be "enough *greater than* it to be able to 'discount' what it says" (1984, p. 55).

Rights and obligations do not have to be incompatible, but they are constructed as such in the water wars. Thus, new insight can perhaps be gained by considering the idea of water rights and obligations as a challenging paradox to be resolved. The need for an appropriate balance of rights and obligations was illustrated clearly in the parable Garret Hardin (1968) referred to as "the tragedy of the commons," where the right to graze in an open pasture allowed herdsmen to bring more and more animals, while thinking only of their own individual costs and benefits, until the pasture becomes overgrazed and no longer sustainable. When (social) obligations are ignored or left for someone else, resource use can only achieve ecological and human objectives by accident and the typical pattern is short-term realization for human objectives with the long-term objective of neither. The scenario is becoming the same for water. Phil Norton, manager of the Klamath Basin National Wildlife Refuges, observed, "To me we're like a guy with five credit cards that are all maxed out. . . . We've overused everything we have. Any bump in the road, and we're in deep" (cited in Milstein, 2001a, p. A7).

With such a perspective for analysis, this chapter identifies the overemphasis on rights and underemphasis on obligations in the discourse that comprises the war over water as problematic from the burlesque frame of reference described by Burke. In doing so, the emphasis on rights versus obligations is considered within the context of two overlapping issues that concern the demands (and difficulties in meeting demands) for water in quantity and in quality. The chapter argues that a well-rounded frame of rights and obligations is not currently offered or encouraged by conflicting interests, but can be viewed as essential to the future use of what fresh water remains to be used. The chapter intends to increase the understanding of rhetoric in environmental controversies by illustrating the manner in which it is shaped by and about stakeholders with a burlesque frame of reference that is partial in the sense that it is not only partisan, but incomplete. The chapter begins with an overview of Burke's burlesque frame of reference, with specific regard to rights and obligations, identifies the method of analysis, applies this critical perspective to the discourse of water wars surrounding Norton's announcement to release irrigation water for the Klamath Project, and then discusses implications of the discourse surrounding the controversy.

ENVIRONMENTAL CONFLICT IN THE BURLESQUE
FRAME OF REFERENCE

In *Attitudes Toward History*, Burke (1984) identified several poetic categories that can be used to analyze symbolic structures that convey meanings and attitudes in response to significant issues and events during a given time period. Among others, these categories include *tragedy, comedy,* and *burlesque.* Symbolic structures in ancient Greek tragedy, for example, communicate meanings and attitudes that stress fatality, resignation, and humility, in the acceptance of human limitations. In the comic frame, Burke suggests that humans turn their limitations and liabilities into assets by changing the rules of hierarchy and social order to make the best out of a system that perpetuates inequality in the name of equality. Unlike the fatal resignation of tragedy, the comic frame offers strategies for living that can heighten the ability to gets things done, in light of human limitations. As such, the comic frame communicates a sense of acceptance with regard to the human condition and provides a well-rounded perspective for confronting life situations. The burlesque frame, with polemical style, conveys attitudes of superiority and rejection from a narrow perspective designed primarily for caricature and debunking.

With regard to all of the poetic categories, Burke feels that symbolic action allows humans to maintain social order and control conflict through a dramatistic process. Drama begins with an act or action in a disordered scene. In other words, something has gone wrong, a transgression of some kind that generates a sense of guilt or rejection that demands attention, but more specifically a need for correction. The burlesque frame of reference divides controversial issues and problems into what Burke describes as black-and-white, all-or-nothing schemas, hence the emphasis on polemic (Burke, 1984; Appel, 1996). In addition to the debunking of a ritual clown or fool, the burlesque converts every "manner" into a "mannerism" and offers "logical conclusions" that reduce all behavior to the absurd (Burke, 1984, pp. 54–55). While comedy controls conflict by stressing positive aspects of the problematic situation through acceptance, burlesque emphasizes rejection with negative characteristics. The overemphasis on the negative distinguishes burlesque from comedy, but nevertheless, this act of rejection also implies some corresponding acceptance of something else, thus the tendency toward extreme partisanship. In all, then, the method of burlesque can be observed paradoxically as "a form of 'comic rejection' or as a negative method of revealing acceptance" that lacks a well-rounded frame, since it is partisan and incomplete (Moore, 1992, p. 112). The critical challenge, therefore, is to identify the burlesque frame when it is at work and then discount properly what it contains.

The partisanship and incompleteness expressed by those who speak in the burlesque frame combine with an all-or-nothing attitude in a rhetorical strategy that overemphasizes a value, belief, or principle at the expense of an unpleasant although necessary opposite value, belief, or principle. This aspect of burlesque, a strategy that has not been considered in previous studies on this frame of reference, is a characteristic feature of the rhetoric surrounding water wars in the Pacific Northwest and perhaps debates concerning natural resources in general. In terms of what Burke (1984) described as "the burlesque genius" (p. 55), apologists in the water wars primarily demand water "rights" without considering adequately the "duties" or "obligations" that go along with them. The overemphasis on rights has given rise to what can be viewed, at least in part, as an antithetical or polemic relationship between rights and obligations, one that privileges rights over obligations.

As a result, rights and obligations with regard to water can be viewed in terms of paradox, and in this relationship the deck seems to be stacked against obligations. In addition, this is a fundamental paradox that is by and large overlooked in the water wars. When addressed within the context of human rights to water, the corresponding human obligations to water can seem to sound absurd. However, in a society and government based on the protection of rights, as in The Bill of Rights, where there is no corresponding Bill of Obligations, environmental advocates are at a great disadvantage when speaking of our duties to nature, to fellow humans, to future generations. The paradox points to a fundamental problem in natural resource debates but it also represents a basic challenge that must be faced. By constructing the war over water in terms of a paradox to be solved, one that stresses the ambivalence of rights and obligations with an admission of limitations, rather than rights above or even without obligations, a more complete frame of reference can be considered that discourages the partisanship and short-term benefits to special interests that are bringing to an end the very source of life itself. To learn from this paradox, the rhetoric of water wars is examined as a discourse advanced by competing interests from a burlesque frame of reference that seeks strategic advantage by stressing rights without due consideration of obligations, which would by definition include an admission of the limitations deemed necessary, but paradoxically eluded, to sustain life.

Following a description of the conflict, this study proceeds with an analysis of selected textual fragments from primary sources in various media accounts that reflect viewpoints of major stakeholder groups in the Oregon water controversy, and examines how this discourse is limited within a burlesque frame of reference. Because the conflict involves questions of irrigation water for Klamath Basin farms, the study focuses primarily on the protest rhetoric of the Klamath Basin farmers and their supporters as major stakeholders. However, the study also considers the views expressed by fishers, Native Americans, power companies, conservationists, and environmentalists, as stakeholders who also stress their rights over obligations.

The discourse analyzed consists of primary source accounts taken from statewide, regional, and local newspapers in Oregon during the time of the water crisis and protests. These newspapers include *The Oregonian* (Oregon's largest daily circulating, statewide newspaper), *The Albany (OR) Democrat-Herald* (a regional newspaper serving Linn and Benton Counties), and *The Herald and Times* (the local daily newspaper in Klamath Falls, Oregon). Quotes and accounts of stakeholders taken from these newspapers in the analysis section of the essay run from May 2001 to April 2003. Although some of the complexity and introspection may be lost by the tendency of newspaper coverage to condense reports, the proliferation of primary-source quotes by actual stakeholders on a local and statewide basis can serve arguably to inform, influence, and exacerbate the conflict as discussed in this study (Moore, 2004; Bowers, Ochs, & Jensen, 1993). News sources were selected on the basis of geographical location, intended audience, and proximity of the conflict with respect to major stakeholders. The depth of the coverage varied accordingly, with the greatest detail and number of sources quoted going to the local Klamath Falls paper. Although much of the coverage overlapped, the greater detail and depth of the local paper revealed the burlesque frame with its partisanship and incompleteness most clearly. Nevertheless, the primary accounts from all of the sources by all stakeholders stress rights over obligations. Finally, the analysis of the water wars in the Klamath Basin begins with a section on the discourse of the farm stakeholders and then follows with a section on the discourse of other competing stakeholders for water in the controversy. Overall, the analysis illustrates how the rhetoric develops dramatistically in the burlesque frame of reference during the water crisis between 2001 and 2003.

THE BURLESQUED WATER WARS

Much more exists on the topic of environmental law and legal rights to natural resource use than can be discussed in this chapter, but after reviewing much literature, a few general comments can be made on the subject of human rights to nature that can serve as context for the analysis. To begin, humans have always held grand thoughts about water as a natural resource and these thoughts have led to grand advancements in civilization by building dams, constructing canals, establishing irrigation, moving rivers, stopping oceans, creating lakes, and greening desert landscapes.[1] As Reisner (1993) observed in *Cadillac Desert*, we now face the disturbing fact that the water supply is limited and defy this limitation by digging deeper wells, removing salt from the ocean water, competing for shrinking rivers, even thinking about towing icebergs from the Arctic Circle to cities with short supplies.[2] In all of this, questions about ownership and rights to water occur, of course, on a worldwide basis, but in the West and the Northwest United States, recent disputes include: Yellowstone National Park and landowners who are tap-

---

[1]For tributes to water and human achievements with it in the Pacific Northwest, see, for example, Egan (1991), White (1995), and Rapp (1997).

[2]The shrinking supply of fresh water in the American West is a growing problem caused by the constant increase in demand from a number of sources and this is discussed at length, for example, by Reisner (1993).

ping into geothermal water supplies that threaten Old Faithful; Montana farmers who accuse North Dakota of stealing their rainwater; and also Wyoming officials who prevented Idaho from seeding clouds over the Grand Tetons in order to improve its snow pack, because the snow would create excessive runoff and overload local dams (Ward, 2002).

A review of recent battles over water indicates that it is difficult, if not impossible, to establish a well-defined set of rights that consider all consequences of actions taken by natural resource consumers. To address environmental problems more broadly, policies could focus on what Hanna, Folke, and Maler (1996) characterized as "property-rights regimes that are designed to fit the cultural, economic, geographic, and ecological context in which they are to function" (p. 4). That is to say, there may be general principles that apply across the board, but specifics will vary from context to context. Although property-rights regimes may be critical to the effective use of the natural environment, they often fail when overwhelmed by pressures from human population growth and increased demand for natural resources, pressures that are identified regularly as problematic in the Northwest water wars. If, as Hanna, Folke, and Maler (1996) explained, American citizens in the form of the federal government, represented by the Bureau of Reclamation, have owner rights to a water system that serves agricultural, industrial, energy, and other needs, those citizens, in the form of the Bureau of Reclamation, have the right, through the passage and enforcement of legislation, to determine the rules for use. In other words, the public decides, and as an arm of the federal government, the Bureau of Reclamation serves the public.

Water rights in and of themselves are ambiguous, perhaps indeterminate, but the federal government took charge of them in 16 western states with the passage of the Reclamation Act of 1902 (Lee, 1980; Pisani, 2002). Since that time, rights have been allotted on the basis of land use. Farmers, for example, who use land for agriculture, are a part of the public but they are not *the* public in and of themselves.[3] As such, they do not have the right to determine water use as owners themselves. However, they do have the same obligations to maintain as the rest of the public. In this way, rights and obligations are the same for all citizens. Whereas water rights are determined by rules established by the public, obligations also include the maintenance of social, not individual or collective, objectives (Hanna et al., 1996). In general terms, such objectives include constraining the rates of use and/or avoiding those uses that would be socially unacceptable. Acceptable and unacceptable uses are determined typically by rules as well, such as the Clean Water Act, the Endangered Species Act, and Reclamation contracts that set quotas on consumption. Nevertheless, when push comes to shove, rights often supersede

---

[3]Marbut (2001) provided an explanation of Oregon's water law, water-right appropriation under Oregon's water code, and the Klamath Basin Reclamation Project. One primary strategy in the Klamath Basin water wars that emerges from the burlesque frame of reference is vilification. That is, competing groups vilify each other, while they characterize their own water rights in an ennobling fashion (Lange, 1993).

obligations and sometimes they do so in an extreme fashion, as in the case of "persons unknown" in the Klamath Basin who opened irrigation canal gates illegally on several occasions without recourse or even reprimand after the region had been deprived of water legally by the Bureau of Reclamation.

Farm Stakeholders in the Burlesque

The Klamath Basin protests represent a good starting point for an analysis of water wars in the burlesque frame of reference. To begin, the farmers take an all-or-nothing position with an overemphasis on water rights without due consideration for reverse social obligations in the form of the Endangered Species Act (ESA) invoked to protect endangered and threatened fish. During a bucket brigade ceremony on May 7, 2001, in which thousands of Klamath Basin farmers and their supporters protested against the ESA by passing water into an irrigation canal, farmer Tim Parks set the stage for burlesque when he stated, "We've got two options: to quit or fight" ("Brigade urges," 2001, p. A9). And Marion Palmer, a World War II veteran who received a Klamath Basin homestead in 1949, added, "Fifty-nine years ago, we were welcomed home as heroes and asked to feed a hungry world. . . . Today we may be reduced to welfare recipients standing in line for rice and cheese" ("Brigade urges," 2001, p. A9). In response to the protest, Oregon Senator Gordon Smith also contributed to the partisanship as he offered an absurd reduction about the effects of the ESA on the farmers by reasoning that "If the government chooses to save the sucker fish, it must not make suckers of Klamath County. . . . We must never forget that it is not OK to say a sucker fish is of more value under the law than a family farm" ("Brigades urges," 2001, p. A9). In these examples, any mitigating circumstances that might put the federal government and the ESA in a better light are suppressed as the cry for water rights drowns any sense of duty. Thus, the protest obliterates the discriminations of the federal government to enforce the ESA and to honor other obligations such as Native American tribal agreements.

Without water provided (though not guaranteed) previously by the federal government, farmers continued to attack the federal government and reduce their situation to either–or terms. After someone violated a federal court order and illegally opened one of the channel's headgates at the end of May in an "apparent" act of defiance against the federal government, Bob Gasser, organizer of the May 7 bucket brigade noted, "I am not worried about the court order. . . . Our lives were destroyed by that order" (Gibson, 2001a, p. A2). After the Klamath Tea Party on July 4, during which time another headgate was opened illegally, one farmer described himself and his situation in the burlesque by noting, "People say we are a bunch of dumb farmers. . . . Well, we are. We are a bunch of dumb farmers who have been put in this position by a bunch of smart politicians. We didn't create this problem, we are just the ones who have been left to deal with it" (Gibson, 2001b, p. A2). Irrigation District Manager, Dave Solem, said he could not blame the farmers for "taking action they believe to be just," adding that "The water is private property that has been taken from them [water users]" (Juillerat, 2001, p. A2).

Others were sympathetic to the farmers as well. Klamath County Sheriff, Tim Evinger, defended the farmers while declaring that "It just appears to me that they are trying to save their lives" ("Klamath residents," 2001, p. A5). Moreover, a Klamath Falls farm implement dealer, Ron Johnson, stated that "It is really unfair to a lot of people who make their livelihood from farming, having everything taken away from them like this" ("Klamath residents," 2001, p. A5). To refer to "lives being destroyed" and "having everything taken away" by the ESA, farmers drove the action taken by the federal government to a logical conclusion that becomes their reduction to absurdity, without considering the fact that there is too much demand for water overall. As Wendell Wood of the Oregon National Resource Council explained, "There has to be a reduction in total water use. Because if the ESA wasn't there, that is still what they'd have to do" (Bragg, 2001b, p. A3). But while Wood points out the misguided logic of *farmers or fish*, the former continue to view the latter as the real problem.

As protests continued in early July, Klamath County Commission Chair, M. Steven West announced that county commissioners were "asking the county board of commissioners to look into creating some kind of ordinance to overturn the federal government in terms of water use" (Dworkin, 2001a, p. A1). It might seem ironic that farmers in a U.S. reclamation project would want to override the federal government, but as West stated, "People are desperate. . . . They're watching their lives in ruin and they're looking for a ray of hope somewhere" (Dworkin, 2001a, p. A1). With desperate lives in ruin, the farmers posted road signs that attacked the federal government in general and the ESA in particular. One sign in front of Wong Potatoes declared, "We want to farm. We need water. Not government aid," and another sign erected nearby displayed two swastikas with the words "Welcome to Oregon, the Dictator state" (Dworkin, 2001b, p. E4). In addition, other road signs read, "Welcome to Klamath Project, largest water theft in history" and "Federally created disaster area" (Attig, 2001, p. F2). As such, the overall sentiment was one of government betrayal. Farmers not only felt that their rights to water had been taken away, but that the federal government had failed to live up to its obligations to provide water for them.

Protesters also justified their actions in partisan terms without due consideration of other factors involved in the water shortage that might justify actions taken by the federal government to withhold water from them. Bob King, an alfalfa farmer, stated that "We're paying these people [federal marshals who shut down a canal headgate opened illegally by protesters] to starve us out" (Young, 2001, p. A1). When asked why he was participating in the protest that involved an illegal opening of canal headgates in mid-July, farmer Barron Knoll explained that "I wouldn't be down here doing what a lot of hippies do to get attention if I hadn't lost all hope. . . . The last stand is the last stand" (Milstein, 2001b, p. A20). Another protester observed that the federal government "made us desperate. . . . And desperate people do desperate things" (Attig, 2001, p. F2). Indeed, as farmer Alvin Cheyne watched one of his prize bulls graze on dead grass, he noted, "People aren't much different than these animals [bulls]. If you push them into a corner, if you give them no place to go but you keep pushing, they're going to come out fighting" (Attig, 2001, p. F1). Even though Cheyne owns a Klamath Project farm, he concluded by saying, "I had never felt that I couldn't trust government. . . . But today I don't trust them for nothing" (Attig, 2001, p. F2).

In general, the protesters agreed that the opening of headgates represented a symbolic act to attract media attention and create national support for local farmers on the basis of a failed obligation on the part of the federal government. Local farmer, Doug Staff, noted that the protests were "proving a point that we care about our rights as Americans" and that he would risk being arrested in order to keep the water flowing (Milstein, 2001c, p. A1). "It has become a national issue and people outside the region are covering it in very volatile and simplistic terms," as Patricia Foulk of the U.S. Fish and Wildlife Services observed, but "To characterize it as suckers vs. farms is doing a great disservice to the Klamath Basin, when what we have is an ecosystem that is stressed beyond its capacity to provide everything we're demanding of it" (Milstein, 2001c, p. A4). Disservice notwithstanding, even local newspapers in Oregon condoned the illegal headgate openings. One editorial from the *Albany Democrat-Herald,* for example, stated that "Farmers in Klamath Falls have a good justification for their defiance of the federal government" and then referred to the Declaration of Independence to argue that "People have an inalienable right to life, liberty and the pursuit of happiness, and 'whenever any form of government becomes destructive of these ends, it is the right of the people' to do something about it. There is nothing in there [the Declaration of Independence] about fish" ("Klamath farmers," 2001, p. A8).

By the time Interior Secretary Norton announced that 75,000 acre-feet of water would be released from Upper Klamath Lake, in the "hope that," as she stated, "this will be viewed by everyone as taking care of the situation" ("Cool comfort," 2001, p. C8), farm protesters had pried apart canal headgates on four occasions and twice created irrigation lines that bypassed canal headgates from Upper Klamath Lake to farms, in what Jeff McCracken of the U.S. Bureau of Reclamation called a "symbolic" attempt to restore the flow of water in the region ("Farmers bypass," 2001, p. A1). Protesters felt a "moment of joy" but viewed the release as too little too late (Brinckman, 2001c, p. A1). Furthermore, one farmer, Jeanne Anderson, noted, "If they [the militant factions] keep holding out for all or nothing, I think we're going to end up with nothing" (Detzel & Barnett, 2001, p. A1). Here, Anderson speaks to the burlesque frame within which the "militant factions" operate. That is to say, she referred to those who characterize the scene from a partisan and incomplete point of view. The "quit or fight" or the "all or nothing" attitude allowed nothing in between and thus eliminated any sense of duty or obligation that goes along with rights for water use.

The water was released on July 24, after new measurements showed that Upper Klamath Lake held more water than scientists projected previously. Nevertheless, there were still no guarantees about future water supplies and tempers continued to flare. For example, as what he described as a situation where "extremists and out-of-control federal agents continue to push," Jack Redfield, a Klamath Falls police officer, told a group of protesting environmentalists that farmers have seen "their entire lives destroyed" and "their frustration will undoubtedly escalate to the point of boiling over," where "the potential for extreme violence, even to the extent of civil war is possible if action is not taken in the very near future to remedy this tragedy" (Bernard, 2001, p. D4). After delivering the speech, Redfield was placed on administrative leave. However, because he was in his uniform when he addressed the crowd, environmentalists claimed that their civil rights were violated and filed a $100,000 claim against the city of Klamath Falls. Andy Kerr, who Redfield singled out by name, explained through a letter written by his lawyer that "In the course of my actions as an environmentalist, I have received death threats on a regular basis, but never by someone in a police officer's uniform" ("Environmentalists claim," 2001, p. D10).

Other Stakeholders in the Burlesque

If Norton's decision to release water could at least be viewed as a symbolic victory for the Klamath Basin farmers, it came at a high price. Because Norton based her decision to release the water on a surprise surplus, she overruled a previous mandate by federal biologists to give any such excess to national wildlife refuges instead, who also claimed rights. Conservationists argued that Norton's decision, along with the Bureau of Reclamation's release of the water, violated the ESA by allowing the wildlife refuges to remain dry during the summer drought. Then, on August 7, the Oregon Natural Resources Council, WaterWatch of Oregon, the Golden Gate Audubon Society, and the Northcoast Environmental Center filed a lawsuit over water rights against the Bureau of Reclamation. On the following day, the Klamath Basin irrigation districts agreed to release enough water, in what one district manager called a "huge sacrifice," to the Klamath Basin National Wildlife Refuges (a major stop for migrating waterfowl and home to the largest population of wintering bald eagles in the lower 48 states) to keep wildlife from dying (Brinckman, 2001d, p. A13). Even with the agreement, the refuges in August and September only received about one-third the amount of water that they normally use. As Bob Hunter, lawyer for WaterWatch of Oregon, observed, with everyone demanding more, there is simply not enough water to meet competing needs (Dworkin, 2001c, p. D4).

In addition to wildlife refuges already mentioned, other demands for water that compete with farms include energy, Native Americans, fishers, and fish (Moore, 2003). After an energy crisis emerged in December of 2000, the Bonneville Power

Administration of Oregon (BPA) announced that a lack of water in the Columbia River Basin was the source of the problem. The shortage was complicated by the threatened and endangered salmon that needed water held at the time by power companies for electricity. As the farmers began to protest in the spring of 2001, the Northwest Power Planning Council announced that electricity has priority over fish and water would be sent to turbines to maintain production of low-cost hydropower. Council chair, Larry Cassidy, called the situation an "emergency" and added that "It's critical we do the best we can to avoid power supply problems" (Brinckman, 2001b, p. A1). Water rights for energy clashed with those of conservationists like Tims Stearns of the National Wildlife Federation, who stated that "We really have bad choices. ... We can either stop, spill, or bankrupt Bonneville. . . . If we continue to allow the river to be viewed only for hydropower, eventually there will be no more salmon" (Brinckman, 2001a, p. B5). With a water shortage for energy and fish, an either–or dilemma, reinforced by an all-or-nothing attitude, surfaced for both power companies and conservationists: either protect fish or produce energy. While farmers protested for water at the end of June, the BPA outraged Northwest tribal officials and fish advocates by declaring that they would not risk power to aid fish migration, because, as Acting Administrator, Steve Wright, stated, "Summer spill would reduce power system reliability to an unacceptably low level" (Cole, 2001, p. D1).

The main concern for water rights by Native Americans involved fish and wildlife habitats. With Upper Klamath Lake serving as home for the endangered Kaptu (sucker), which holds deep meaning for the Klamath Tribes, and threatened coho salmon, tribe members stood firm on water rights protected by treaty agreements with the federal government. Allen Foreman, chair of the Klamath Tribes, explained that "the demand for water in the Klamath Basin has been allowed to exceed supply. ... When the government invited farmers and veterans of World War I and II to move into the Basin and suggested water would be available, the government did not tell the farmers about tribal water rights" (Bragg, 2001a, p. A3). Water rights for the Klamath Tribes have been upheld in court repeatedly and Foreman added that abolishing the ESA "simply will not change the tribal trust responsibility, nor will this fix the problems that exist today" (Bragg, 2001a, p. A3). In a lawsuit decision on Klamath Basin water rights in March of 2002, a U.S. District Judge in Portland reiterated that water rights of the Klamath Tribes supersede all others in the basin and that they are entitled to all water necessary to support a healthy habitat for fish and wildlife. In response to the ruling, Foreman asserted, "The court could not have been more clear and direct in reconfirming the rights that we have maintained all along. . ." (cited in Milstein, 2002a, p. A10).

Fishers share concerns about fish habitat and enter the water wars for rights on behalf of what they consider to be a dwindling industry. According to *The Oregonian*, many commercial fishers believe the federal government was right to stop the irrigation water in the Klamath Basin. One fisher, Tom Stockley, referred to the farmers as "water robbers" and noted that "Commercial fisherman have given up, given up, and given up. . . . I think it's time for someone else to give up something" (Brinckman, 2001e, p. A6). If there is not enough water to support salmon runs, then there will be fewer and fewer fish left to migrate out into the ocean for fishers to catch. In March of 2002, while courts were upholding tribal water rights, fishers threatened to sue the federal government for violation of the ESA if they resumed full irrigation deliveries to the Klamath Basin in the spring. Commercial fishers filed a 60-day notice to sue after President Bush created a special task force to ensure that irrigators in the Basin will never have their water cut off again by the ESA. Todd True of Earthjustice explained that "What we are trying to flag for the Secretary of Interior is that the fish in the Klamath Basin still need water" (Bernard, 2002a, p. A6). Glen Spain of the Pacific Coast Federation of Fisherman's Associations observed, "The administration, I think, is deliberately creating a train wreck. . . . Commercial fishermen are not going to lay down and let the bureau roll over them" (Bernard, 2002a, p. A6).

Finally, the battle over water in the Klamath Basin also raised a contingency issue that is often overlooked, if not taken for granted in the Pacific Northwest, that being water quality.[4] As it became scarce, farm water was also withheld to dilute pollutants that feed algae blooms that are deadly to protected fish, in what is known as the Klamath Drain. According to a state report, the Drain contains the worst quality water in Oregon. It "resembles a stagnant pond, complete with fluorescent green patches of mold floating on mats of decaying algae" and it is "too polluted to do anyone, or any fish, much good" (Milstein, 2001a, p. A1).[5] In addition to the pollution from agriculture, industry contributes increasingly to the decline in water quality in the Northwest and as a result, legal battles have increased as well. For example, in May of 2001, when farmers in the Klamath Basin were protesting, the Blue Heron Paper Company in Oregon City agreed in a legal settlement to lower the amount and the temperature of wastewater pollution discharged into the Willamette River in order to renew their wastewater permit. The mill discharges eight to ten million gallons of wastewater a day above the Willamette Falls that reach temperatures of 94 degrees (Hunsberger, 2001, pp. C1, C7). As of April 2003, more than 13,000 miles of rivers and streams in Oregon do not meet the federal clean water standards and "Oregon rates the worst in the nation for how far behind it is in renegotiating and approving expired industrial wastewater discharge permits" (Cole, 2003, p. B3). Stephanie Hallock, director of the Oregon Department of Environmental Quality, noted that there is no real environmental agenda in the state at present, and feels "like the little boy with his finger in the dike" (Cole, 2003, p. B1). Former state Senator Ted Hallock, who

---

[4]Efforts have been made since 2001 to reduce water pollution and improve drainage by combining cropland and wetlands. The Klamath Basin Ecosystem Restoration Office helped to pay for a wetlands project on a farm in the Basin in the summer of 2003 (Darling, 2003).

[5]For a discussion of water-quality problems as well as a review of local, state, and federal water-quality regulations, including the Clean Water Act, see Vigil (2003). For a discussion of water-quality problems in the Pacific Northwest, see chapter 3, "Troubled Waters," in Barker (1993).

served as the director before his daughter, admitted, "I suffer for her. . . . This is a merciless time" (Cole, 2003, p. B3).

When Interior Secretary Norton wanted everyone to agree that she had taken care of the water crisis in the Klamath Basin by releasing a surprise surplus of water for irrigation, she only received more criticism for doing so a year later, when a large die-off of an estimated 30,000 salmon occurred in the lower Klamath River in late September, from what activists called flaws in the region's water diversion policy (Milstein, 2002c, p. A1). Conservationists and fishers sued the federal government immediately after the die-off was discovered and within 3 days, Bush administration officials pledged to release more water for fish, even though they would not admit that water diversions were to blame. In early October, Native Americans later threatened to sue for the government's violation of tribal treaty agreements. Sue Masten, chair of the Yurok Tribe, explained, "The government promised to protect the resources we depend on for our very survival, and that's not something that should be taken lightly" (Cole, 2002, p. A17). At the same time, scientists determined that the fish died from bacterial and parasitic diseases that attacked the gills, resulting in suffocation. The diseases flourish when water temperatures rise and fish are crowded together by low flows (Bernard, 2002b, p. C3). Could this calamity in the Klamath indicate possible consequences for the future of water rights that are based in rhetorical demands for *I, me, mine*?

CONCLUSIONS: ON THE LACK OF OBLIGATIONS IN THE BURLESQUE

With obligations lacking, an overemphasis on rights in the rhetoric of water wars from a burlesque frame of reference emerges from all stakeholders and concerned parties in the attempt to serve the interests of I, me, mine. Even conservationists, who, by definition want to keep from losing or wasting natural resources, stress their rights to receive, not to protect water. Otherwise, wildlife, and even our beloved bald eagle, will not survive. The charge holds, that which we cherish will perish. Farmers, fishers, industrialists, energy resource managers, all say the same thing. As Burke suggests about the burlesque frame, the competing interests divide the problematic issues that concern water rights into black and white, binary categories. The farmers may be the most extreme with their bellicose discourse about "ruined" lives by the federal government and the ESA, but fishers are quick to blame farmers for "destroying" their industry, energy managers assail fishers for "loss" of power, and so on (Moore, 2003).[6] The result is a doomsday discourse

---

[6]Assessing total losses in the Klamath Basin during the summer of 2001 proved to be a difficult and complex task. As farmers and community members expressed great concern about their planted fields and their future, local community members also expressed a sense of loss in their way of life and a sense of betrayal by government. The social and psychological damage to community members is therefore recognized with all due respect. According to Jaeger (2003), in one comprehensive, 400-page study con-

of rights lost by a particular group. It is an all or nothing characterization for the maintenance of rights, be they inalienable or otherwise. As such, this finger-pointing discourse in the burlesque frame of the water wars surfaces as a defining characteristic aimed at the protection of clashing and individual rights held by all competing interests.

Rather than being cast in a negative light, obligations are essentially avoided as topics for discussion. Instead, stakeholders cast each other in a negative light when they exhort their cases concerning the loss of rights. In the burlesque frame, which stresses rejection and the negative, stakeholders who argue for their rights to water are also accused of depriving the rights of others. This can be considered as a form of "Mirror and Matching" strategy Lange (1993) identified as "Vilify/Ennoble" that has been modified in the burlesque frame to overemphasize vilification (p. 248) in an attempt to protect the rights that are presumed by each group to be rightfully theirs to protect. As a rejection frame, the burlesque certainly invites this mirror and matching strategy, and as each of the interest groups vilify each other, they do so on the basis of a perceived greater right that is in itself an ennobling strategy. In the water wars, the notion of duties or obligations, such as constraining rates of use and avoiding uses that are socially unacceptable, mentioned by Hanna et al. (1996), imply a negative (or non) condition that imposes an unattractive limit on use and restricts freedom. Stakeholders essentially avoid the topic of obligations as they vilify and cast each other in a negative light. However, to the extent that waters rights of others are cast negatively, an obligation to others would be as well. In the burlesque frame of rejection, such obligations do not emerge. In fact, only water rights for each group seem to matter. In this way, the emphasis on rights implies a rejection of obligations, even though obligations are not cast specifically in a negative light in the same way that competing groups emphasize the negative when they vilify each other to preserve rights.

Many solutions to the water crisis have been offered and some even call for what would amount to an increase in obligations on the part of users, such as restoring wetlands as natural filters to reduce pollution. Retiring farmland to reduce water demand, improving both logging and grazing practices to reduce erosion

---

ducted by Oregon State University and the University of California, Berkeley, irrigation curtailment carried a high economic cost as well, however, a large portion of that cost was shouldered by taxpayers ($45 to $47 million). The aggregate, net losses on the Klamath Project was estimated to be between $27 and $46 million. Several groups also suffered economic losses, such as farm workers, tenant farmers, sharecroppers, and agricultural input suppliers. Landowners suffered considerable hardship as well, but all groups benefitted at least to some extent from government emergency programs. In some cases, payments exceeded direct losses by individuals.

Although local citizens complained throughout the summer that the media oversimplified the crisis, the extent to which lives were "ruined," as repeated throughout the summer by farmers and their supporters, can also be questioned. Jaeger (2003) pointed out that nearly all of the decline in employment in 2001, which grew only 3.3 percent from the previous year, was due to factors that were unrelated to the irrigation curtailment and could be traced specifically to a contraction in the lumber and wood products sector and the construction sector.

and pollution, and removing dams to assist imperiled fish have also been suggested. However, the overwhelming concern lies with the acquisition of more water for competing groups. This would include building more reservoirs, drilling deeper wells to boost supplies during dry years, reforming the ESA to ease pressure on local economies, and settling and enforcing overlapping water rights to ensure that everyone gets what they are entitled to get. In Oregon there are no state statutes that specify who owns water rights. Authority over water has always been decided by case laws. Nevertheless, the conflict was further exacerbated in May of 2003, when the Oregon State House of Representatives passed a water-rights bill declaring that the person who has the title to property described by the water right owns that right ("Water-rights," 2003, p. A6). Proponents say the bill will essentially guarantee landowners' private-property interests in their water rights, while opponents warn it "could have devastating effects on irrigation districts, conservation projects and agricultural lifestyle of central Oregon" ("Water-rights," 2003, p. A6).

The problem with obligations in the water wars, and perhaps resource conflicts in general, is that they not only imply a negative, in this case a shortage of clean, fresh water, but also stress, as Burke (1984) noted, "a formal admission of strictures" (p. 56). This formal admission can be stated paradoxically in the following way: Concerned parties in the water wars have an obligation not to exercise (that is to say, limit or narrow) their rights. This obligation will only last as long as there is water over which to battle, for only through the dispossession or even disappearance of water will we be free of our obligations that concern it. In "The Tragedy of the Commons," Hardin (1968) argued for the recognition of such an obligation as a necessity when he stated, "Every new enclosure of the commons involves the infringement of somebody's personal liberty" (p. 1248). Although few if any complain about a loss based on infringements from the distant past, Hardin (1968) contended, "It is the newly proposed infringements that we vigorously oppose; cries of 'rights' and 'freedom' fill the air. But what does freedom mean?" (p. 1248). Freedom, Hardin (1968) responded by quoting Hegel, "is the recognition of necessity" (p. 1248).

Hardin's (1968) views on the commons can function dramatistically to call Oregon's water warriors to a new understanding of resource shortages. For example, his views on the need for "Mutual Coercion Mutually Agreed upon" can serve as a possible comic corrective for the selfish, burlesque approach taken by Oregon's water war advocates (p. 1247). Hardin's tragedy of the commons speaks to the need for individuals and nations to look beyond their short-term, private interests and work together to deal with issues such as overpopulation that involve everyone. Hardin referred to responsibility as "the product of definite social arrangements," such as a mutually agreed-upon coercion that keeps downtown shoppers from hoarding public parking spaces by introducing parking meters and traffic fines to insure limited and short-term use (1968, p. 1247). In Burke's terms, Hardin's mutually agreed-upon coercion functions as a comic corrective, because it involves a formal admission of strictures based on the necessity for public responsibility that attempts to bring the relationship between rights and obligations into greater balance. This ironic perspective can help stakeholders see beyond the simplistic reductions of their own rhetorical depictions, for as Hardin (1968) noted, "Individuals locked into the logic of the commons are free only to bring on universal ruin; once they see the necessity of mutual coercion, they become free to pursue other goals" (p. 1248).

With its lack of obligation to others, the logic of the commons offers a partisan and incomplete frame of reference similar to that of the burlesque in the Oregon water wars. The necessity of mutual coercion identified by Hardin also resonates with Burke's claim that it is necessary to recognize the burlesque as partisan and incomplete in order to rise above it whenever it may appear. With a well-rounded frame of reference based on both rights and obligations, like Hardin's concept of mutual coercion, all stakeholders in the Oregon water wars would be in a position to benefit by discounting properly what the incomplete partisanship has to offer in the burlesque frame of reference and by agreeing to an appropriate form of mutual coercion that could function as a comic corrective for the current overemphasis on rights. By rising above the limits of partisanship in this way, stakeholders can therefore become greater than it.

REFERENCES

Appel, E. (1996). Burlesque drama as a rhetorical genre: The Hudibrastic ridicule of William F. Buckley, Jr. *Western Journal of Communication, 60*, 269–284.

Attig, R. (2001, July 15). Klamath Basin's desperate days. *The Oregonian*, pp. F1, F2.

Barker, R. (1993). *Saving all the parts: Reconciling economies and the Endangered Species Act.* Washington, DC: Island Press.

Bernard, J. (2001, July 27). FBI asked to investigate incident involving Klamath Falls policeman. *The Oregonian*, p. D4.

Bernard, J. (2002a, March 2). Fishermen warn of lawsuit over Klamath water. *Albany (OR) Democrat-Herald*, p. A6.

Bernard, J. (2002b, October 6). Fish deaths: Worst fears come true. *Albany (OR) Democrat-Herald*, p. C3.

Bowers, J., Ochs, D., & Jensen, R. (1993). *The rhetoric of agitation and control* (2nd ed.). Prospect Heights, IL: Waveland Press.

Bragg, J. (2001a, June 17). A public meeting. *Herald and Times* (Klamath Falls), pp. A1, A3.

Bragg, J. (2001b, July 3). Feds turn water off. *Albany (OR) Democrat-Herald*, pp. A1, A3.

Brigade urges farms over fish. (2001, May 8). *The Oregonian*, pp. A1, A9.

Brinckman, J. (2001a, March 13). Salmon second to power grid, BPA says. *The Oregonian*, p. B5.

Brinckman, J. (2001b, March 28). Salmon may lose to power. *The Oregonian*, pp. A1, A12.

Brinckman, J. (2001c, July 26). Klamath water eases tensions. *The Oregonian*, pp. A1, A14.

Brinckman, J. (2001d, August 9). Klamath irrigation districts will release water to refuges. *The Oregonian*, p. A13.

Brinckman, J. (2001e, September 4). Farmers aren't first left high and dry. *The Oregonian*, pp. A1, A6.

Burke, K. (1984). *Attitudes toward history* (3rd ed.). Berkeley: University of California Press.

Cass, P., & Miner, R. (1993). *The historical Tualatin River Basin* (Tualatin River Basin Water Resources Management Rep. No. 3). Corvallis, OR: Water Resources Institute & Oregon State University.

Cole, M. (2001, June 30). BPA won't risk power to aid fish migration. *The Oregonian*, pp. D1, D4.

Cole, M. (2002, October 6). With deep ties to fish, tribes mourn die-off. *The Oregonian*, pp. A17, A20.

Cole, M. (2003, April 14). Stephanie Hallock: Swimming in cuts, her job is to make the water cleaner. *The Oregonian*, pp. B1, B3.

Cool comfort in Klamath. (2001, July 25). *The Oregonian*, p. C8.

Darling, D. (2003, July 8). Owners turn farmland into wetlands. *The Oregonian*, p. B2.

Detzel, T., & Barnett, J. (2001, July 22). Klamath farmers taking their case to Capitol Hill. *The Oregonian*, pp. A1, A21.

Dworkin, A. (2001a, July 6). Farmers' fight for water intensifies. *The Oregonian*, pp. A1, A9.

Dworkin, A. (2001b, July 9). Klamath becoming new dust bowl. *The Oregonian*, pp. E1, E4.

Dworkin, A. (2001c, September 5). Klamath wildlife gets turn for water. *The Oregonian*, pp. D1, D4.

Egan, T. (1991). *The good rain: Across time and terrain in the Pacific Northwest.* New York: Alfred A. Knopf.

Environmentalists claim speech by policeman violated civil rights. (2001, September 5). *The Oregonian*, p. D10.

Farmers bypass Klamath headgate. (2001, July 16). *Albany (OR) Democrat-Herald*, p. A1.

Gibson, K. (2001a, July 1). Water is flowing into A canal. *Herald and Times* (Klamath Falls), pp. A1, A2.

Gibson, K. (2001b, July 5). Klamath tea party. *Herald and Times* (Klamath Falls), pp. A1, A2.

Hanna, S., Folke, C., & Maler, K. (1996). Property rights and the natural environment. In S. Hanna, C. Folke, & K.-G. Maler (Eds.), *Rights to nature: Ecological, economic, cultural, and political principles of institutions for the environment* (pp. 1–12). Washington, DC: Island Press.

Hardin, G. (1968). The tragedy of the commons. *Science, 162*, 1243–1248.

Hunsberger, B. (2001, May 30). Paper mill plan may keep salmon out of hot water. *The Oregonian*, pp. C1, C7.

Jaeger, W. (2003). What actually happened in 2001? A comparison of estimated impacts and reported outcomes of the irrigation curtailment in the Upper Klamath Basin. In *Water allocation in the Klamath Reclamation Project 2001: An assessment of natural resources, economics, social, and institutional issues with a focus on the Upper Klamath Basin* (pp. 265–283). Corvallis, OR and Berkeley, CA: Oregon State University & University of California. Also available at http://eesc.orst.edu/klamath/

Juillerat, L. (2001, July 5). Protest criticism is mild, but some fear escalation. *Herald and Times* (Klamath Falls), pp. A1, A2.

Klamath farmers have justification. (2001, July 16). *Albany (OR) Democrat-Herald*, p. A8.

Klamath residents defy ban, reopen irrigation gates. (2001, July 5). *Albany (OR) Democrat-Herald*, p. A5.

Lange, J. (1993). The logic of competing information campaigns: Conflict over old growth and the spotted owl. *Communication Monographs 60*, 239–257.

Lee, L. (1980). *Reclaiming the American west: An historiography and guide.* Oxford, England: Clio Press.

Marbut, R. (2001). Legal aspects of Upper Klamath Basin water allocation. In *Water allocation in the Klamath Reclamation Project, 2001: An assessment of natural resource, economic, social, and institutional issues with a focus on the Upper Klamath Basin* (pp. 75–90). Corvallis, OR and Berkeley, CA: Oregon State University & University of California. Also available at http://eesc.orst.edu/klamath/

Milstein, M. (2001a, May 9). High and dry in the Klamath: Water quality, future murky. *The Oregonian*, pp. A1, A7.

Milstein, M. (2001b, July 15). Klamath head gates peacefully reclosed. *The Oregonian*, pp. A1, A20.

Milstein, M. (2001c, July 16). Klamath Basin protest takes new route. *The Oregonian*, pp. A1, A4.

Milstein, M. (2002a, March 2). Bush creates Klamath Basin task force. *The Oregonian*, pp. A1, A10.

Milstein, M. (2002b, June 18). West wages 100-year war over water. *The Oregonian*, p. A4.

Milstein, M. (2002c, September 24). Dead fish tied to policy flaws. *The Oregonian*, pp. A1, A4.

Moore, M. (1992). "The Quayle quagmire": Political campaigns in the poetic form of Burlesque. *Western Journal of Communication, 56*, 108–124.

Moore, M. (2003). Making sense of salmon: Synecdoche and irony in a natural resource crisis. *Western Journal of Communication, 67*, 74–96.

Moore, M. (2004). Eulogy for Tobe West: On the agitation and control of a salvage-rider timber sale. In S. L. Senecah (Ed.), *The environmental communication yearbook, Vol. 1* (pp. 33–56). Mahwah, NJ: Lawrence Erlbaum Associates.

Pisani, D. (2002). *Water and American government: The reclamation bureau, national water policy and the west, 1902–1935.* Berkeley: University of California Press.

Rapp. V. (1997). *What the river reveals: Understanding and restoring healthy watersheds.* Seattle: The Mountaineers.

Reisner, M. (1993). *Cadillac desert: The American West and its disappearing water.* Hammondsworth, Middlesex, England: Penguin Books, Ltd.

Shively, D. (1993). *Landscape change in the Tualatin Basin* (Tualatin River Basin Resources Management Rep. No. 6). Corvallis, OR: Water Resources Institute & Oregon State University.

Vigil, K. (2003). *Clean water: An introduction to water quality and water pollution control.* Corvallis, OR: Oregon State University Press.

Ward, R. (2002). *Water wars: Drought, flood, folly, and the politics of thirst.* New York: Riverhead Books.

Water-rights bill clears House, heads for Senate. (2003, May 25). *Albany (OR) Democrat-Herald,* p. A6.

Weiss, R. (2003, March 6). Fresh water reserves shrinking, U.N. says. *The Oregonian,* p. A7.

White, R. (1995). *The organic machine: The remaking of the Columbia River.* New York: Hill & Wang.

Young, A. (2001, July 15). Marshals close canal headgate in K. Falls. *Albany (OR) Democrat-Herald,* pp. A1, A8.

CHAPTER TWO

Articulating "Sexy" Anti-Toxic Activism on Screen: The Cultural Politics of *A Civil Action* and *Erin Brockovich*

Phaedra C. Pezzullo
Indiana University

Environmental movement discourse in the United States increasingly makes use of what is and what is not "sexy" in a variety of ways.[1] The Sierra Club's Corporate Accountability Campaign (n.d.), for example, skeptically notes how *sexy* serves as an important facet of a constellation of terms brought together in the marketing of bottled water: "Advertising for bottled water suggests that drinking water in plastic can make you thin, sexy, healthy, affluent, and environmentally responsible" (n.d., Web site). In referring to declining recycling trends, Steve Kullen of America Recycles Day reluctantly states in an interview with the Natural Resources Defense Council (NRDC): "I hate to use the term, but recycling is clearly not as sexy as it was" (Schueller, 2002, Web site). Responding to the appearance that they work on sexy issues, Greenpeace refutes critics who "say that Greenpeace picks 'soft and sexy' issues that are 'easy to sell' to the public," instead claiming that the "truth is not so much Greenpeace picking 'sexy' issues, ... but that—once Greenpeace had picked up an issue—sufficient attention was drawn to it in a creative way to change the public's perception of it, and to make it popular" and, by implication, sexy (Parmentier, 2003, Web site).

Regardless of the ambivalence many environmentalists seem to feel toward resisting, embracing, or becoming sexy, the value of celebrity endorsements is not

---

[1]Adhering to APA format guidelines, I have taken the word "sexy" out of quotation marks throughout this chapter, except in the title, its first use, when I'm quoting another source, and this footnote. Part of the goal of this chapter, however, is to denaturalize the many ways we reference the term and to map the various significations that are articulated together through this signifier.

far behind in this debate. Chris Collins (2001), an associate editor of Salon.com, noted, "Our 'product' is the planet—not that sexy. Look at the more successful outlets: MTV doesn't lure its demographic with spotted owls, does it? . . . The way I see it, you catch more flies with hot beats and sizzlin' celebs" (Web site). Not all agree. One university student, for example, challenges the logic behind the environmental movement's use of celebrities to appear sexier:

> In our culture, being environmentally conscious is just terribly unsexy. Sure, there have been a handful of attempts made to "sexify" the environmental movement. Some notable examples include Pamela Anderson's ads for PETA (People for the Ethical Treatment of Animals) and work done by celebrities such as Woody Harrelson advocating a vegan lifestyle. But what is really needed is a fundamental shift in values, away from luxury and excess and towards environmentally sustainable living. (Calla, 2003, Web site)[2]

This disagreement over celebrity endorsements is illustrative of the broader ambiguity indicated by environmentalists toward the idea of sexy: Yes, the linking of environmentalism with sexy stars brings increased attention, but does it help or harm environmental efforts to be associated with what or who is sexy?

As a way of beginning to answer this question, this chapter examines two Hollywood films in which celebrities perform as environmentalists. *A Civil Action* (*CA*; 1998), directed by Steven Zaillian, was based on the National Book Award-winning, best seller of the same title, which chronicled "the true story" of a lawyer's struggle to advocate on behalf of the toxically contaminated community of Woburn, Massachusetts. *Erin Brockovich* (*EB*; 2000), directed by Steven Soderbergh, was based on the "real-life" story of the woman for whom the film was named, who helped establish a legal case for the toxically contaminated community of Hinkley, California. Both were nominated for and won numerous awards,[3]

---

[2]This quote raises several areas of related debate. In contrast, I perceive the animal rights movement and the environmental movement as separate, but sometimes overlapping struggles. Given the high profile ad campaigns of PETA, the controversy over the role of sexiness has been debated in popular venues, including *Ms.* magazine. The use of naked stars in PETA ads, as Olson and Goodnight (1994) noted in passing, attempt to rearticulate what is sexy through star endorsements: "[c]elebrities appearing in anti-fur materials designed to shatter the image of fur as smart and sexy" (p. 264). The debate over whether or not such sexy portrayals of stars in animal rights campaigns are necessarily objectifying to women, however, goes beyond the scope of this chapter.

[3]*CA* was nominated for two Oscars (Best Actor in a Supporting Role and Best Cinematography) and won Best Supporting Actor from the Screen Actors Guild, Boston Society of Film Critics Awards, and Florida Film Critics Circle Awards, as well as the USC Scripter Award and U.S.A. Political Film Society's Human Rights Award (Awards for *Civil Action*, n.d., Web site). *EB* was nominated for several Oscars (including Best Actor in a Supporting Role, Best Director, Best Picture, and Best Writing, Screenplay for the Screen) and won Best Actress in a Leading Role from the Oscars, the Golden Globe, the BAFTA Film Awards, Blockbuster Entertainment Award, Bogey Award, Broadcast Film Critics Association Awards, and MTV, in addition to the Florida Film Critics Circle Award for Best Director, the U.S.A. Environmental Media Award for a Feature Film (Awards for *Erin Brockovich*, n.d., Web site).

fared well at the box office,[4] and continue to reverberate culturally.[5] Although they are not the first or sole representations of toxins or anti-toxic activism on screen, *CA* and *EB* are significant because they catapulted anti-toxic activism into the realm of "the popular," or what Stuart Hall (1994) called "the arena of consent and resistance" (p. 466). Millions of people took notice of these two films and their stars and, thus, for better or for worse, resisting toxic pollution became sexy to discuss, even if only momentarily.

Before proceeding, a tentative definition of sexy might be helpful to clarify exactly what is at stake in its use. By sexy, I am referring to a term that has taken on a double life. As *The Oxford English Dictionary (OED)* observes, "sexy" (2005) initially was defined as that which is "concerned with or engrossed in sex" and/or is considered "sexually attractive or provocative." Yet, the *OED* (2005) also noted that in 1978, *Rolling Stone* first referenced "a 'sexy' (TV for a good story) news idea" (Web site). In other words, sexy sometimes refers to anything or anyone that can create a "buzz" or garner excitement in the media. These two related, but distinct meanings make analyzing sexy a precarious task. Not only does one need to take into account the ways in which a sense of sexiness is created through taboo and sometimes elusive categories such as sex and sexual attraction, but one also should attend to how such themes influence the now extensive media complex in regards to what becomes deemed as attention worthy.

Beyond these ephemeral and culturally constructed definitions, it is even more challenging to analyze the discursive and cultural weight of sexy because even these explanations remain woefully inadequate for the task of signifying what sexy suggests. I believe, therefore, that sexy is most productively appreciated by cultural and rhetorical critics as an affective *structure of feeling*. Raymond Williams (1961) initially defined "structure of feeling" as both the "definite" and "delicate" "culture of a period" (p. 48). Its "very deep and very wide possession, in all actual communities" is important because "it is on that communication depends" (p. 48), and it is "generationally distinct" (p. 49). Despite their cultural significance, structures of feeling are hard to identify, even when one is living in their presence. As Williams (1977) elaborated, these "articulation[s] of presence" (p. 135), "although they are emergent or pre-emergent, they do not have to await definition, classification, or rationalization before they exert palpable pressures and set effective limits on experience and on action" (p. 132). Thus, it is useful to struggle with naming their effects even though we might be unable to delimit their parameters.

---

[4]*CA* is estimated a total gross of approximately $57,000,000 and *EB* of approximately $125,600,000 (Box Office Mojo; Download date: March 10, 2001; http://www.boxofficemojo.com/movies).

[5]Both films are available for home rentals. The popularity of *EB* helped launch a *Lifetime* television series hosted by Erin Brockovich, who is currently writing her second book, among other activities. Although relatively unknown prior to the films, both Schlichtmann and Brockovich continue to work on legal cases that make national headlines (e.g., Avril, 2002; Breslau & Welch, 2001; Orecklin, 2002; Peterson, 2002; Schlichtmann, 1999; Staudinger, 2003).

Importantly, Lawrence Grossberg (1992) elaborated on Williams' concept by introducing a keener appreciation of the role of "affect" as "the nature of the concern (caring, passion) in the investment . . . the way in which the specific event is made to matter. Too often," he notes, "critics assume that affect—a pure intensity—is without form or structure. But it too is articulated and disarticulated . . . through social struggles over its structure" (pp. 82–83). Appreciating the labeling of sexy as an affective structure of feeling, therefore, enables us to consider how environmentalism not only is made to mean, but perhaps more importantly "is made to matter" within specific contexts. "Affective organizations," as Grossberg (1992) emphasized, "are a complex and contradictory terrain, but one that the Left ignores at its peril" (p. 87).

The work of both Williams and Grossberg also reminds us of the importance of *articulation theory* in examining the cultural politics of linking two elements together (e.g., environmentalism and sexiness) and in forging alliances between social movements.[6] Drawing on Ernesto Laclau's (1977) theory of articulation, Hall defines articulation concisely as both "to utter, to speak forth, to be articulate" and "the form of the connection that can make a unity of two different elements, under certain conditions . . . a linkage which is not necessary, determined, absolute, and essential for all time" (cited in Grossberg, 1996, p. 141). Ernesto Laclau and Chantal Mouffe (1985) also note that when these linkages occur, the elements themselves are modified as a result of this process (p. 105).[7] It seems less useful, therefore, to hold onto a stringent definition or list of what or who is sexy, as opposed to identifying and examining how *sexy* functions culturally, politically, and historically from the standpoint of articulation.

This chapter is motivated in part by a desire to consider the possibilities and limitations of articulating environmental concerns (e.g., toxically polluted water) to a broader audience through a sexy star. Because sexy is an ephemeral and contextually based label, I want to examine these two films as heightened performances of this structure of feeling in these times. Although *CA* and *EB* ostensibly are

---

[6]Laclau and Mouffe (1985) claimed that their collaborative work on articulation is occasioned by a need to rethink "the theoretical and political bases on which the intellectual horizon of the Left was traditionally constituted" due to the "proliferation of struggles" such as "the rise of the new feminism, the protest movements of ethnic, national and sexual minorities, the anti-institutional ecology struggles waged by marginalized layers of the population, the anti-nuclear movement, the atypical forms of social struggle in countries on the capitalist periphery" (p. 1). "Most promising," Judith Butler (1997) reiterated of/for the Left, "are those moments in which one social movement comes to find its condition of possibility in another" (p. 269). Similarly, Kevin Michael De Luca (1999), argued: "Articulation is a way of understanding how, in a postmodern world with neither guarantees nor a great soul of revolt, diverse groups practicing an array of micropolitics can forge links that transform their local struggles into a broad-based challenge to the existing industrial system" (p. 82).

[7]Bobo (1988/1998) elaborated: "When an articulation arises, old ideologies are disrupted and a cultural transformation is accomplished. The cultural transformation is not something totally new, nor does it have an unbroken line of continuity with the past. It is always in a process of becoming" (p. 316).

stories about environmental struggles, it is my belief that as sexy films they also tell important stories about gender and sexuality, which always (already) should be understood as raced and classed categories. Conversely, in focusing much of my readings on how sexy identities are constructed via the protagonists in these films, I build an argument about the ways these portrayals may or may not challenge the social structures that perpetuate environmental degradation.

Following this introduction, this chapter expands on the theoretical assumptions underpinning my chosen focus on popular films, stars, and articulation theory. Because *Environmental Communication Yearbook* is a multidisciplinary forum, I draw on communication and media scholars to emphasize the importance of critically examining films as environmentally relevant texts. Then, I stress the importance of Richard Dyer's theory of "star analysis" to the ways I am reading these two films and how they articulate sexy with environmentalism. Finally, I revisit articulation theory. More specifically, I put forth the argument that analyzing *CA* and *EB* offers an opportunity to expand our understanding of the different modalities through which articulation may occur.

Subsequent to my theoretical detour, I provide critical readings of each film. It is important to analyze both *CA* and *EB* within one chapter, I believe, because of their similarities and their differences. On one hand, these films share commonalities such as narrative themes, critical acclaim, and relative financial success. As such, repetition of these patterns should help to indicate the persistence of the attention worthiness of certain cultural politics. On the other hand, reading these films against each other also enables a rich opportunity to explore how sexy can be articulated within environmental narratives in different ways depending on the sexual portrayals of their lead stars and the roles they perform.[8]

To elaborate, probably less surprising is the desire to read *EB* as a film about a sexually attractive woman, because the female lead was promoted overwhelmingly as "the perfect role model for the new millennium" (Shamberg, 2000), and women long have been perceived as sexual objects on and off screen. Conversely, although film critics frequently linked *EB* with *CA* at the time of the former's release,[9] *CA* did not receive such explicit attention as sexy. Nevertheless, I plan to show how both films invite critical analysis of their articulations of sexy as sexual insofar as they emphasize the sexuality and gender of their protagonists as pivotal to framing their narratives.[10] As cultural critics long have emphasized, it is often

---

[8]The rationale for this approach resonates with Sefcovic's (2002) argument for a "contrastive rhetorical reading" (p. 330) of the films *On the Waterfront* and *Salt of the Earth*.

[9]See: Berardinelli, 2000; Ebert, 2000; Trainer, 2000; Travers, 2000; and Turan, 2000.

[10]*EB* emphasizes sex also, particularly in a scene between Roberts/Brockovich and a resident who just has had her breasts removed in addition to the earlier removal of her uterus. The latter asks the former if she still qualifies as "a woman" (to which Roberts/Brockovich affirms, "Sure you are. Yeah, you're actually a happier one because you don't have to worry about maxi pads and underwire"). *CA*, however, doesn't have any direct dialogue about male biology and, therefore, I am not highlighting sex as a category in this analysis as much as gender and sexuality, which both films focus on.

the dominant social group that remains "unmarked" in discourse and, therefore, it is not surprising that *EB* has received more explicit sexual recognition as a film about a woman than *CA* has as a film about a man. This chapter, thus, aims to correct this popular sexist oversight by marking both films as gendered and sexualized stories of their lead characters in order to consider the limits and the possibilities of the relationships they portray between environmentalism and sexiness.

In addition, analyzing both films within one chapter enables us to consider two different ways in which sexiness can be articulated to achieve environmental goals. The narrative trajectory of *EB* is that of a success story whereas *CA* is not. By illustrating how the narrative of "success" in *EB* is predicated on sexy sexuality, whereas the narrative of "failure" in *CA* is predicated on a fall from what is sexually sexy, we further layer our understanding of sexy by considering its relationship with "success" and "failure" in environmental stories. Like most "dreams come true," I argue that *CA* and *EB* provide mixed results politically.[11]

FILMS, STAR ANALYSIS, AND ARTICULATION THEORY

For environmentalists, as Mark Meister and Phyllis M. Japp (2002) wrote, popular culture is important to analyze insofar as it can "situate humans in relation to natural environments, create and maintain hierarchies of importance, reinforce extant values and beliefs, justify actions or inaction, suggest heroes and villains, [and] create past contexts and future expectations" (p. 4). Because environmentalists do not exist in vacuums, popular culture is a telling site of analysis for environmental communication scholars. "Films," Jean Retzinger (2002) specified, "offer their viewers far more than a straightforward mirror of social reality; they serve as well as expressions and storehouses of public dreams and myths" (p. 47).[12]

---

[11]As a result of the specific question driving this chapter, many facets of these two films are deflected from our attention. Kenneth Burke (1966) noted when writing about terministic screens: "any such screen necessarily directs attention to one field rather than another" (p. 50). Perhaps the most notable deflections in this chapter's reflections are: the portrayal of water in both films, the importance of the co-stars (including Robert Duvall, James Gandolfini, Albert Finney, and Marg Helgenberger), and a detailed comparison of what "really happened" on the ground versus the filmic narratives chosen. The subsequent readings, therefore, make no promise of being exhaustive. These absences and the unmentioned ones that you as readers no doubt will discover, hopefully, will fuel further discussions about these two films and the ways we can critically engage their cultural politics and environmental impacts.

[12]Likewise, Hall (1981) suggested, when engaging media, "we have to 'speak through' the ideologies which are active in our society and which provide us a means of 'making sense' of social relations and our place in them" (p. 32). Mapping the politics enabled in and by film, therefore, is worthwhile for those invested in unpacking environmental connections between ideology and resistance. Hence, Douglas Kellner (1995) observed, "[m]edia spectacles demonstrate who has power and who is powerless" (p. 5).

In this sense, films—especially those as popular as *CA* and *EB*—are a significant medium of communication about both our public dreams and nightmares.

Although all films arguably have some relevance to social change, as struggles based on "true life stories," *CA* and *EB* might best be appreciated as part of the more specific category of the *social problem film genre.*[13] Understood as such, *CA* and *EB* offer particularly rich texts for environmental communication research, enabling "us to understand better what the environmental movement means to society, what values it emphasizes, and what motives it affirms" (Peterson, 1998, p. 385). Since *CA* and *EB* focus on struggles against toxic polluters, their texts and responses to them provide compelling opportunities to explore contemporary culture's attitudes about toxic pollution and antitoxic advocates.

This chapter's critical readings of films are heavily influenced by Richard Dyer's (1979) theory of stardom. What Dyer's insightful writings about stars emphasize is the ideological functions of these people and the roles they perform on screen. Due to a series of factors, including the emergence of mass media and industrial production, the historical phenomenon of modern stars did not begin until the early 1900s. Dyer argues that, at a fundamental level, when modern performers achieve "star" status, their perceived persona influences how an audience may or may not identify with a cinematic character. "Stars . . . collapse this distinction between the actor's authenticity and the authentication of the character s/he is playing" (p. 21). This phenomenon of blurring between actors' performances on and off screen has increased the possibilities of branding or marketing stars as a form of capital investment, defining cultural norms through their embodiment of characters on screen, and inviting audience identification with the narratives of films.

One of the primary criticisms of the star system, according to Dyer (1979), is that it suppresses "notions of human practice, achievement, [and] making the world" by commonly portraying social issues "in terms of the individual versus society/the mass" (p. 23). Given the elite status of stars and those that produce the films in which they appear, many have pointed out that stars tend to serve a conservative function, reinforcing dominant values. Dyer (1979), however, cautioned us against a hermetically sealed perspective of hegemony: "My own belief is that the system is a good deal more 'leaky' than many people would currently main-

---

[13]Waller (1987) observed, "this genre usually employs a set of recurring elements . . . for instance, sympathetic, innocent victims, embodiments of social authority and professional knowledge, a familiar, contemporary setting, and an attitude of justifiable outrage and/or pity" (p. 4). This category is identified as one of the earliest film genres. According to Sloan (1988), "early risk-taking silent filmmakers saw their new medium as one that could both entertain and, in due course, instruct. They . . . reflected the traditional American belief that once social wrongs were exposed to the people, the people would see to it that they were righted" (p. 16). Rapping (1992) argued social problem films historically have "dealt seriously with issues of the day that called into question the promise of the American Dream and showed the tragic implications of its failed vision" (p. 9). The narrative assumptions of social problem films, therefore, both reflect and shape social movements and cultural change.

tain. In my view, to assert the total closure of the system is essentially to deny the validity of class/sex/race struggles and their reproduction at all levels of society and in all human practices" (p. 25). Despite the traditional rhetorical force stars wield, therefore, oppositional and negotiated readings remain possible and even likely.[14]

Even with the role of audiences in co-constructing the meaning of texts, Dyer (1979) argued that stars and their performances remain important to analyze because their privileged positions enable them a "political significance" whether it be the "overt stands of a John Wayne or Jane Fonda, or the implicit political meanings of a Bette Davis or a Marlon Brando" (pp. 7–8). Perhaps most importantly for this chapter, the ideological significance of stars is established in the ways that stars on and off screen are constructed as heroes or antiheroes as they negotiate the political tensions of their times. As Dyer (1979) contended, "stars embody social values that are to some degree in crisis" (p. 25). In the plots of *CA* and *EB*, the crisis undoubtedly is the threat unregulated or unobserved corporate polluters pose to the environment or, more specifically, healthy public water supplies. Critically reading the portrayals and receptions of the leading stars in *CA* and *EB*, therefore, provides a rich opportunity to explore the contemporary cultural crisis of toxic water pollution and environmentalism.

As stated in the introduction, this chapter particularly is interested in responses to this crisis that *articulate* sexiness to environmentalism or environmentalists. Although *articulation* previously has been defined as a process, what remains generally unexplored is the nature of this process or the specific ways in which this linkage occurs. Can it be that all articulations are the connections of two elements that then are modified? Do these modifications always occur in equal proportions? This analysis of *CA* and *EB* offers a preliminary answer to these questions by illustrating the distinct possibilities of at least two modalities of articulation. The first is the one I already summarized and is most commonly discussed (i.e., when two elements are articulated in a way that enables something new and alters the elements themselves). This sense of articulation can be found in the joining of the environmental movement and the civil rights movement to form, at least in part, the environmental justice movement. More than merely combining the agendas of the older movements, the newly articulated environmental justice movement fosters new alliances and struggles for previously underrecognized concerns, such as lead paint poisoning in inner-city housing.

As a means to imagine more than this one modality of articulation, I believe *CA* and *EB* illustrate the possibility of another way that two elements may be linked, what I want to call *eclipse*. By examining how sexy and environmentalism is articulated in these films and their reception by film critics, this essay highlights how one element may come to overshadow another through the process of

---

[14]For more on his theory of dominant, oppositional, and negotiated readings, see Hall (1973/1980).

articulation. More than merely modifying both elements in some relatively equivalent fashion, this practice ends up concealing one element in the process of outshining another and, in turn, risking the viability of an alliance. Opportunities for environmentalists to articulate their causes to sexiness, therefore, may seem an important gesture to increase the salience or news-worthiness of environmental concerns; yet, such attempts may warrant more deliberate and self-reflexive discussion, given the possibility of eclipse.

To illustrate my point, I want to turn to the films, but in reverse chronological order (i.e., *EB*, then *CA*). I begin with the film more frequently and readily identified with sexuality and gender, as a means for identifying some of the sexy markers that thus far have been ignored in the other. Then, I return to *CA* with these criteria to illustrate how both films reinforce and challenge perceptions of sexiness in environmental communication.

"SHE BROUGHT A SMALL TOWN TO ITS FEET AND A HUGE COMPANY TO ITS KNEES"

Since her initial debut in the late 1980s in films such as *Mystic Pizza* (1988) and *Steel Magnolias* (1989), Julia Roberts' consistent star appeal to mass media audiences has been undeniable. It is well known that Roberts is the highest paid female actor in the world today. This notoriety comes, at least in part, because Roberts' economic success is touted as an accomplishment for women, who generally are paid less and are offered more limited roles than their male counterparts.

Roberts' stardom makes her sexy by both criteria established previously: The films in which she stars tend to become sexy or attention worthy topics of conversation and her attractiveness repeatedly is attributed as central to her appeal. Dyer (1979) noted a primary challenge for female stars is their having been pigeonholed into characters who must balance "both sexy and pure and ordinary" (p. 26). Roberts' persona continually navigates these extremes, particularly in discussions about her body. As Hilary Radner (2002) pointed out in her analysis of *Pretty Woman* (1990), although the producers of the film tried to hide the body doubles used for Roberts, the media relished publicly acknowledging that Roberts herself was more ordinary than the sexy woman portrayed in the film. "The following headline is characteristic: 'Those sexy curves super-star Julia Roberts flaunts in her Pretty Woman poster aren't hers!' . . . Even Julia Roberts is not in and of herself adequate to her own image, which must be created through an excess of bodies" (pp. 73–74). Similarly, many giggles were inspired by the suggestion that it took significant effort to make Roberts appear to have noticeably large breasts for her performance of Brockovich—instigated further when Roberts herself lovingly quoted her boyfriend at the time on *Oprah* suggesting, "it takes a whole village to raise that cleavage" ("Julia Roberts," 2000).

Although many once considered *Pretty Woman* Roberts' defining role, *Erin Brockovich* undoubtedly trumped that impression. As film critic David Edelstein (2000) noted: "The movie is all Julia, all the time." Film critic Mick LaSalle (2000) further suggested "Erin Brockovich is the apotheosis of every woman Roberts has played." Evidence of this linkage between the actress and her role and, thus, to the person for whom the role was based, may be found in the second tag line that was added to the film, after hearing the overwhelmingly positive reception of Roberts' performance: "Julia Roberts is Erin Brockovich."

Despite its predictability, the absurdity of the complete conflation between the actress and the character Roberts performs is perhaps most profound for the political economics of the film. For *EB*, Roberts was reported to have received between $20 and $30 million (Cameron, 2000; Travers, 2000). This factor alone also should help to temper any utopian sense of the film's message regarding environmental justice because, toward the end of the film, Roberts/Brockovich[15] hands a settlement check of $5 million to a resident stating: "It's enough for whatever you could ever need or whatever your girls need or your girls' girls need. It'll be enough." Are we supposed to believe that all of the needs of this resident combined with the next two generations of her entire family are only one quarter to one sixth the amount that Roberts alone "needs" every few years? Does this statement not epitomize the skewed value system of our culture in which someone can be paid four to six times as much money to perform a role on screen as a woman dying from toxic pollution legally deserves for the premature ending of her life? How does the belief that a monetary check, however large, might provide someone with all of her "needs" reinforce the corporate logic that supports a cost–benefit analysis assessing the worth of her, her family, and her community solely in dollars and cents?

In this same scene, the resident reacts by crying, hugging Roberts/Brockovich, and expressing her gratitude: "Oh, God, Erin. Thank you so much. I don't know what I would have done without you." This response raises a second point regarding what *EB* communicates about environmental change: In the film, the person who decides this is a case worth pursuing and the one that is credited with winning the case is the one for whom the film is named and the actress who portrays her: Roberts/Brockovich. Although she is poor and White, like most of Hinkley, Brockovich is not from Hinkley. Of all the stories of environmental injustice to be told, then, it is at least worth asking what it means for Hollywood to choose one in which a White outsider is portrayed as willing to risk everything to help "save" a community that is not her/his own. As environmental justice scholar and activist, Robert D. Bullard, wrote in 2003: "We are still waiting for an EJ [environmental justice] movie by and about people of color" (Web site).

---

[15]Given Dyer's argument regarding the collapsing between stars and the roles they perform and the reiteration of this pattern in the films which are the subject of this chapter, I subsequently will refer to the screen characters as "Roberts/Brockovich" and "Travolta/Schlichtmann" to represent this blurring.

Kelly J. Madison (1999) poignantly noted the damaging politics indicated by this broader trend in film history:

> From *Cry Freedom* in 1987 to *Amistad* in 1998, the mass dissemination of historical "anti-racist" narratives that marginalize African and African American agency, thereby highlighting white "heroism," mark whiteness in crisis, resolve the crisis through a paternalistic white supremacist co-optation of anti-racist struggle, and provide a re-legitimating historical fiction supportive of the white backlash against equality. (p. 400)

The choice of foregrounding predominantly European American communities that are "toxically assaulted"[16] might raise questions of environmental pollution; but, can the environmental movement afford to continue to ignore questions of race?

Even if *EB* is a relatively "accurate" portrayal of Brockovich's life during that time, the medium of film pushes us to ask the performative "what if": What if, instead, Hollywood chose to depict a story of a community that primarily organized itself? Or what if a story of a community of color were chosen? Would knocking door to door become complicated by racial or linguistic differences? What if the hero wasn't one individual ("I don't know what I would have done without you"), and instead were a group of community members? What if that group of community activists was a collective of people of color?

Whether or not one chooses to entertain these questions, it nevertheless is important to remember that "based on a true story" still entails lots of editing and interpretation. In both the opening and the closing scenes, it is important to note that *EB* begins and ends with Roberts/Brockovich. When asked how she related to the film *EB*, for example, Brockovich herself unsurprisingly admits: "I don't associate with the name as much as the event. It would feel great to say I'm a lone ranger and did everything myself, but the people of Hinkley and judges and so many other people helped the case come to fruition" (cited in Orecklin, 2002, p. 8). This inaccurate reduction of a collective struggle with an individual one, as was acknowledged by Dyer earlier, is typical when producers hone in on the leading star as the major marketing point of a film; but, it is not typical of off-screen, grassroots environmental struggles.

Despite the environmental storyline, *EB* overwhelmingly was received as a story about sex. Film critic, Roger Ebert (2000), wrote, "*Erin Brockovich* is *Silkwood* (Meryl Streep fighting nuclear wastes) crossed with *A Civil Action* (John Travolta against pollution) plus Julia Roberts in a plunging neckline."[17] As one of

---

[16]Layne, 2001, p. 26.

[17]In *Silkwood* (1983), directed by Mike Nichols, Meryl Streep (as Karen Silkwood) doesn't wear a plunging neckline; however, she does flash her naked breasts at one of the many co-workers who finds her attractive, and Kurt Russell frequently appears shirtless. Sexuality, including Cher's role as a lesbian, plays a predominant role throughout the film's plot, as well.

the few willing to criticize *EB*, Ebert (2000) notably stated, "the costume design sinks this movie" because "dressed so provocatively in every single scene, she upstages the material. If the medium is the message, the message in this movie is sex" (Web site). To crudely translate, Ebert suggests that no reasonable viewer could be expected to pay attention to the substance of the film with an attractive, scantily dressed woman on screen. If nothing else, this response should indicate that environmentalists need to care about the gendered and sexualized politics of the film in addition to the ways those portrayals are and are not articulated to what is sexy. To further unpack how this process of articulation occurs, I now turn to the film itself. Beyond making the struggle of Hinkley "newsworthy" to a broader audience, this reading illustrates how *EB* is a sexy film insofar as it portrays Roberts'/Brockovich's sexual life as central to the storyline.

First, Roberts'/Brockovich's dress is foregrounded as a trope both visually and in the dialogue. Consistently, she is shown in clothing fitting tightly overall, often in bold prints, with short hemlines, low necklines exposing cleavage, and made of inexpensive materials. In the following exchange with her boss Ed Masry (Albert Finney), her dress is questioned directly:

Masry: Where's Anna?

Brockovich: Um, she's out to lunch with the girls.

Masry: I have to open a file. Real estate thing, *pro bono*. You know how to do that, don't you?

Brockovich: Yeah, yeah, I've got it.

Masry: You're a girl.

Brockovich: Excuse me?

Masry: Why aren't you out to lunch with the girls? You're a girl.

Brockovich: I guess I'm not the right kind.

Masry: Oh. Ah, look, now you may want to now that you're working here, you may want to, uh, rethink your wardrobe a little?

Brockovich: Why's that?

Masry: Well, I think, uh, some of the girls are a little uncomfortable because of what you wear.

Brockovich: Is that so? Well, it just so happens, I think I look nice. And as long as I have one ass instead of two, I'll wear what I like, if that's all right with you. [Pause]. You might want to rethink those ties.

Masry's stumbling comment that Roberts/Brockovich might be alienating the other "girls" in the office because of her wardrobe is quickly rebuffed by the seemingly confident Brockovich/Roberts. The script suggests that if the other women are uncomfortable with her dress, it's a statement about their problems, not any fault of hers. The cause is ambiguous: Perhaps the women in the office are classist and feel her clothes mark her status as economically "beneath" them, from the "wrong side of the tracks"? Or maybe her obvious economic status is too painful a reminder of their economic "class"? Perhaps they feel the exposure of her body threatens the safety of their own bodies by creating a more sexual working environment? Or perhaps not being as attractive by mainstream standards, they are "just jealous" of the Pretty Woman?[18]

In addition to her dress, Roberts'/Brockovich's body is not only part of, but is central to the plot of *EB*. When asked by her boss why she believes she can obtain copies of essential documents for their legal case, for example, Roberts/Brockovich responds in one of the most quoted lines from the film: "They're called boobs, Ed." More than mere sarcasm, the audience knows at this point in the film that Roberts/Brockovich did obtain the documents by flirting with the young man who worked at the water board (by emphasizing her cleavage, assuring him "I'm divorced," etc.). Further, the audience is informed in a prelude to Roberts'/Brockovich's only "sex scene" that her beauty contest victory as Miss Wichita is vital to her self-image: "You are living right next door to a real live beauty queen. I still have my tiara. And I thought it meant that I was going to do something with my life. That it meant I was someone."

Despite the power Roberts/Brockovich appears to derive from her appearance, *EB* also suggests the continued need to debunk a prevailing cultural prejudice that "sexy looks" are mutually exclusive from "brains." For example, her excitement in obtaining 634 affidavits and a copy of the vital document linking PG&E corporate with PG&E Hinkley prompts Kurt Potter (a male European American "partnering" lawyer, played by Peter Coyote) to ask in disbelief, "Wha-ha-how did you do this?" Roberts/Brockovich sarcastically responds by stating: "Well, um, seeing as I have no brains or legal expertise . . . I just went out there and performed sexual favors. 634 blow jobs in five days." In response, Masry and an unidentified African American male legal professional snicker as the two other lawyers (Potter and his European American female assistant) stare dumbstruck.

In addition to challenging the assumption that sexy women cannot be intelligent, Roberts/Brockovich seems to struggle with a pattern of men who desire her body, but who don't care about her long-term happiness. In the first scene between Roberts/Brockovich and George (Aaron Eckhart), her new neighbor and primary love interest, he asks for her number. She responds angrily:

---

[18]Throughout, Roberts/Brockovich performs complicated relationships with other women, which I believe is a telling measure of one's commitment to feminism. Roberts/Brockovich is hostile to many of her female co-workers, though this dynamic arguably is presented as a "chicken versus egg" dynamic. Recall her snapping at a larger colleague whom she felt was staring at her, "Bite my ass, Krispy Kreme!" The only women Roberts/Brockovich appears comfortable around are those in Hinkley, generally poor European American, working-class mothers who appear around her age and closer to her body size than the other "professional" women highlighted in the film. In short: Roberts/Brockovich is portrayed as feeling comfortable around women most like herself.

Brockovich: Which number do you want, George?

George: George, now, I like the way you say that. George. Well, how many numbers you got?

Brockovich: Oh, I got numbers coming out of my ears. For instance, ten.

George: Ten?

Brockovich: Yeah, that's how many months my baby girl is.

George: You got a little girl?

Brockovich: Yeah, yeah. Sexy, huh? How about this for a number: six. That's how old my other little daughter is. Eight is the age of my son. Two is the times I've been married and divorced. Sixteen is the number of dollars I have in my bank account. 850-3943 is my phone number. And with all the numbers I gave you, I'm guessing zero is the number of times you're going to call it.

Loud, foul-mouthed, poor, twice divorced and a single mother of three, Roberts/Brockovich is portrayed as a survivor perhaps above all else. *EB* begins with a job interview in which Roberts/Brockovich suggests that as a mother, she has gained a wide array of knowledge and skills. Throughout the film, her desire to provide economically for her children is portrayed as driving her sense of urgency for work and increased compensation for her labor.[19]

When George asks her to quit her job or find a new man, Roberts/Brockovich chooses the job because, she claims, "for the first time in my life, I've got people respecting me. Up in Hinkley, I walk into a room and everybody shuts up to hear if I've got something to say. I've never had that before, ever." In addition to respect, she notes how the salary provides for her children and that her deadbeat ex-husbands left without a care for her or her children's happiness. Thus, *EB* remains consistent in its message from beginning to end: It's not too late for Roberts/Brockovich to realize that she does not need to place a man first in her life. Because men will come and usually go (particularly in the case of the sexy Brockovich/Roberts), she decides to place her career and her sense of obligation to support her children financially above all other needs or desires.[20]

---

[19]According to Nolan (2003): "The film defines Erin as sacrificing mom on a grander scale than ever before . . . she's mothering a whole damn town (a justification that, alas, will not fly for most working mothers)" (p. 37).

[20]There is ambiguity with regard to who is "right" here, insofar as George's stated reason for leaving is that Roberts/Brockovich allegedly hasn't done or said one nice thing for or to him in six months. As far as I know, there is no feminist who claims treating one's partner poorly is a good thing. Whether one perceives Roberts/Brockovich to be admirable or not in this moment, however, the choice itself is one with which feminists long have struggled for decades and have defended as a choice that women have the right to make. The dream, of course, is that a woman could "have it all," if that is what she wants; but, *EB* suggests that there is no shame in the decision to choose one's career first, whether or not the man is a handsome Harley Davidson rider who likes kids and is emotionally accessible.

In the end, there seems to be neither regret nor a sense of desperation on behalf of Roberts/Brockovich. As the updates at the end of the film inform audiences to the tune of Sheryl Crow's "Every Day is a Winding Road," the case became "the largest class-action lawsuit in history" and life moved on to the next seven cases pending. The film is a celebration of individual triumph and the hope that more "Davids" can defeat even more "Goliaths," despite all odds. Although it does not claim toxic assaults on our environment are over, *EB* serves as a reminder that the American Dream is alive and flourishing for those of us who want to believe in it. And feeling sexy is part of that dream.

"JUSTICE HAS ITS PRICE"

CA's protagonist is performed by one of cinema's biggest stars. As one of the most consistent actors of what Schatz (1992) called "the New Hollywood," John Travolta has starred in at least one major blockbuster every 5 years since the mid-1970s (e.g., *Saturday Night Fever*, 1977; *Grease*, 1978; *Staying Alive*, 1983; *Look Who's Talking?* and its sequels, 1989, 1990, 1993; *Pulp Fiction*, 1994; *Face/Off*, 1997). "Since he first rocketed to acclaim as a sexy, whippet-hipped heartthrob in the 1970s" (Cawley, 2003, p. 38) to twisting "his hips in *Pulp Fiction*" (Smolowe & Leonard, 1999, p. 210), Travolta has been portrayed as a sexy, though perhaps sometimes unconventional, leading man. Further, from dancing to disco music to becoming involved in independent filmmaking, Travolta's role choices often have made certain practices sexy. Yet, Travolta's roles to greater and lesser degrees have defied the overall trend of most male movie stars in the past three decades.[21]

In many ways, his career has been foreshadowed by the role in which he first became a cinematic star, disco dancer Tony Manero, in *Saturday Night Fever*. "The movie," Susan Bordo (1999) argued, "might poke affectionate fun at him, but it also admires him. A hero-narcissus—a very new image for postwar Hollywood" (p. 200).[22] Arguably, Travolta has maintained this trend of a more physically and emotionally available male lead for much of his career. "A star like John

---

[21]Since the Leftist movements of the 1960s, films have tended to reflect "defensive reaction[s] to traditional masculine failure," "a new homophobic scrutiny," and the need for men "to reclaim their manhood" (Kimmel, 1996, pp. 287, 288, 289). Jeffords (1992), for example, argued that male film roles generally depicted "hard-bodied male action heroes . . . [in] the eighties" (p. 197) and the "unloved" White family heroes in the 1990s (p. 206). Travolta's career reflects a set of possibilities for a man that differs from these norms. As someone who tends not to portray stoic, hard-bodied, or heroic figures, Travolta "hasn't made his reputation playing professional men" (Turan, 1998).

[22]Bordo (1999) recognized two conventions that were broken by Travolta in his first starring film role. First, she claims, "[n]ever before *Saturday Night Fever* had a heterosexual male movie hero spent so much time on his toilette" (p. 198), primping his hair and appearance in front of a mirror. Second, she notes: "Travolta was also the first actor to appear on-screen in form-fitting (if discretely black) briefs" (p. 199). In addition to these shifts in representation, as Steve Neale (1983/2000) pointed out, Travolta is presented in a rare instance of "feminization" given its status as a musical, "the only genre

Travolta (*Saturday Night Fever, Urban Cowboy, Moment by Moment*)," E. Ann Kaplan (1983/2000) argued, "has been rendered the object of woman's gaze and in some of the films (i.e., *Moment by Moment*) placed explicitly as a sexual object to a woman who controlled the film's action" (pp. 128–129). At least until *CA*, Travolta usually has embodied nontraditional male roles in film or, at minimum, has exceeded more prevalent masculine cinematic depictions through acting as the object of desire in song and dance.[23]

Travolta's vulnerability physically often parallels an emotional and/or intellectual vulnerability marked by a series of roles as a working-class man in love with a wealthier and/or smarter woman (e.g., *Saturday Night Fever, Grease*, and *Look Who's Talking*). Although it is possible to consider this *Welcome Back Kotter* clueless charm as something akin to the younger Keanu Reeves' simple charisma, I would argue that Travolta's history of performing "inner" vulnerability linked to a working-class identity also suggests a willingness on the part of audiences to believe in the capitalist myth that men who are poor must naturally be more physical rather than cerebral.

In *CA*, however, Travolta isn't a working-class hero, he doesn't sing, and he barely dances. Yet, unlike *EB*, it is important to note that *CA* is not the story of a hero in any romantic sense of the term. John Travolta/Jan Schlichtmann does not win the court case around which *CA* centers. He alienates all of the co-workers in his firm. He begins economically well-off and then proceeds to lose all of his material wealth. In the end of the film, he is alone, attempting to build a new life for himself. The only way we can imagine this role as even an antihero is insofar as his story has repercussions for the community of Woburn and, therefore, we find ourselves perhaps reluctantly rooting for his side. This role represents a challenge that Travolta describes in an interview:

> I didn't like Schlichtmann, and I wondered how the hell I was going to play him. But I didn't like the character I play in *Pulp Fiction* very much when I read it first, and I didn't really know how to play a lot of characters I've been offered, because people often ask me to solve unpleasant characters. . . . With this guy, I thought if I played a greedy, self-centered guy who was just unaware that he was that way, if I could really pull that off, maybe the audience would stay with it. ("Travolta's Law," 1999, p. 64)

Considering how Travolta's star persona as a more vulnerable man interacts with the story of Schlichtmann, one might imagine that an audience's familiarity with,

---

in which the male body has been unashamedly put on display in mainstream cinema in any consistent way" (p. 263). Not only was Travolta himself placed on display as a performer, but also as the object of female desire. Further, the film portrays Travolta crying, a rare display emotional vulnerability for a man on screen at the time.

[23]Despite his consistent commercial success, most claim Travolta's acting career has had a "comeback" in terms of substance and quality since *Pulp Fiction* (1994). These characteristics of his star persona in 2000, interestingly, seem to have remained relatively consistent.

and lack of threat from, Travolta would make his performance of Schlichtmann more amicable or sexy—though perhaps less believable—than another casting choice.

Environmentally, like *EB*, *CA* may be applauded for the attention it garners for antitoxic politics and might be worth questioning for the narrative alone that it represents. Yet, in addition to choosing another predominantly European American community and portraying the story of an individual White outsider attempting to "save" the locals, *CA* places the voices of the residents even farther in the background than *EB*. Again, environmentalists might be wise to consider how this portrayal might marginalize collective efforts. I will return to this point. For now, to identify the gendered and sexual politics portrayed in this analysis of *CA*'s leading star, I draw on the two themes suggested by my critical reading of how sexiness is performed in *EB*: the politics of appearance and one's eligibility romantically.

Throughout the film, Travolta/Schlichtmann appears to embody conventional masculine values and traits. The film begins and ends with a voice-over of Travolta/Schlichtmann quoting statistics and odds in a tone of authority and certainty. The opening scene depicts Travolta/Schlichtmann dressed in a well-tailored suit, walking down a hallway, wheeling a disabled European American man whom the reader is invited to see as "the perfect victim":

> It's like this: a dead plaintiff is rarely worth as much as a living, severely maimed plaintiff. However, if it's a long, agonizing death as opposed to a quick drowning or car wreck, the value can rise considerably. A dead adult in his twenties is generally worth less than one who is middle aged. A dead woman less than a dead man, a single adult less than married, black less than white, poor less than rich. The perfect victim is a white male professional, forty years old, at the height of his earning power, struck down in his prime. And the most imperfect? Well, in the calculus of personal injury law, a dead child is worth the least of all. (*A Civil Action*, 1999)

Immediately, this cold, calculating logic draws on a traditionally masculine, "objective" science of human worth. Translating life and death into formulas with which one may gamble is, as the film's tagline provokes us to consider, "the price of justice."

After winning this so-called "perfect" case by settling out of court for $2 million, Travolta/Schlichtmann is represented as "on top of the world." Set to the song "Hard Workin' Man" (performed by Captain Beefheart), a montage of images—from champagne flowing to scenes of Boston Harbor—follows. Travolta/Schlichtmann then appears on a radio show where he states:

> Personal injury lawyers have a bad reputation. They call us ambulance chasers, bottom feeders, vultures who prey on the misfortune of others. Well, if that's true, why do I lie awake at night worrying about my clients? [Screen cuts to him dancing with a woman.] Why does their pain become my pain? I wish I could find some way not to

empathize. It would be a lot easier. [Screen cuts to him picking out silk ties and scarves at an upscale clothing store.] (*A Civil Action*, 1999)

The insincerity of Travolta/Schlichtmann is blatant: This man who is drinking champagne, laughing with friends, and buying expensive accessories does not appear to be in pain at all. By mainstream standards, he is performing a sexy life for a man, including happiness, camaraderie, money, power, persuasive moves, and smooth talking. Similar to that of *EB*, Travolta's/Schlichtmann's physical appearance is portrayed as a telling reflection of his character: flashy, proud, and perhaps a bit too obsessively tailored. Later in the film, we see that his car, a black Porsche 928, with which he receives multiple speeding tickets, matches this persona.

For anyone mistakenly caught up in the lure of all of this certainty and glamorous lifestyle, the tone of the film quickly changes as a resident of Woburn, Anne Anderson, calls into the radio show to ask Travolta/Schlichtmann: "Hey, I have an idea. Why don't you come up to Woburn one of these days and actually meet some of those people who's pain is your pain? [Pause.] Jan?" Travolta/ Schlichtmann hangs his head in response before the film cuts to his office the next morning.

In this scene and throughout the film, working on the legal case for Woburn is portrayed as the antithesis of feeling sexy, or to borrow a phrase from Captain Beefheart, "bad to the bone." In the conclusion, Travolta/Schlichtmann writes the EPA about the possibility of the agency picking up the case, because he has lost his "gambling spirit." An explicit statement of self-reflection from the protagonist, this letter is worth quoting at length:

I have the evidence, but no longer the resources or *the gambling spirit* to appeal the decision of the Beatrice case. I have no money, no partners, and, as far as I can tell, no clients anymore. The Woburn case has become what it was when it first came to me, an orphan. I am forwarding it on to you, in all its unwieldiness, even though you might not care to adopt it any more than I did at first. But, if you do decide to take it on, I hope you'll be able to succeed where I failed. If you calculate success and failure as I always have, in dollars and cents divided neatly into human suffering, the arithmetic says: I failed completely. What it doesn't say is if I could somehow go back, knowing what I know now, knowing where I'd end up if I got involved with these people, knowing all the numbers, all the odds, all the angles, I'd do it again. (*A Civil Action*, 1999, emphasis added)

Measured against even Travolta's/Schlichtmann's own standards at the beginning of the film, losing one's money, partners, clients, and spirit is not a happy ending. The question prompted in the end is why Travolta/Schlichtmann would "do it again." Is it because he remains committed to the faults that led to his downfall and, therefore, still wants to prove he was/is "right"? Or is it because he has learned that those characteristics (greed and self-centeredness) are not desirable?

In either case, the audience—environmentally minded or otherwise—is invited to value the lesson and to recognize it is worth learning.

When *CA* was received in gendered terms, it was in the film's insistence on portraying Travolta/Schlichtmann as an antihero, an image some film critics found refreshing in contrast to films based on books of the best-selling "legal thriller" novelist, John Grisham (Berardinelli, 1998). Film critic Cindy Fuchs (1999), for example, argued, "Travolta's Schlichtmann [is] a welcome antidote to John Grisham's heroic lawyer plots, where Tom Cruise or Matt Damon is noble through and through, and incidentally saves/gets the girl. Schlichtmann is frankly shown to be self-centered and obsessive" (Web site). Ebert (1998) further enforced this impression of the film's gender politics: "*Civil Action* is John Grisham for grownups. Watching it, we realize that Grisham's lawyers are romanticized hotshots living in a cowboy universe with John Wayne values" (Web site). In contrast to the mainstream masculine hero of legal narratives that pervade our popular "law and order" culture, Travolta/Schlichtmann is depicted as an undesirable man and lawyer. His stubbornness to act as an individual who doesn't need others is portrayed as his Achilles' heel rather than his strength. This depiction denaturalizes the normative portrayal of masculinity today, what Katz and Earp (1999) called the "tough guise." Travolta/Schlichtmann is not sexy.

Part of being perceived as sexy, as earlier definitions imply, is having or the possibility of having sex. As Fuchs (1999) noted, the film *CA* does not portray Schlichtmann "getting the girl." This is, in fact, one of the primary points of departure between the film and the book, because the latter delves into two of Schlichtmann's romantic relationships (Harr, 1995). Instead, the film discards his sexual affairs and, alternatively, references *Boston Magazine's* listing of most eligible bachelors. As the film begins, Schlichtmann is on the list and in the second to last written update offered at the end, the film notes: "It took Jan several years to settle his debts, but only one year to fall off Boston's Ten Most Eligible Bachelor list." Unlike Roberts/Brockovich, therefore, Travolta/Schlichtmann is portrayed not merely as single in the end of the film, but sexually ineligible, without prospects. Further, this status as an undesirable bachelor is juxtaposed with another "cost" of his environmental efforts—debt. Here, the personal cost of dedication is starkly portrayed. *CA* represents the protagonist's work ethic as an overly inflated ego's obsession, symptomatic of the stereotypical male workaholic, rather than an empowering means to build one's self-esteem.

The tormented sexual and environmental struggles highlighted in *CA* appropriately resonate with the last song choice on the soundtrack, the Talking Heads' cover of the Al Green song, "Take Me to the River." The River metaphorically serves as both a biblical reference to the possibility of baptism and rebirth (as we are told that Schlichtmann now only practices environmental law) and a place where temptation may reside (as we are told that Schlichtmann remains single and broke at the end of the film). Further, as the film's environmental narrative pivots around the health of the water in Woburn, it would not be difficult to imagine the song as an enthymematic challenge to consider what or who is the ambiguous "you" referenced in the lyrics that admit "I don't know why I love you like I do/All the troubles you put me through" and beg the question: What might "Take me to the river/Drop me in the water" entail today?

IMPLICATIONS

In a study of residents in Alabama, a seemingly frustrated scholar documents some of the reasons why local communities often are unwilling to oppose corporate polluters, noting that, due to "their highly sympathetic portrayals of grassroots environmentalism, recent commercially successful films such as *A Civil Action* and *Erin Brockovich* have no doubt contributed to misconceptions about the efficacy of collective action against corporate polluters" (Moberg, 2002, p. 377). This scholar's criticism of the films (also encapsulated in his essay's title: "Erin Brockovich Doesn't Live Here")[24] reminds us that *CA* and *EB* may place unrealistic expectations on the part of other communities and advocates regarding the sacrifices people are willing to make and the successes that may be obtained. Though I would agree based on my own experiences, *CA* and *EB* also enable much more.

This chapter began by pointing out that and who is or is not sexy is a major constraint to the environmental movement today. Although more work needs to be done to trace the ways this arrangement functions within U.S. culture, this chapter provides an initial scholarly attempt at taking the label sexy seriously as an affective structure of feeling with consequences for our times.

To elaborate, in accounting for the gendered and sexualized politics of these environmentally based films, this analysis illustrates how *EB* and *CA* articulate anti-toxic activism with sexiness. At a basic level, both Roberts and Travolta are considered sexy superstars and, therefore, following star analysis, the respective portrayals of Brockovich and Schlichtmann articulate these "true stories" to sexiness for mainstream audiences. In addition, the politics of appearance I have identified in these films illustrate how gender and sexuality are central to the reception and commercial success of these environmental narratives.

The significance of Travolta's/Schlichtmann's and Roberts'/Brockovich's eligibility to the storylines of the films further emphasizes how desirability can frame environmental struggles. As noted earlier, it was strategic on the part of the screenwriters to depict both Travolta/Schlichtmann and Roberts/Brockovich as romantically single at the conclusion of these two films. In *CA*, I have argued that this decision is consistent with its overall story of an antihero. Film critic James

---

[24]Though appreciating the attention-getting value of such a title, it is disappointing that an academic would obscure the point that Brockovich didn't ever live in Hinkley herself (hence, the need for so many car rides, hotel rooms, and babysitting).

Verniere (1999) noted, "Without a love interest or a confessor, one of the consequences is we never warm up to him" (Web site). In *EB*, though Roberts/
Brockovich is not portrayed as desperate, this decision unnecessarily reinforces
the false perception that women must choose between careers (in environmentalism) and love.

Yet, as noted earlier, these films do differ in significant ways. As far as the
broader cultural politics of gender performed by *CA* and *EB*, *CA* offers a clearer
message. As an antidote to Grisham heroes, the film indirectly promotes the value
of collective action and challenges traditional masculine traits of cockiness and
self-obsession. As an ecofeminist perspective might suggest, this questioning of
the ideologies of domination, objectification, and ownership helps us as a culture
become more reflexive about our everyday practices.[25] How did Schlichtmann's
desire to be in control, to ignore the wishes of the Woburn community members,
to take for granted the labor of the rest of the people working in his law firm, and
to want to protect his pride end up hurting people whom he claimed to legally
represent? How does the film suggest that, in the end, the type of man that
Schlichtmann was is indicative of the pervasive attitude in our culture that continues to enable and condone a headstrong production of toxic chemicals without
care for who is hurt along the way? The moral lesson for the film's antihero is, perhaps above all else, that his initial value for what dominant hegemonic culture
might consider sexy in heterosexual men (i.e., valuing dollars over lives, gambling
odds over initiating careful planning, and aiming for individual success over collective achievements) is precisely the value system that contributes to the toxic
culture in which we live. As such, the sexy and environmental themes of the film
seem inextricably interlinked.

Although analyzing *EB* reinforces the linkages between sexy and environmental themes, the film's overriding visual, aural, and narrative message regarding gender is more ambiguous. Overall, *EB* illustrates how sexually attractive
and tenacious women can reclaim a sense of self-esteem and take important
steps to improving this world; yet, this narrative resonates with two contrary political movements. In a positive sense, it echoes third-wave feminists who "con-

---

[25]Despite some exceptions, many ecofeminists have emphasized that it is less useful to perceive
ecofeminism as a commitment to the belief that women are essentially "closer" to nature (see, e.g.,
Bullis, 1996; Taylor, 1997; Warren, 1997). Carolyn Merchant (1996) outlined a history of the various
branches of ecofeminism (see also Di Chiro, 1998). She emphasizes that ecofeminists, like all feminists,
are motivated by a variety of reasons and employ an array of preferred tactics, though "all [have] been
concerned with improving the human/nature relationship" (p. 5). Overall, ecofeminists do claim that
"important connections exist between the treatment of women, people of color, and the underclass on
the one hand and the treatment of nonhuman nature on the other" (Warren, 1997, p. 3). For example,
Carol Adams (1990) argued that U.S. culture's sexualization and objectification of "meat" (or "dead
animals") reinforces treating women like meat, defined solely as sex objects to be consumed; thus, she
argues that feminists would do well to resist this ideology, in part, by becoming vegetarians. By definition, therefore, ecofeminism is a movement of articulation, one that defines the identification, assessment, and challenge of connections between social movements and issues as central to their struggle.

tinue to build upon a feminist legacy that challenges the status quo, finds common ground while honoring difference, and develops the self-esteem and confidence it takes to live and theorize one's own life" (Walker, 1995, p. xxxv). In a more troubling sense, *EB* just as plausibly could reiterate the general position of "power feminists," who encourage women to resist labels of *victim* and to claim sole responsibility for improving their situation in this world. As critics of this so-called branch of feminism point out, the problem with this perspective is that it all too often "works" for a limited number of women, those who tend to be "White, upperclass, successful, well educated, and physically attractive" (Wood, 2003, p. 79). Although the former philosophy of feminism offers possibilities for alliances with environmental struggles, the latter reinforces a damaging conception. Namely, although Brockovich is not upperclass or well educated in any formal sense, the appropriation of her traditionally understood sexual appeal for "power" is an option available to far too few women for that performance alone to make her an empowering environmental heroine. The challenge remains, therefore, to encourage audiences to locate *EB* within a third-wave paradigm, rather than one of power feminism. With this perspective, we can appreciate Brockovich as a woman who has faced major constraints in life and has overcome them through her own resourcefulness *and* the help of others (such as the community members who risked sharing their lives with her, Ed Masry, who provided her with a job in the first place, and the babysitter who looked after her children while she was away at work). Clearly, regardless of the perspective an audience member chooses, any gendered reading of *EB* is far from peripheral to the cultural politics of its environmental storyline.

Despite the dual cultural meanings of sexy as both sexual and attention worthy, this chapter has aimed to keep the contested and messy cultural politics of judging sexiness a definite and delicate task. Although there is a recognizable structure to this feeling in our times, I also believe that scholarly attempts to engage what is sexy remain challenging. Doing so requires something akin to a leap of faith: a belief not only that feeling sexy or not matters deeply to politics today, but also that the effects or implications of this belief are worth examining beyond whether or not we agree with the articulations constructed by a dominant, hegemonic society.[26] In addition to identifying and interpreting the sexy dynamics portrayed by the stars of *CA* and *EB*, therefore, this chapter also has provided further opportunity to theorize articulation. More specifically, as indicated earlier, these two films provide an occasion to examine a risk of articulation: when environmental struggles may be eclipsed by the affective structure of feeling sexy.

---

[26]Feminists have long debated what is or is not an empowering response to sexual practices (e.g., whether or not to engage in sadomasochism, heterosexual sex, or miscegenation) and sexual representations (e.g., debates over pornography, Barbie, and supermodels). Given the complexity of these wide-ranging conversations, critically examining the moral or political worth of sexy itself goes beyond the scope of this chapter. This chapter, instead, assumes that because sexy matters to environmental politics, it is worth exploring how.

There were many signs that *CA* stimulated attention for environmentalism. The Sierra Club "piggybacked" on the film's buzz by releasing a report entitled, "Cancer, Chemicals, and You" (Greenwire, 1999). High schools initiated new education projects related to toxic pollution, water, and the environment (Talbot, 1999). News stories were published in every major newspaper in the country and broadcast by other media sources such as National Public Radio and Court TV. This type of attention was fostered by both Travolta and Schlichtmann. Travolta stated: "If people leave the theater more enlightened about the situation in our environment we're in good shape"; whereas Schlichtmann said: "It's through popular culture that we communicate. We get marginalized by not being in popular culture. The greatest entertainment is that which touches us in some way, that offers a chance for enlightenment. Here's the opportunity" (cited in Greenwire, 1999).

Despite these positive implications, the catalyst for environmental change focused on Travolta/Schlichtmann such that Anne Anderson, a resident of Woburn who inspired one of the largest supporting roles in the film (performed by Kathleen Quinlan), protested prior to the film's box office release: "If they're [Disney] going to use our characters and our story, then they certainly should ask our permission and look for input from us so that it's accurate" (cited in Mueller, 1997, p. 4). By placing even an antihero in the foreground of this story instead of residents, the sexiness of the film's protagonist risks eclipsing the people for whom the negative environmental impact is most felt.

Doubling *CA*'s profits at the box office, *EB* further attracted attention to the toxic pollution of water. In addition to generating more media coverage in general, *EB* brought in larger audiences and increased attention within markets more often targeted at women. Yet, epitomized by the tagline, "Julia Roberts is Erin Brockovich," I would argue that *EB* dangerously eclipses Brockovich with Roberts by overarticulating the film with the story on which it is based, such that the sexy buzz surrounding this successful Hollywood film risks completely overshadowing the environmental politics. Roberts, for example, forgot to thank Brockovich in her Oscar reception speech. Although she publicly apologized, it seems that "if Roberts had things in perspective, it would not have been possible for her to forget" (Tacey, 2001, p. 17). The absurdity of this eclipse persists. A recent CNN (2003) report on a current Brockovich case asserts: "The claims are a precursor to lawsuits Masry and Brockovich, subjects of the popular film *Erin Brockovich*, expect to file" (Web site). If the film is named after Brockovich, wouldn't it seem obvious that we would remember her name? Why must we be reminded continually of the film? The danger of eclipsing Brockovich in the process of articulating her with Roberts is that, at minimum, it disregards the importance of social location, specificity, and political critique.

Admittedly, both films could have been more representative in their depictions of antitoxic struggles and, therefore, this chapter has highlighted the problems of focusing on the personal stories of individual European American outsiders rather than on the collective narratives of multicultural residential communities that are resisting the effects and practices of toxic pollution. This narrative pattern of simplifying cultural change through the story of one person in order to dramatize the unfulfilled promise of the American Dream (or, in *EB*'s case, to dramatize the fulfillment for Roberts/Brockovich personally) is typical of the types of stories that are chosen to be depicted in the social problem film genre (Waller, 1987). Yet, although *CA* and *EB* are limited tales that offer resolution to unresolved social problems and, as such, to greater and lesser degrees, risk eclipsing the environmental messages that might be heard otherwise, the films also should provoke us to continue to analyze the ways in which the powerful medium of film and the influence afforded by a Hollywood budget may raise awareness of environmentalism on a scale and in a manner rarely possible.[27] If the environmental movement is going to persist in defining itself as aligned with or in opposition to what is sexy, we would do well to examine further the possibilities and limitations of such articulations.

As articulation theory continues to gain traction in academic research, this chapter's analysis indicates that we need to be more specific about the nature of the linkages we are studying. If connecting an environmental cause in a popular text or persuasive campaign to someone or something may lead to an overshadowing of the cause itself, environmental communication scholars and practitioners would do well to consider not just to whom or what they want to link their causes, but also *how*. Further theorization of additional modalities of articulation also seems warranted.

In summary, *CA* and *EB* are important sites of cultural struggle because they have become a part of the current popular imaginary about anti-toxic activism. As a result, environmentalists have an opportunity to reexamine what is entailed in implicating their cause in a structure of feeling sexy. It is important to take seriously this task of laying claim to and reflexively considering the processes of articulation that shape the presence of entertainment and education if we are to increase the salience and the popularity of environmentalism. Engaging how environmentalism intersects with discourses of gender, sexuality, race, and class fosters not only a more complex rendering of public life, but also a more compelling one that promises the possibility of making environmentalism matter to an even broader coalition for progressive social change.

ACKNOWLEDGMENTS

An earlier version of this chapter was presented at the annual convention of the National Communication Association, Miami, Florida, 2003. The author would like to thank her colleagues at Indiana University, Rachel Hall, and Ted Striphas for their feedback and encouragement.

---

[27]This echoes Gray's (1995) study of *Roots*, in which he suggests that despite the primary narrative's focus on one particular slave's experience, its success had significant cultural effects.

REFERENCES

Adams, C. J. (1990). *The sexual politics of meat: A feminist–vegetarian critical theory.* New York: Continuum.

Amistad. (1998). Steven Spielberg (Director). DreamWorks SKG and Home Box Office (HBO).

Avril, T. (2002, January 23). Children with cancer, siblings will get millions in Toms River settlement. *The Philadelphia Inquirer,* p. A01.

Awards for *Civil Action.* (n.d.). [online]. *Internet Movie Database Inc.* Retrieved March 10, 2003, from http://us.imbd.com/Name?Travolta,+John

Awards for *Erin Brockovich.* (n.d.). [online]. *Internet Movie Database Inc.* Retrieved March 10, 2003, from http://us.imbd.com/Tawards?0195685

Berardinelli, J. (1998). [online]. *A Civil Action*: A film review. *James Berardinelli's ReelViews.* Retrieved December 18, 2002, from http://movie-reviews.colossus.net/movies/c/civil_action.html

Berardinelli, J. (2000). [online]. *Erin Brockovich*: A film review. *ReelViews.* Retrieved December 18, 2002, from http://movie-reviewscolossus.net/movies/e/erin_brockovich.html

Bobo, J. (1998). 'The Color Purple': Black women as cultural readers. In J. Storey (Ed.), *Cultural theory and popular culture: A reader* (2nd ed., pp. 310–318). Athens, GA: University of Georgia Press. (Original work published 1988)

Bordo, S. (1999). *The male body: A new look at men in public and private.* New York: Farrar, Straus & Giroux.

Breslau, K., & Welch, C. (2001, February 5). Another civil action. *Newsweek, 137*(6), 48–49.

Bullard, R. D. (2003). [online]. Crowning women of color and the real story behind the 2002 EJ [Environmental Justice] Summit. *Environmental Justice Resource Center,* Atlanta, GA. Retrieved December 8, 2003, from http://www.ejrc.cau.edu/SummCrowning04.html

Bullis, C. (1996). Retalking environmental discourses from a feminist perspective: The radical potential of ecofeminism. In J. G. Cantrill & C. L. Oravec (Eds.), *The symbolic earth: Discourse and our creation of the environment* (pp. 123–150). Kentucky: University Press of Kentucky.

Burke, K. (1966). *Language as symbolic action: Essays on life, literature, and method.* Berkeley: University of California Press.

Butler, J. (1997, Fall/Winter). Merely cultural. *Social Text, 25*(3&4), 265–277.

Calla, M. (2003, November 13). [online]. Environmentalism requires real effort. *The Silhouette.* Retrieved December 6, 2003, from http://www-mus.mcmaster.ca/sil/oped/o31113evn.html

Cameron, M. (2000, March 26). Julia's a woman in $30m. *Nationwide News Party Ltd., Sunday Mail,* p. 31.

Cawley, J. (2003, April). John Travolta flying high. *Biography, 7*(4), 38–42.

A Civil Action. (1998). Steven Zaillian (Director). Touchstone Pictures.

CNN. (2003, April 29). [online]. Brockovich takes on Beverly Hills High. Retrieved April 29, 2003, from http://cnn.com/2003/TECH/science/04/29/california.brockovich.reut/index.html

Collins, C. (2001, May 4). [online]. U sexy mother nature. *Grist magazine.* Retrieved December 6, 2003, from http://www.gristmagazine.com/imho050401.asp

Cry Freedom. (1987). Richard Attenborough (Director). Marble Arch Productions, Inc. and Universal Pictures.

De Luca, K. M. (1999). *Image politics: The new rhetoric of environmental activism.* New York: Guilford Press.

Di Chiro, G. (1998). Environmental justice from the grassroots: Reflections on history, gender, and expertise. In D. Faber (Ed.), *The struggle for ecological democracy: Environmental justice movements in the United States* (pp. 104–136). New York: Guilford Press.

Dyer, R. (1979). *Stars.* London: British Film Institute.

Ebert, R. (1998). [online]. *A Civil Action. Chicago Sun-Times.* Retrieved December 18, 2002, from http://suntimes.com/ebert/erbert_reviews/1999/01/010801.html

Ebert, R. (2000). [online]. *Erin Brockovich. Chicago Sun-Times.* Retrieved December 18, 2002, from http://suntimes.com/ebert/erbert_reviews/2000/03/031703.html

Edelstein, D. (2000, March 17). [online]. The riot girl next door: Julia Roberts is a larger-than-life hell-raiser in Erin Brockovich. Retrieved December 18, 2002, from http://slate.msn.com/?id=77437

Erin Brockovich. (2000). Steven Soderbergh (Director). Jersey Films.

Face/Off. (1997). John Woo (Director). Douglas/Reuther Productions, Paramount Pictures.

Fuchs, C. (1999, January 7–14). [online]. *A Civil Action:* He's not your usual Hollywood lawyer—but then the melodrama kicks in. *Philadelphia City Paper.* Retrieved December 18, 2002, from http://www.citypaper.net/movies/c/civilaction.shtml

Gray, H. (1995). *Watching race: Television and the struggle for "blackness."* Minneapolis: University of Minnesota Press.

Grease. (1978). Randal Kleiser (Director). Paramount Pictures.

Greenwire. (1999, January 8). [online]. *A Civil Action:* Travolta takes on toxic torts. *Environment and Energy Publishing, LLC.* Retrieved March 7, 2003, from Lexis/Nexis Environmental. http://web.lexis-nexis.com/envuniv/printdoc

Grossberg, L. (1992). *We gotta get out of this place: Popular conservatism and postmodern culture.* London: Routledge.

Grossberg, L. (1996). On postmodernism and articulation: An interview with Stuart Hall. In D. Morley & K.-H. Chen (Eds.), *Stuart Hall: Critical dialogues in cultural studies* (pp. 131–150). New York: Routledge.

Hall, S. (1980). Encoding/decoding. In Centre for Contemporary Cultural Studies (Ed.), *Culture, media, language: Working papers in cultural studies, 1972–79* (pp. 128–138). London: Unwin Hyman. (Original work published 1973)

Hall, S. (1981). The whites of their eyes: Racist ideologies and the media. In G. Bridges & R. Brunt (Eds.), *Silver linings* (pp. 28–52). London: Lawrence & Wishart.

Hall, S. (1994). Notes on deconstructing 'the popular.' In J. Storey (Ed.), *Cultural theory and popular culture: A reader* (2nd ed., pp. 442–453). Athens, GA: University of Georgia Press.

Harr, J. (1995). *A civil action.* New York: Vintage Books.

Jeffords, S. (1992). The big switch: Hollywood masculinity in the nineties. In M. Collins, H. Radner, A. Preacher Collins, & J. Collins (Eds.), *Film theory goes to the movies: Cultural analysis of contemporary films* (pp. 196–208). London: Routledge.

Julia Roberts. (2000, March 10). *The Oprah Winfrey Show* [Transcript]. Livingston, NJ: Burrelle's Information Services.

Kaplan, E. A. (2000). Is the gaze male? In E. A. Kaplan (Ed.), *Feminism and film* (pp. 119–139). Oxford: Oxford University Press. (Original work published 1983)

Katz, J., & Earp, J. (1999). *Tough guise: Violence, media, and the crisis in masculinity.* S. Jhally (Director). Media Education Foundation.

Kellner, D. (1995). *Media culture.* London: Routledge.

Kimmel, M. (1996). *Manhood in America: A cultural history.* New York: The Free Press.

Laclau, E. (1977). *Politics and ideology in Marxist theory: Capitalism, fascism, populism.* London: New Left Books.

Laclau, E., & Mouffe, C. (1985). *Hegemony and socialist strategy: Towards a radical democratic politics,* 2nd edition. London: Verso.

LaSalle, M. (2000, March 17). [online]. Ready for her close-up: True-story role of flashy, crusading law clerk was made for Julia Roberts. *San Francisco Chronicle.* Retrieved December 18, 2002, from http://www.sfgate.com/cgi-bin/article.cgi?f=/c/a/2000/03/17/DD78120.DTL

Layne, L. L. (2001). In search of community: Tales of pregnancy loss in three toxically assaulted U.S. communities. *Women's Studies Quarterly, 1 & 2,* 25–49.

Look Who's Talking. (1989). Amy Heckerling (Director). MCEG Productions and TriStar Pictures.

Look Who's Talking Too. (1990). Amy Heckerling (Director). Big Mouth Production and TriStar Pictures.

Look Who's Talking Now. (1993). Tom Ropelewski (Director). TriStar Pictures.

Madison, K. J. (1999, December). Legitimation of crisis and containment: The "Anti-Racist-White-Hero" film. *Critical Studies in Mass Communication, 16*(4), 399–416.

Meister, M., & Japp, P. M. (2002). *Enviropop: Studies in environmental rhetoric and popular culture.* Westport, CT: Greenwood Press.

Merchant, C. (1996). *Earthcare: Women and the environment.* New York: Routledge.

Moberg, M. (2002). Erin Brockovich doesn't live here: Environmental politics and 'Responsible Care' in Mobile County, Alabama. *Human Organization, 61*(4), 377–389.

Moment by Moment. (1978). Jane Wagner (Director). Robert Stigwood Organization (RSO).

Mueller, M. (1997, August 16). Cellucci takes heat from Woburn families over movie bill. *The Boston Herald*, p. 4.

Mystic Pizza. (1988). Donald Petrie (Director). Night Life Inc. and Samuel Goldwyn Co. (Production).

Neale, S. (2000). Masculinity as spectacle: Reflections on men and mainstream cinema. In E. A. Kaplan (Ed.), *Feminism and film* (pp. 253–264). Oxford: Oxford University Press. (Original work published 1983)

Nolan, M. (2003, Fall). Unmarried . . . with children. *Bitch: feminist response to pop culture, 22*, 35–41.

Olson, K. M., & Goodnight, G. T. (1994). Entanglements of consumption, cruelty, privacy, and fashion: The social controversy over fur. *Quarterly Journal of Speech, 80*, 249–276.

Orecklin, M. (2002, May 20). 10 questions for Erin Brockovich. *Time, 159*(20), 8.

Parmentier, R. (2003, September 25). [online]. Then & now: Iceland, Greenpeace, and whales, Part I. *Greenpeace.* Retrieved October 18, 2004 from http://www.greenpeace.org/international_en/features/details?item_id=319905&campaign_id=&print=1

Peterson, L. (2002, July). Erin Brockovich: The real-life sequel. *Biography, 6*(7), 66–72.

Peterson, T. R. (1998). Environmental communication: Tales of life on earth. *Quarterly Journal of Speech, 84*, 371–393.

Pretty Woman. (1990). Gary Marshall (Director). Touchstone Pictures.

Pulp Fiction. (1994). Quentin Tarantino (Director). A Band Apart, Jersey Films, and Miramax Films.

Radner, H. (2002). Pretty is as pretty does: Free enterprise and the marriage plot. In M. Collins, H. Radner, A. Preacher Collins, & J. Collins (Eds.), *Film theory goes to the movies: Cultural analysis of contemporary films* (pp. 56–76). London: Routledge.

Rapping, E. (1992). *The movie of the week: Private stories, public events.* Minneapolis: University of Minnesota Press.

Retzinger, J. (2002). Cultivating the agrarian myth in Hollywood films. In M. Meister & P. M. Japp (Eds.), *Enviropop: Studies in environmental rhetoric and popular culture* (pp. 45–62). Westport, CT: Greenwood Press.

Saturday Night Fever. (1977). John Badham (Director). Paramount Pictures and Robert Stigwood Organizations (RSO).

Schatz, T. (1992). The new Hollywood. In M. Collins, H. Radner, A. Preacher Collins, & J. Collins (Eds.), *Film theory goes to the movies: Cultural analysis of contemporary films* (pp. 8–36). London: Routledge.

Schlichtmann, J. I. (1999, March). To tell the truth. *American Bar Association Journal, 85*(3), 100.

Schueller, G. H. (2002). [online]. Wasting away: Is recycling on the skids? *Natural Resources Defense Council.* Retrieved October 18, 2004, from http://www.nrdc.org/onearth/02fal/recycling1.asp

Sefcovic, E. M. I. (2002, September). Cultural memory and the cultural legacy of individualism and community in two classic films about labor unions. *Critical Studies in Media Communication, 19*(3), 329–351.

Sexy. (2005). [online]. *The Oxford English Dictionary Online, 2nd ed., 1989.* Retrieved December 6, 2003, from http://www.letrs.indiana.edu/cgi-bin/oed-idx

Shamberg, C. S. (2000). [online]. *Erin Brockovich:* Behind the scenes, getting started. *Universal Studios.* Retrieved December 18, 2002, from http://www.erinbrockovich.com/scenes.html

Sierra Club Corporate Accountability Campaign. (n.d.). [online]. Corporate water privatization, bottled water campaign: What are the issues? *Sierra Club.* Retrieved October 18, 2004, from http://www.sierraclub.org/cac/water/bottled_water/

Silkwood. (1983). Mike Nichols (Director). 20th Century Fox and ABC Motion Pictures.

Sloan, K. (1988). *The loud silents: Origins of the social problem film.* Urbana: University of Illinois Press.

Smolowe, J., & Leonard, E. (1999, March 15–22). John Travolta. *People, 51*(10), 210–213.

Staudinger, C. (2003, August 5). Expert advice. *Woman's Day, 66*(13), 21.

Staying Alive. (1983). Sylvester Stallone (Director). Paramount Pictures.

Steel Magnolias. (1989). Herbert Ross (Director). Rastar Films and TriStar Pictures.

Tacey, E. (2001, April 2). Pampered star's omission unforgivable. *South China Morning Post,* p. 17.

Talbot, D. (1999). 'Civil' session with attorneys, author awes high school kids. *The Boston Herald,* p. 25.

Taylor, D. E. (1997). Women of color, environmental justice, and ecofeminism. In K. J. Warren (Ed.), *Ecofeminism: Women, culture, nature* (pp. 38–81). Bloomington, IN: Indiana University Press.

Trainer, J. (2000, March 19). [online]. Movie review: *Erin Brockovich. Normal Guys Movie Reviews.* Retrieved December 18, 2002, from http://normalguyreviews.com/review.asp?id=25

Travers, P. (2000). [online]. *Erin Brockovich*: The Rolling Stone review. *Rolling Stone.* Retrieved December 18, 2002, from http://www.rollingstone.com/reviews/moview/printer_friendly.asp?mid=73000

Travolta's Law. (1999, April 10). *The Irish Times,* p. 64.

Turan, K. (1998, December 25). [online]. *A Civil Action*: Legal entanglements. *LA Times.* Retrieved December 18, 2002, from http://www.calendarlive.com/movies/reviews/cl-movie981225-6.story

Turan, K. (2000, March 16). [online]. *Erin Brockovich*: The smile wins the day. *LA Times.* Retrieved December 18, 2002, from http://www.calendarlive.com/movies/reviews/cl-movie000316-43.story

Urban Cowboy. (1980). James Bridges (Director). Paramount Pictures.

Verniere, J. (1999). David, et al., V. Goliath; *Civil Action* bears witness to lawyer's transformation, but doesn't do justice to families' fight against the system [Movie review]. *The Boston Herald,* p. 55.

Walker, R. (1995). *To be real.* New York: Anchor Books.

Waller, G. A. (1987, Spring). Re-placing *The Day After. Cinema Journal, 26*(3), 3–20.

Warren, K. J. (1997). Taking empirical data seriously: An ecofeminist philosophical perspective. In K. J. Warren (Ed.), *Ecofeminism: Women, culture, nature* (pp. 3–20). Bloomington, IN: Indiana University Press.

Williams, R. (1961). *The long revolution: An analysis of the democratic, industrial and cultural changes transforming our society.* New York: Columbia University Press.

Williams, R. (1977). *Marxism and literature.* Cambridge, England: Oxford University Press.

Wood, J. T. (2003). *Gendered lives, 5th edition.* Belmont, CA: Wadsworth.

CHAPTER THREE

From Dualisms to Dialogism: Hybridity in Discourse About the Natural World

Tracy Marafiote
University of Utah

Emily Plec
Western Oregon University

> *In all ethical problems, we must consider the rules for community formation, but in environmental disputes, we must additionally understand how the disputants construct their views of the natural or nonhuman worlds. One group will view nature as a warehouse of resources for human use, while an opposing group will view human beings as an untidy disturbance of natural history, a glitch in the earth's otherwise efficient ecosystem. Between such extremes, there are any number of conventional or idiosyncratic constructions of the person–planet relation.*
> —Killingsworth & Palmer, 1992, p. 4

Scholars of environmental communication have fruitfully explored the ways in which civic discourses about environmental issues shape personal and public perceptions of those issues, reveal anthropocentric biases, frame the meaning of particular events, and encourage (or discourage) human action (e.g., DeLuca, 1999; Opie & Elliot, 1996; Ulman, 1996). Such studies add much to our understanding of the complexity of the ideological and discursive formations that constitute public ways of communicating about the environment. We should also attend to the ways private, colloquial, or vernacular expressions of human–nature relationships reflect and shape knowledge, attitudes, and behavior, for it is these discourses that "gird and influence local cultures first" (Ono & Sloop, 1995, p. 20).[1] Peterson and

---

[1] A number of communication scholars have recently explored the concept of the *vernacular* with different emphases but with an abiding focus on the discourses that emerge from and appeal to everyday people (see, e.g., Hauser, 1995, 1999; Ono & Sloop, 1995, 1999) "within local communities" (Ono & Sloop, 1995, p. 20). To expand our understanding of these and other types of private or informal en-

Horton (1998), for instance, illustrate the salience of this latter focus for the management of environmental disputes by attending to landowner discourses as an important part of larger conversations about land use.

Forms of informal or colloquial discourses about the natural world may share many features with dominant, civic environmental discourse, but they also demonstrate the complexity and contradiction of much linguistic consciousness.[2] For example, the natural world may be described as a resource for human consumption or a setting for outdoor entertainment, and, at the same time, as a site or living system that is, or should be, untouched by humans. Discourse about the natural world, like discourse generally, is composed of fragments of multiple social texts (McGee, 1990), ideologies (Hirschkop, 1989), or styles, and is, therefore, heteroglossic (Bakhtin, 1975/1996). Contrasting heteroglossia, Glover (2000) argues that most Americans' perspectives on environmental issues are influenced by official discourses such as public hearings and mass-mediated government accounts, and that these "official languages are by nature monologic" (p. 46); their intrinsic diversity is repressed in favor of coherence or unification (Bakhtin, 1975/1996; Hirschkop, 1989). Drawing upon Russian theorist Mikhail Bakhtin's concept of *heteroglossia*, Glover illustrates the ways in which official environmental discourses tend toward a monologic framework that obscures the diversity and suppresses, rather than elicits, the dialogic potential of any utterance.

> For Bakhtin, language, and particularly the language of the novel, is a rich and complex dialogue of voices, each voice having its own unique language arising out of its own ideology, culture, and social stratum. . . . Environmental issues, of course, elicit very diverse views on the world and its resources, ecosystems, and populations; and these diverse views are expressed through a variety of discourses and rhetorical approaches. (Glover, 2000, p. 37)

In contrast to the posited monoglossia of civic expression, various discourses and rhetorical perspectives are apparent within the heteroglossic colloquial utterance. The utterance, for Bakhtin, is complete in itself, whether it is an advocate's sigh, a politician's speech, or a series of nature stories. In the same moment, the

---

vironmental discourses used by and among "everyday people," we encourage the examination of multiple forms of colloquial discourses: conversational, informal ways that "everyday people" communicate both verbally or in written form (see Foertsch, 1998).

[2]The juxtaposition of informal and civic discourses is most clearly outlined by Ono and Sloop (1999, 2002), who argue that vernacular discourses, in particular, represent dominant and/or outlaw perspectives, depending upon the context, whereas civic discourses are those crafted to impart information to particular groups. They also suggest that outlaw logics that circulate widely in the civic sphere tend to become dominant logics. As an encompassing category, informal or colloquial discourses may include the vernacular, and may represent outlaw or dominant logics.

utterance is never fully autonomous, but is always in relation to another utterance (Danow, 1991). The particular sociohistorical context, therefore, is significant in interpreting any discourse. The variety of social influences and perspectives articulated in discourses taken together constitutes a *pastiche* composed of various cultural elements as well as personal experience. Examining a single utterance as a pastiche of fragmentary texts allows us to consider the "internal dialogism of discourse," which "inevitably accompanies the social, contradictory historical becoming of language" (Bakhtin, 1975/1996, p. 330).

As environmental communication scholars increasingly attend to the dialogic features of private or colloquial expression, theoretical perspectives that can guide the explication of symbolic environmental consciousness are necessary. In this chapter, we develop a dialogic perspective grounded in Bakhtin's (1975/1996) theorizing of *hybridity*, the combining of distinct socioideological voices within an utterance, and extend it to colloquial environmental discourse in order to broaden our understanding of human relationships with the natural world. Whereas Glover (2000) explores the possibilities of dialogism for civic environmental communication, and Hauser (1995, 1999) and Ono and Sloop (1995, 1999) argue the ideological significance of informal, private voices, our goal is to put such theoretical agendas together in order to examine the dialogic dynamics of verbal environmental communication at the individual level. Such an approach can illuminate the communicative processes that ground social and civic environmental communication—those processes undertaken by the individual consciousness: How do everyday women and men use socioideological language(s) to describe and explain their understanding of, and relation with, the natural world? We believe that the heteroglossic qualities of language, the multivocality of any particular utterance, and the sometimes complicated and contradictory quality of human thought exceeds the grasp of foundational environmental communication theories. We therefore explore this question using the linguistic theory of hybridity, and pursue the ways it can contribute both to understandings of environmental communication discourse and to environmentalist goals.

Bakhtin's conceptions of dialogic hybridity, both organic/unintentional and conscious/intentional, are primarily developed within *Discourse in the Novel* (1975/1996). While evocative, his discussion of organic (or unintentional) hybridity, the basis for the first part of this chapter, is somewhat sparsely detailed in comparison to his delineation of the conscious (or intentional) hybrid. The conscious form of the hybrid, referenced in the last section of this chapter as a foundation for a discursive environmental strategy, is "artistic" and "systematic," as opposed to the organic hybrid, which is "mute and opaque," yet "profoundly productive" (Bakhtin, 1975/1996, p. 360). The productive potential of hybridity has received significant attention within literary and postcolonial cultural criticism as a means through which to account for colonizers' discursive constructions of the "Other," as well as a model of cultural exchange in which different identities may converge into forms that contest dominant cultural power (Bhabha, 1995; Young, 1995).[3] Outside of these forms of criticism, however, hybridity has been minimally developed as an analytic concept; we believe that its utility in understanding the complexity and implications of various discursive forms contributes a critical theoretical lens to environmental communication research. Consequently, this chapter focuses on developing hybridity as a theoretical framework—one which, in its analytical form, may be utilized effectively when complemented with current theories within environmental communication.

The chapter's first section lays a foundation for this effort by reviewing the capacity of some existing theories of environmental discourse and consciousness. The current environmental models include Evernden's (1992) discussions of the nature/culture dualism and materialist and idealist monisms, and Herndl and Brown's (1996) triad of "centrisms": anthropocentrism, ethnocentrism, and ecocentrism. We find that these prominent theoretical lenses provide foundational, yet primarily monoglossic, frameworks for identifying specified discourses within the individual utterance.

After briefly examining the ways in which these perspectives attend to certain elements of discourse about the natural world, we turn in the second section to theoretical concepts that allow us to examine the intersections of discursive elements that are not fully accounted for by monologic frameworks. Our approach, like Glover's (2000) and Hauser's (1995), is ultimately grounded in the literary theorizing of Bakhtin, whose dialogic perspective helps us account for the polyvocality within discourses about the natural world. Specifically, we utilize Bakhtin's concept of *organic hybridity*, the unintentional mixing of distinct voices, linguistic consciousnesses, or syntaxes within an utterance (Bakhtin, 1975/1996). Bakhtin (1975/1996) asserts, "in essence, any *living* utterance in a *living* language is to one or another extent hybrid" (p. 361, emphasis in original). Pursuing this contention, we develop a dialogic approach grounded in organic hybridity, which begins to account for the ambiguity, complexity, or ambivalence within utterances, and functions as a complementary addition to theories of environmental communication.

In the third section of the chapter, we briefly demonstrate the theoretical potential of hybridity by applying it to examples of colloquial environmental discourse drawn from a previous study. These utterances, also briefly used in the first section to illustrate the existing theories of environmental discourse, were collected as part of a study of nearly 100 open-ended, self-report questionnaires completed by three groups of college students in order to explore their percep-

---

[3]The introduction of hybridity to postcolonial cultural criticism is widely credited to Homi Bhabha's (1994) *Location of Culture*. In his (also widely cited) text, *Colonial Desire: Hybridity in Theory, Culture, and Race*, Young (1995) asserts, "Bakhtin's intentional hybrid has been transformed by Bhabha into an active moment of challenge and resistance against a dominant cultural power" (p. 23; see also Ashcroft, Griffiths, & Tiffin, 2000).

tions of humans' relationships with the natural world.[4] These discursive illustrations are not intended to represent generalizable views of the environment, but rather provide examples of the types of layered, heteroglossic discourses about the natural world that we attempt to account for within the theoretical framework of hybridity.

In the chapter's final section, we address the significance of hybridity as an environmental communication theory, and suggest directions for future research. In particular, we propose strategic hybridity as an approach grounded in Bakhtin's conscious form of hybridity that mobilizes divergent values in the interests of environmental advocacy. Building on organic hybridity, strategic hybridity is a rhetorical and applied communication strategy that allows us to identify and utilize intersections and overlapping constructions among divergent environmental discourses with the aim of advancing the goals of environmental movements.

THEORIES OF ENVIRONMENTAL COMMUNICATION

Several notable theories of environmental communication provide foundational explanations of discourse about the natural world. These theories identify critical forms of environmental discourse that reveal particular socioideological perspectives, and may reflect certain characteristics of dominant, civic environmental discourse. Included among these theoretical frameworks are nature/culture dualisms as well as a different and more expansive approach found in the concepts of *idealistic* and *materialistic monisms* (Evernden, 1992). Also, Herndl and Brown's (1996) concepts of *ethnocentrism, anthropocentrism,* and *ecocentrism* provide perspectives that help explain a broader spectrum of attitudes and discourses about

---

[4]These discursive illustrations are drawn from our study entitled "College Students' Perceptions of Humans and Nature: An Inquiry into Identification with the Natural World," in which 93 open-ended, self-report questionnaires from three academic areas have been analyzed: Business (32), Parks, Recreation and Tourism (28), and Writing (33). The project's goals include explorations of students' perceptions of humans' relations with the natural world and how analyses of these perceptions could inform academic, civic, and environmental discourses. Utterances quoted in this chapter are cited by department (B-Business; PR-Parks, Recreation and Tourism; WP-Writing Program) and survey number; for instance, WP.a2 indicates survey number two from the "a" group of Writing Program respondents.

We believe that, despite obvious limitations, students provide a unique opportunity to examine a wide range of educational interests and goals, which may indicate a broad array of social ideologies, perspectives, and discourses, particularly those relevant to humans' relations with the natural world. A breakdown of the solicited demographic information includes: 40 female and 48 male respondents (of 93) identify their sex. Although ranging from 18 to 44 years old, all but one of the participants stating an age (75 of 93) are under age 30, and nearly three quarters are under age 24. Of those identifying a religious affiliation (53 of 93), nearly three quarters name a Christian religion (e.g., Baptist, Catholic). More than half of these respondents (29 of 53) identify themselves as members of The Church of Jesus Christ of Latter Day Saints. Additional information about questionnaire data is available from the authors upon request.

the environment. In this section, we review these three theories' important contributions, as well as their limitations, for understanding discourse about the natural world.

The Nature/Culture Dualism

Binary constructions such as the nature/culture dualism have been problematized within environmental communication literature (as well as within other theoretical areas) due to their reductive, potentially damaging, and inherently value-laden nature (Curtin, 1997; Evernden, 1992; Oelschlaeger, 1991; Wells & Wirth, 1997). Yet, the influence of such dualisms is pervasive in current, as well as historical, Western conceptions of the natural world. Evernden (1992) explores the origin and influence of the nature/culture dualism in some depth, concluding that although it has undergone major ideological transformations, its impact on human experiences and perceptions of the natural world remains relevant and undeniable.

Examining colloquial discourse through the lens of dualisms provides an understanding of the pervasiveness of perceptions of the natural world as exclusive of humans and human constructs, and vice versa. This separation is illustrated both implicitly and explicitly in descriptions of the natural world that may define it as "the trees, plants, animals, waters, rocks ... the universe" (PR.b2), "that sphere which does not include man or man-made things" (WP.a2), or "everything that man did not build. Like the mountains, forests, and etc." (PR.a5). Evernden (1992) explains the lingering discursive power of the nature/culture dualism, stating that "unless we have this absolute separation, we cannot claim unique qualities that justify our domination of the earth" (p. 96). The assertion is especially poignant given the ubiquity of voices expressing this nature/culture distinction; whereas the distinction historically positions humans, as reasoning agents, as having dominion over the environment, we maintain our control over the natural world only by attentively monitoring and reifying the categorical borders.

Although the recognition of the prevalence of the nature/culture dualism does have significant implications for environmental communication research, the binary also poses difficulties for making sense of multifaceted human/nature relations. The conclusions reached from this approach alone are incomplete in that they cannot account for diverse, even ambivalent, socially and historically situated voices. Rather, viewed through the lens of the nature/culture dualism, colloquial utterances appear primarily monoglossic, that is, as "language whose natural [heteroglot] tendency is repressed or obstructed by some external force" (Hirschkop, 1989, p. 5). Put another way, recognizing that private, informal discourses express views of humans as separate from the natural world is an informative perspective in examining perceptions of the natural world; it is also, however, insufficient to account for the complexity of environmental discourses.

Materialistic and Idealistic Monisms

In further examining the implications and evolution of the nature/culture dualism, Evernden (1992) asserts that we are in fact moving into a "postdualistic" era. This is a cultural setting in which environmental consciousness is dominated by two competing, seemingly holistic, "monisms": modern materialism and modern idealism. Of the two, the predominant outlook, he argues, is the materialistic monism. The monism of modern materialism represents a view that foregrounds humanity, and human interests and needs, as the controlling historical force that encompasses and subordinates nature. Conversely, the idealistic monism views the natural world as the primary unifying force that encompasses humans and their constructs as secondary elements within the natural order (Evernden, 1992). These models shift views of humanity and the natural world from locations at the ontological poles of a binary, to allied, interconnected forces. This perspective is illustrated, for example, in the comment: "all aspects of my existence are interactions with the natural world, for is anything really 'unnatural'?" (B.b6). In discourse viewed through monistic lenses, such relational views and questions are illuminated.

The theoretical perspective of monisms differently frames views of the explicit and implicit relations between humanity and the natural world. A dilemma arises, however, when we embrace one or the other monism, as the framework insists that we must either accept that nature is a subcategory or construction of humanity, or that humans and all human constructs are a subcategory within the larger genre of nature. As Evernden (1992) notes, "the attempt to explain all by reference to either of the polar monisms appears to lead to doomed attempts to explain the whole by reducing it to a part, with the usual consequences" (p. 95). In other words, rather than resolving dueling dualistic discourses, the modern monisms unify the ontological poles of nature and culture through reducing the binary to a domination of one over the other. Ultimately, then, the notion of postdualistic monisms leads to many of the same human/nature hierarchical dilemmas and distinctions as the dualisms themselves.

The monisms do offer further insight into particular perspectives revealed in environmental discourses, specifically those encompassing interconnected views of a human/nature relationship. Alone, however, they do not offer a theoretical lens through which we might see beyond the either/or monologue of the nature/culture dualism. Although the monisms differ from the dualisms in that they allow for the identification of discourses that express a connection between nature and culture, they do not account for discourses that espouse that relationship as mutual, egalitarian, or variable. On one hand, within this monologic frame, the monisms encourage more thoughtful understandings of potentially monoglossic discourses—as "the product of the dialogical struggle between opposing tendencies" (Crowley, 1989, p. 37)—as opposed to two cleanly distinctive ontological categories. On the other hand, instead of providing a broader or more fluid alternative to a dualistic view, the monisms in effect reproduce a monologic framework that offers little room for investigating overlap, contradiction, and complexity within discourses.

A Triad of "Centrisms"

To complement the dualistic and monistic perspectives discussed by Evernden, and to help us further theorize the multiplicity in environmental discourse, we turn to Herndl and Brown's (1996) triangular mapping of environmental discourse. In their introduction to *Green Culture*, Herndl and Brown assert that discourses about the natural world cannot be fully understood through simple dichotomies. Because a single text can function in multiple ways, it may be more effectively examined using rhetorical approaches that attempt to account for its diversity. An effort to facilitate analyses of environmental discourses that might consider, for instance, cultural and institutional influences, the "centrisms" identify three prevailing orientations within environmental discourse: anthropocentric/scientific, ethnocentric/regulatory, and ecocentric/poetic (Herndl & Brown, 1996). The triadic model provides broader, yet more focused, lenses through which to theorize heteroglot discourse about the natural world.

One point of Herndl and Brown's (1996) discursive triangle represents anthropocentric, or scientific, discourse. This is discourse that espouses a human- or culture-centered perspective and "locates the human researcher as outside and epistemologically above nature" (p. 11). Further, anthropocentric/scientific discourse positions the natural world as an object of knowledge, framing colloquial utterances that posit, for instance, that the natural world "is not only nature itself, but the behavior [*sic*] and educational extent to which it is known and used" (PR.b4).

In its privileging of humanity, the anthropocentric view can be understood as corresponding in some ways with the culture component of the nature/culture dualism as well as with the materialistic monism, as binary thought patterns often indicate a greater valuing of one element in a pair over the other (Wells & Wirth, 1997). For example, the dualism frequently invokes what has been referred to as a masculine bias, privileging scientific over situated ways of knowing (Curtin, 1997; Wells & Wirth, 1997). As Cantrill (1996) notes, outside of "ardent environmentalism . . . our culture still supports a traditional paradigm of growth, progress, and anthropocentrism" (p. 85). On one hand, an anthropocentric/scientific lens allows the examination of the human-centeredness identified within the dualistic frame from a different viewpoint. On the other hand, anthropocentric ways of knowing are often limited in their potential depth and complexity because they, like dualisms and monisms, rely upon either/or conceptions of humanity and nature, which may encourage both a privileging of human culture and a converse subordination of the natural environment.

A second element of the triad is ethnocentric, or regulatory, discourse, which "regards nature as a resource . . . to be managed for the greater social welfare" (Herndl & Brown, 1996, p. 10). This element of the model frames utterances that position the natural world as a resource or setting for human use, action, recreation, or pleasure. Ethnocentrism is a legacy of Gifford Pinchot's utilitarian conservationism (Herndl & Brown, 1996) that, like those forms of discourse highlighted through the lens of the materialistic monism, positions the natural world as controlled by and therefore subordinate to human culture; as such, it also reflects the dualistic view of the nature/culture relation seen in anthropocentric discourses. An instance of such assessments is seen in the confident assertion that "the world has been placed here for my benefit. It is a way for me to learn, grow and gain experience. There are so many beautiful things to see and new technology to be discovered in our world" (B.b15). This utterance represents a perception of human dominance over the natural world, and a privileging of human need and advancement. Declarations of approval not withstanding, the natural world here is positioned as "other," fulfilling some human need or desire.

Utilizing ethnocentric discourse as a theoretical frame provides a means to further identify and examine particular discursive constructions of the natural world that focus upon its uses in relation to humanity; views that could not be understood through the dualistic lens only. As with anthropocentric discourses, however, this discourse focuses upon the relation of humanity and the natural world in either/or terms and, by itself, limits the possibilities of identifying complex voices and views within utterances.

Finally, ecocentric, or poetic, discourse represents the third element of the triad. This is a form of discourse that brings into focus "the beauty, value, and emotional power of nature" (Herndl & Brown, 1996, p. 12), and which typically positions humans as part of the natural world. This can be seen in assertions that: "the natural world contains all events" (B.b2), is "the world in which we all live" (PR.a22), and is "everything we can feel, touch, taste, hear and see" (B.a15), including both "urbanization" (PR.b25) and "dirt, soil, trees, water, air, food, honesty, peace" (B.a8). Conversely, however, ecocentric expression may occasionally be located within utterances indicating a distinction between humans and nature. Ironically, this dualism may be evident in discourse containing "aesthetic or spiritual responses" to nature's beauty or emotional value (Herndl & Brown, 1996, p. 12), such as this statement: "The natural world teaches to me the concepts of power, beauty, deception, amazement, miracles. In the end I yearn for a more natural life in the natural world where I am directly a part of the ecosystem" (PR.b1). While this speaker idealizes nature, he also perceives himself as separate from it.

Regardless of how this relationship is understood, in its valuing of nature, ecocentrism intersects with both the idealistic monism and the nature side of the nature/culture dualism. As such, ecocentric discourse shows less evidence of the anthropocentrism highlighted by the other points of the triangular, centrism model; instead, these voices illustrate discourse that "seeks to locate human value in harmonious relation to the natural world" (Herndl & Brown, 1996, p. 12).

By identifying foundational perspectives or forms of discourse within a single utterance, Evernden's theories of the nature/culture dualism and materialistic and idealistic monisms, as well as Herndl and Brown's model of centrisms, provide significant models for examining discursive constructions of the natural world. The primary limitation of these concepts is their tendency toward monoglossic interpretation, while, we argue, environmental discourses often contain the intermingling of different languages or belief systems, and as such are heteroglossic. Herndl and Brown (1996) recognize, however, the likelihood of multiple forms of discourse within an utterance, noting for instance that "nature writing often . . . connect[s] scientific knowledge to a spiritual sense of nature and beauty. . . . to pathos as well as to reason" (p. 12). To begin to account for such discursive conjuncture, the centrism model builds upon the insights of the dualisms and monisms in offering, with its triadic form, a more complex framework for examining discourse than these former theoretical approaches. The model elaborates options for identifying different social discourses as well as their categorical intersections, although it provides little explanatory power for the overlap of the three centrist discourses, or for the presence of voices unaccounted for by the three centrisms.

Herndl and Brown (1996) concede that their model "identifies only the dominant [discursive] tendencies," and recognize that the discourses framed by this triad "are not pure" (p. 12). It is, in fact, the very heteroglossic impurity of discourse—which cannot be fully accounted for within the above theoretical perspectives—that we further elaborate using Bakhtin's (1975/1996) theory of dialogue, and specifically, his concepts of *heteroglossia* and *hybridity*. Bakhtin's notions of heteroglossia and the utterance help us to theorize contemporary cultural, discursive practices, and to disclose the ways in which discourses are contextualized by historical conditions (Danow, 1991).

INTERNAL DIALOGISM AND HYBRIDITY

As noted earlier, Bakhtin's understanding of language, speech, and the novel provides insights into the intricate discursive intertextuality of utterances. Pechey (1989) points out that dialogue, as we commonly think of it, is an "epiphenomenon of a generalized inner dialogism of all discourse" (p. 46). Thus, the multivocality of inner speech projects outward, expanding the range of voices and perspectives, contributing to the pastiche of social discourse and, as Glover (2000) notes, enriching the possibilities of public environmental participation. As opposed to an understanding of dialogue as occurring externally between persons, groups, or texts, it is this inner dialogism—particularly as it is manifested in the private or colloquial utterance—that is largely unaccounted for in studies of environmental discourse.

Dialogue and Heteroglossia

Bakhtin labels one element of internal dialogism "internally persuasive discourse," a process marked by the interplay of distinct socioideological voices combined with "one's own word" (Bakhtin, 1975/1996, p. 345). Internal discourse or dialogue influences the development of individual consciousness from an array of social ideologies. Rather than dissolving into a coherent and monologic perspective, these discourses engage in struggle within the speaker's consciousness for ideological dominance. Bakhtin explains that a person's consciousness is based in these competing socioideological languages, and "language, for the individual consciousness, lies on the borderline between oneself and the other" (p. 293). In this way, discourse is never completely one's own, but is always heteroglossic—both one's own *and* other. The study respondent, for example, who states "I like to go . . . exploring but it's getting harder to do so. You can't go off trail, for you'll destroy plant life. We are becoming very limited in what we are able to do" (PR.b13) illustrates the interactions between public discourses about, as well as personal experience with, the natural world. In these brief sentences, she acknowledges and verbalizes civic and environmentalist discourses of environmental protection ("do not go off trail") interspersed with concern over limitations of personal pleasure and action, perhaps influenced by voices disparaging limitations on access to public land. Bakhtin (1975/1996) explains that prior to being appropriated by this speaker, these words were not "neutral and impersonal," but were spoken by others and "populated—overpopulated—with [their] intentions" (p. 294). The presence of such distinct socioideological languages influencing and populating environmental discourses characterizes one conception of heteroglossia (Bakhtin, 1975/1996; Danow, 1991).

Heteroglossia is the interplay among various social discourses within a national language (Bakhtin, 1975/1996). This common definition, however, is only partially complete. An equally essential characteristic is the sociohistorical context of these voices, languages, or utterances. Heteroglossia, Holquist (1996) explains, is a "function of a matrix of forces." That is, it is located in the intertextuality of "a particular set of conditions—social, historical, meteorological, physiological—that will ensure that a word uttered in that place and time will have a meaning different than it would under any other conditions" (p. 429). Consequently, heteroglossia is not simply the presence of various socioideological voices, but the presence of a *particular set* of socioideological voices and conditions. For the colloquial environmental discourses sampled later in this chapter, those heteroglot voices and conditions include the speakers' demographics and social and cultural environments.

A closely related conception of heteroglossia highlights the contextualized, un-repeatable moment and place as the intersection of centripetal and centrifugal tensions or forces (Holquist, 1996). *Centripetal forces*, which strive to centralize meaning, illustrate the utility of strategic essentialisms[5] while *centrifugal tendencies*, which decenter and disrupt meaning, undermine such essentialisms with a cacophony of multilanguaged dialogue (Bakhtin, 1975/1996; Danow, 1991). These ongoing tensions among social discourses are critical "generative forces of linguistic life" (Bakhtin, 1975/1996, p. 270), and provide rhetorical spaces for the examination and engagement of environmental issues.

By examining the heteroglossic, centrifugal forces that challenge centralized (centripetal) monologic discourse, we can begin to see the interaction of multiple socioideological discourses within single utterances. As Bakhtin (1975/1996) explains:

> A language is revealed in all its distinctiveness only when it is brought into relation-ship with other languages, entering with them into one single heteroglot unity of so-cietal becoming. Every language . . . is a point of view, a socio-ideological conceptual system of real social groups and their embodied representatives. (p. 411)

Because the heteroglossia of socioideological languages is formed by the process of hybridization (Morson & Emerson, 1990), those distinct voices of the "heteroglot unity" are then brought into play in singular utterances as hybridity.

From Heteroglossia to Hybridity

Both drawing on and (re)producing the heteroglossia of language, competing socioideological discourses come into play as hybrid constructions, or the merg-ing of two voices or belief systems, in the arena of the utterance (Bakhtin, 1975/ 1996; Holquist, 1996). The competing languages of heteroglossia "do not *exclude* each other, but rather intersect" (Bakhtin, 1975/1996, p. 291, emphasis in origi-nal). When these intersections occur without conscious reflection within the ut-terances of individual speakers, organic hybrids, "new socially typifying 'lan-guages,' " are fashioned (p. 291). These are sites in which new forms for seeing the world, new forms of consciousness, are born. In this way, in addition to being the source of hybridity, the heteroglossia of language is formed by the processes of hy-bridization, although this unconscious combining of languages is seldom appar-ent to the speaker (Morson & Emerson, 1990).

Just as "language is heteroglot from top to bottom" (p. 291), Bakhtin (1975/ 1996) asserts that all language is "to one or another extent hybrid" (p. 361).

---

[5]Essentialism is commonly viewed as a problem of language and ideology because it represents a reductionist tendency. Some theorists and critics, however, advocate strategic essentialisms as tactical identities through which we may secure power (e.g., Spivak, 1988).

Hybridity occurs in multiple, versatile manifestations; here, we focus on the categories of organic/unconscious and intentional/conscious hybridity. A purposeful merging of socioideological voices, based in the conscious form of hybrid discourse, is developed within the final section of this chapter in which we propose the notion of strategic hybridity as an environmental strategy. The organic or unintentional hybrid, which we locate and develop in the next section in examples of colloquial environmental discourse, occurs when speakers unconsciously mix two voices, linguistic consciousnesses, belief systems, or even syntaxes, in a single utterance in order to make sense of an experience that may elicit ambivalent beliefs, thoughts, or reactions (Bakhtin, 1975/1996; Morson & Emerson, 1990); it is, as characterized by Young (1995), "a living heteroglossia" (p. 22).

Given its inherent multiplicity, the organic hybrid provides a foundation for identifying and analyzing the complexity of private or colloquial discourses about the natural world, in particular, the implications of individual and internal dialogues that are not adequately accounted for solely by applying the theories of environmental communication already reviewed. Building on the insights gleaned through dualisms, monisms, and centrisms, hybridity brings to environmental communication a means through which to examine the polyvalent contradictions and complexities present within individuals' utterances. More specifically, in drawing on heteroglossic and organic hybrid notions of dialogue in order to theorize environmental discourse, we posit that, in conjunction with the aforementioned theories of environmental communication, hybridity's explanatory potential *exceeds* the presence of only two voices, linguistic consciousnesses, or socioideological positions within the utterance, as it is defined by Bakhtin. In surpassing this definitional limitation, organic hybridity accounts more fully for a range of heteroglossic presences within environmental discourse.

From this expanded framework, rather than identifying evidence of *either* nature *or* culture, *either* materialistic separation *or* idealistic connection, *either* anthropocentrism, ethnocentrism, *or* ecocentrism, we seek evidence of all of these, as well as other heteroglot voices that may be present in discourse about the natural world. From a Bakhtinian perspective, limiting discursive possibilities to one of two options is fundamentally problematic. Such restrictions serve, centripetally, to centralize meanings and subdue heteroglossia (Bakhtin, 1975/1996; Danow, 1991; Morson & Emerson, 1990). This move usually serves the interests of monologic discourses, such as the official languages examined by Glover (2000), which silence voices that do not belong to dominant, public-discourse communities. In contrast, as we develop it here, organic hybridity acknowledges and searches for the multiplicity of discourses and socioideological views within linguistic utterances about the environment. Hybridity, then, whether organic or conscious, acts as a centrifugal force in decentering and disrupting limited, monoglossic discourse or dualistic meanings, and helps us to locate occurrences of linguistic and ideological disjunctures.

ORGANIC HYBRIDITY IN DISCOURSE
ABOUT THE NATURAL WORLD

As expressed, *organic hybridity* is an unconscious process in which various socio-ideological discourses are combined by a speaker in a single utterance. Conversely, implementing hybridity as a conceptual, analytic framework allows the environmental communication researcher to "work backward" to deconstruct the utterance. This interpretive process can facilitate the identification of differing socio-ideological voices in the utterance in order to supplement understandings of how persons make sense of the natural world, as well as experiences and issues related to it. It may also allow identification of those social discourses that are most successful in the heteroglossic struggle for internal, linguistic, ideological dominance (Bakhtin, 1975/1996), and therefore most influential in how persons understand and talk about the natural world. Further, because it seeks to situate the speaker or author of the utterance within particular sociohistorical contexts, hybridity is not only an interpretive apparatus but also provides an explanatory framework.

Throughout the theoretical discussions in this chapter, we positioned the move from dualisms to dialogism as located along a continuum from less to more complex theoretical lenses, and from a theoretical lens that frames utterances as monoglossic to one that brings into view the multivocality of discourse. We insist, however, that the "more complex" approach of hybridity does not mitigate the significance of the insights that can be gleaned from the other theories, but in fact embraces and is usefully informed by them. In this section, we illustrate the utility of hybridity and how it is supplemented through the use of the theoretical lenses of dualisms, monisms, and centrisms. We believe that the incorporation of a variety of frameworks into hybridity effectively complements Bakhtin's conception of heteroglossia in that it brings together multiple theoretical voices for the purpose of gaining a more meaningful understanding of the multiple languages present in environmental discourses.

Discourse Samples

For Bakhtin (1975/1996), the languages that are present and dominant within speakers' utterances are informed by their sociohistorical subject positions. Likewise, every utterance is "located at the intersection of a number of circumstances: nation or region, class, gender, profession, age, historical moment, leisure or labour" (Hirschkop, 1989, p. 21). The demographics of the speakers whose voices are to be excerpted, therefore, are pertinent for a rich interpretation, not only of the forms of discourse present in an utterance, but also of the significance of these discourses for environmental researchers and advocates. In considering these expressions of hybrid discourse, we emulate Bakhtin's concise conceptual illustrations by referencing "passages" from the participants' utterances in which hybridity is "immediately comprehensible out of context" (Morson & Emerson, 1990, p. 330).

The study from which these colloquial discourses are drawn is one part of our efforts to further map some of the contours of environmental discourse. In that project, we solicited college students' definitions of "the natural world," and descriptions of their interactions and relationships with it. The discourse samples cited (by department and survey number) are excerpted from the responses of these students, who were enrolled in courses in three distinct departments at a large, western university. In addition to area of study, other social demographic factors influence these discourses. For instance, although the gender distribution is reasonably balanced, the participants are largely young adults (under age 24). Furthermore, of the respondents stating a religious affiliation, nearly three quarters identify as Christian and over one half cite membership in a conservative Christian religion centered in the predominantly Republican state in which the university is located. In sum, participants in the study are primarily young, Christian, American college students from a university located in the western United States; the utterances excerpted in this chapter closely reflect the study's overall demographics.

In addition to these demographics, the respondents' geographical and political locations, ages, and religious orientations, along with popular discourses about the environment, all shape their discursive utterances in relation to the natural world. For example, relevant issues could include conservative political and social rhetoric pertaining to water rights, U.S. dependence on oil, or stewardship of and access to public lands, as well as a history of the sagebrush rebellion and Western ideologies of rugged individualism. Besides regional languages, national socio-ideological discourses on each of these issues as well as on human relations with the environment are all potential influences on the consciousness of and voices uttered by these speakers. In this way, whether organic or conscious, hybridity invokes the articulation of self and society. With the particular convergence of social forces influencing the forms of discourse present in any utterance, these specifics are significant for the analysis to come. Consequently, utilizing hybridity as an interpretive framework in order to more fully account for the complexity in the colloquial environmental discourse of speakers or authors, such as these future consumers and potential advocates, can provide critical insights for environmental movements.

Discourse Analysis

Hybridity refers to the merging of multiple speech styles, socioideological languages, or forms of linguistic consciousness, so its presence in organic discourse may be revealed in various ways. It may be as simple as the voicing of a word that "belong[s] simultaneously to two [or more] languages, two [or more] belief systems that intersect" in the speaker's utterance (Bakhtin, 1975/1996, p. 305). Or, hybridity can be as complex as the juxtaposition of opposing values within a single utterance. Like heteroglossia, these conceptualizations of hybridity underscore the presence of multiple social discourses within any language, and consequently, within the consciousness of any individual speaker.

Although hybridity is present to some extent in all language, its specific composition and the extent to which it is evident varies. As seen here, the convergence of a range of languages or belief systems may overtly occur in instances in which a speaker's primary language is modified by the presence of another language (Morson & Emerson, 1990). For example, the study's respondents reveal the internalization of public, governmental discourses evidenced by statements such as "I try to 'take only pictures, leave only footprints.' I'm a big advocate for not littering" (PR.a1), or "I also keep to the saying tread lightly" (WP.a5). The repetition of slogans such as "take only pictures . . ." and "tread lightly," as well as numerous references to recycling and to not littering,[6] reveal the power of dominant, civic languages to influence popular views of the natural world, particularly for members of a group that have grown up with these expressions as part of an international vocabulary surrounding outdoor experiences. As Bakhtin (1975/1996) stresses, "all words have the 'taste' of a profession, a genre, a tendency, a party, a particular work, a particular person, a generation, an age group, the day and hour. Each word tastes of the context and contexts in which it has lived its socially charged life" (p. 293).

Hybridity not only acknowledges the presence of two or more discursive voices, it encourages the identification of these languages as sociohistorically situated in order to better interpret the socioideologically influenced perspectives of the speaker. Compared to the distinct presence of civic languages in the previous paragraph, a more subtle instance of the confluence of discrete languages is illustrated here:

> I consider myself as an integral, although rather a minor, part of the natural world. I feel that as I spend time getting to know the world in which we live I receive strength and understanding from the powers which formed her. My most incredible experiences have been on or near water flowing. I seem to find unbelievable mystery and power within and beneath the water. . . . [The natural world is] all the beautiful and intricate creations that have been created by an even more beautiful and intricate Creator. (WP.a8)

Highlighting the articulation of self and society in the discursive expression, the primary language in this young man's speech aligns with his self-identification as a

---

[6]It is important to note that the cited illustrations, although containing necessarily unique combinations of socioideological languages based upon the speakers' particular subject positions, are not unusual or atypical discourses or views on the natural world expressed within the study; each example can be further supported by multiple respondents voicing similar languages.

member of the locally predominant conservative Christian religion; elements of religious discourse run throughout the utterance, for instance, presenting a female earth created by (a presumably masculine) God. Still, the heteroglossia upon which he unconsciously draws is constituted in the "co-existence of social-ideological contradictions" (Bakhtin, 1975/1996, p. 291). This is evident as the language of religion converges with discourse that may be influenced by his out-door activities of river-running, skiing, and backpacking; ecocentric discourse, particularly in relation to water, that expresses the value and emotional power of the natural world (Herndl & Brown, 1996).

In the organic hybrid, the various socioideological languages present in the ut-terance rarely remain fully distinct, rather, as they are merged, each language is in-fluenced by the others (Bakhtin, 1975/1996). The Judeo–Christian tradition, an influence on the majority of these participants, has historically promoted either domination or stewardship of the environment, as well as the attendant percep-tion of humans as separate from and often superior to the natural world (Bram-well, 1989; Killingsworth & Palmer, 1992; Opie & Elliot, 1996). Consequently, al-though this speaker expresses a view of nature as "a spiritual or transcendent unity" (Herndl & Brown, 1996), its apparent stature is limited to surpassing that of humans. He also values the natural world as a subordinate construct of his "Creator"; a perspective still in line with common Christian views of the environ-ment. Solely classifying his speech as either religious or ecocentric, therefore, would not fully account for the qualifications on each of these perceptions. For in-stance, his expression does not include the element of ecocentric discourse that of-ten positions humanity as a part of the transcendent natural world and human value in relation to it (Herndl & Brown, 1996). Rather, the moderating presence of the voice of Christianity locates both humans and nature as subordinate to a more powerful Creator. Hybridity often occurs as the combining of socioideological voices in order to help the speaker make sense of different experiences, such as this river guide and kayaker's fascination with water's "unbelievable mystery and power." As an interpretive framework, hybridity may also provide a unique un-derstanding of the ways multiple socioideological languages may merge with and modify religious discourse in relation to the environment.

In addition to identifying divergent discursive voices, hybridity enables investi-gation of the implications of discordant values present within the single utterance. Because the outward manifestation of respondents' internal dialogues is a blend-ing of diverse socioideological perspectives, its natural heteroglot tendencies may be expressed as seemingly contradictory views and beliefs. An illustration of such ambivalence is seen in the utterance of a Parks, Recreation, and Tourism major, which begins with a description of the natural world common among the respon-dents: "to me the natural world is that *without manmade utilities or objects*. Where you are one with nature and you can only use nature, etc." (emphasis added). This statement contains the ubiquitous expression of the nature/culture dualism, voic-ing a larger, social opposition between nature and civilization/humanity. She continues, however, " 'the natural world' is what the world naturally is! Vegetation of all sorts, animals, and *even humans*" (emphasis added; PR.b13). Here, in opposition to frequently expressed understandings of the nature/culture dualism, she conveys a perspective of humans as a component of nature. Implicit in her discourse is a distinction between humans—as well as vegetation and animals—as first-order natural constructs, and human artifacts as second-order *un*natural creations. This is a convergence of views that is not adequately accounted for by the monologic framework of the nature/culture dualism; instead, it is better explained as an example of the materialistic monism, which understands humans and the environment as unified, yet foregrounds humanity as the dominant, controlling, historical force (Evernden, 1992). Corresponding with the view of dominant humanity, this utterance also contains ethnocentric discourse in the comment "where you are one with nature and you can only use nature," that positions the natural world as a resource or setting for human use (Herndl & Brown, 1996).

This young woman is the same speaker whose concerns over the limitations of activities in nature were cited in the previous section, "Internal Dialogism and Hybridity." Interestingly, although she states that her interaction with the natural world "is somewhat limited," similar to those quoting the environmental slogans ("take only pictures . . . ," etc.), she voices the languages of public, governmental discourse related to the environment in her statement: "you can't go off trail, for you'll destroy plant life" (PR.b13). Such links between the particular situatedness of a speaker and the content of her discourse further illustrate the significance within hybridity of the intersections of speakers' social subject positions. In this case, for instance, it could be meaningful to environmental advocates to identify that persons who are less involved with the natural world may still be influenced by discourses about nature—yet may also perceive environmental objectives such as not leaving trails as restrictive. On one hand, this speaker's familiarity with this discourse may be an element of her membership in a generation that has not experienced U.S. environmental issues prior to the existence of the Environmental Protection Agency (EPA) and significant environmental regulation; consequently, such guidelines may be commonplace to her. On the other hand, she is a resident of a region in which conservative political and social discourses challenging any perceived limitations on access to public lands are relatively commonplace.

As conceptualized here, hybridity may be simple or complex; the discursive intersection may occur within a word, two speech styles, various socioideological languages, or divergent belief systems. Or it may simultaneously appear in all of these forms within a single short utterance. Examined through the lens of hybridity, the following utterance, quoted in full, reveals the organic merging of each of these possibilities. Similar to the young woman's description of nature, this respondent defines the natural world as "the world as is, created naturally, without interference of any sort by humans." He further contends, "I like nature, I appreciate nature. I do not take it for granted. I love being far away, with only raw nature surrounding me. Hunting, fishing, back to nature. That is one way for me to get peace of mind and really relax. Unless I am starving or dying" (B.b22). In his initial description of the natural world, this young man repeats a perspective that is shared with a majority of the study participants; a dualistic view of nature as necessarily distinct from humans, human constructs, or human impact. This general tendency is further demonstrated at the level of the word. Bakhtin (1975/1996) asserts, "it frequently happens that even one and the same word will belong simultaneously to two languages, two belief systems that intersect in a hybrid construction—and consequently that word has two contradictory meanings, two accents" (p. 305). The use, in this utterance, of the expressions "raw nature" and "back to nature" infer a view of the natural world as an escape, perhaps a place that is pure, "far away" and untainted. These words also, however, invoke definitions of primeval wilderness—a "raw" place that is unprocessed, unrefined, base—and of a site distinct from culture or humanity. As such, this discourse espouses an anthropocentric perspective that centers humans or culture (Herndl & Brown, 1996).

The depth of this anthropocentric worldview is further demonstrated in the last phrase of the utterance, "Unless I am starving or dying." Although it can be understood in relation to the rest of the statement in several ways, one interpretation is that it betrays recognition—or perhaps anxiety—that despite human efforts to dominate it, the natural world cannot be fully controlled by humans (Evernden, 1992; Nash, 1967). This view of the environment provides further evidence of the nature/culture dualism, which, as noted earlier in this chapter, Evernden (1992) asserts is necessary for humans to claim dominance over nature. The speaker's expressed appreciation of nature notwithstanding, the presence of anthropocentric discourse encourages a reading of this phrase as concern over the consequences of a reversal of the taken-for-granted centrality of humans; situations in which humanity does not dominate nature can result in fearful or disastrous outcomes. It is not unlikely that such expression reflects both national and local conservative political and religious discourses of relations with the environment, as well as Western ideologies of the rugged, allegedly autonomous, individual.

The hybrid convergence of belief systems in this participant's utterance occurs most explicitly within the intersection of Herndl and Brown's (1996) "dominant tendencies . . . of environmental discourse" (p. 10), not only the anthropocentrism already identified, but ethnocentrism and ecocentrism as well. Bakhtin (1975/1996) noted that because we are surrounded by different socioideological views, expressed through different languages, "consciousness finds itself inevitably facing the necessity of *having to choose a language*" (p. 295, emphasis in original), all of which "are specific points of view on the world" (p. 291). Despite its thread of anthropocentrism, this utterance contains explicit expressions of ecocentric appreciation of nature: "I like nature, I appreciate nature. I do not take it for granted. I love being far away, with only raw nature surrounding me. . . . That is one way for me to get peace of mind and really relax." Nevertheless, the young man's admiring voice is moderated—in some of the same phrases—by the ethnocentric notion of the natural world as a resource, a site for escape or recreation and a source of psychological diversion: "I love being far away. . . . Hunting, fishing . . . That is one way for me to get peace of mind and really relax." The possibility of multiple meanings within the same sentence, or even within the same words, is an element of hybridity that invites analyses of the complexity of utterances that reflect the multiple internal discourses of any speaker (Bakhtin, 1975/ 1996). Like dualism and anthropocentrism, instances of ethnocentric discourse are, not surprisingly, pervasive among the study's respondents. This is a group whose cultural experiences may be dominated by popular media, through which images of greenwashed corporations and products, along with representations of the natural world as yet another commodity that can be packaged and purchased, circulate widely (Corbett, 2002). Ultimately, anthropocentric and ethnocentric views of the environment, even among those who appreciate and advocate for it, are the norm. Whether "from an ecocentric or biocentric perspective," Oelschlaeger (1991) argues, even "preservationism remains anthropocentric, since human interests are the ultimate arbiters of value" (p. 292).

The intersection of ethnocentric and ecocentric discourses about the natural world expressed in this utterance may reflect a fairly common socioideological environmental discourse: romanticism. Herndl and Brown (1996) begin *Green Culture* with a discussion of this form of rhetoric that both romanticizes the natural world and has human interest at its core. These common "romantic vision[s]" may cloud the ability to see the natural world as "a social responsibility, to think of nature as a scientific, an economic, or an institutional construction or problem" (p. 7). Instead, such discourse allows individuals to celebrate the self and their personal relationship with the natural world, an individualized notion that may prevent human actors from seeing the larger implications of their interactions with ecological systems. Indeed, many study respondents indicate some idealization of the natural world, and, at the same time, focus on the human agent acting in the environment, thus recentering humanity and culture. In spite of honest appreciation of the natural world, this unmitigated human-centered (and again, anthropocentric) perception may reinforce a detachment from meaningful and mutual dialogue, interrelation, or interdependence (Capra, 1996; Rogers, 1998). Identifying and interrupting such views is a goal of our conceptualization of organic hybridity combined with strategic hybridity, a rhetorical environmental strategy discussed in the following section.

Hybridity accounts for social and historical contexts, and identifies not only the interplay, but also the implications of heteroglot voices present in colloquial discourses about the natural world. Like all hybrid utterances, the discursive fragments presented here are written over with multiple languages. As evidenced in our participants' responses, the external expression of a speaker's internal discourses may blend dominant socioideological views together in ways that allow persons to make sense of their (potentially conflicting) experiences, ideologies, needs, and desires. The convergence of languages and social ideologies, however, may be manifested in ways that lie outside of the reach of current prevailing theories of environmental discourse. Examining these discourses as organic hybrid constructions that often exceed only two voices, linguistic consciousnesses, or socioideological positions offers a means through which to further explore their significance. Further, such analysis reveals hybridity as a centrifugal force that both decenters meaning and elaborates on monoglossic explanations of individuals' views of and relations with the natural world. The movement between coherent monologic discourses, on one hand, and sometimes ambivalent, heteroglossic discourses on the other, illustrates the utility of a theoretical perspective that assumes multiplicity and addresses incommensurability.

IMPLICATIONS AND CONCLUSIONS

We believe that organic hybridity as conceptualized in this chapter contributes a critical theoretical lens to environmental communication research that accounts more fully for the range of heteroglossic contradictions and complexities present within individuals' utterances about the natural world. Hybridity provides a means through which to examine the presence and implications of various, internally competing, socioideological languages as persons discursively make sense of experiences and issues related to the natural world. In addition, utilizing hybridity as an interpretive framework may facilitate the identification of those social discourses that are most influential for particular, sociohistorically situated, persons or groups. Ultimately, we believe that hybridity's explanatory potential can provide critical insights for environmental researchers and advocates.

In the previous section, we examined colloquial discourse for the presence of unconsciously merged languages and belief systems; specifically, organic hybrids. We expect that hybridity will also be useful in the analysis of civic, public, or other forms of discourse, and can be as effectively utilized to identify the voices within conscious or intentional hybrids. Future research grounded in Bakhtin's dialogic approach might consider the distinction between these forms of hybridity; at what point, for instance, does a speaker's choice between two languages cease to be organic and become, instead, an intentional hybrid? How might a hybrid analysis of dominant or civic voices differ, in its process or conclusions, from that of private or colloquial discourses? How does the researcher distinguish between organic and intentional hybridity in discourse? As a theoretical model, hybridity encourages the pursuit of these and other questions about the presence and implications of dialogic constructions. As Bakhtin (1975/1996) notes, "our ideological development is just such an intense struggle within us for hegemony among various available verbal and ideological points of view, approaches, directions and values" (p. 346).

Strategic Hybridity

Although each of these proposed research questions offers a productive avenue for future considerations of hybridity, we believe that one of the more fruitful possibilities draws directly from the understandings gained through the analysis of everyday persons' colloquial discourse, as demonstrated. As a heuristic, hybridity posits a broad continuum of possibilities for understanding colloquial discourse about the natural world. Here, as opposed to the organic quality of hybridity as an analytic framework, we expand Bakhtin's (1975/1996) notion of conscious or intentional hybridity to that of *strategic hybridity*. Whereas conscious/intentional hybridity is the inclusion of another language within a speaker's primary language (Bakhtin, 1975/1996), strategic hybridity is a purposeful, calculated implementation of hybridity as a rhetorical strategy in which goal-oriented voices articulate compound discourses, belief systems, or socioideological positions, pursuing the possibility of multiple discursive identifications. Put another way, strategic hybridity is the deliberate joining of a variety of possible voices persuasive to a particular audience. As an environmental tactic, strategic hybridity can be seen as a discursive strategy that subverts binary logics, and embraces the multiplicity of language and dialogism by allowing listeners to hold multiple, even ambivalent, beliefs about the natural world in order to encourage critical action and reflection. We believe that the strategic hybrid can provoke the dialogic imagination and provide useful rhetorical spaces for the development of environmental consciousness.

In the calculated utilization of hybridity, speakers for environmental organizations might seek to include information, attitudes, or concepts from multiple social discourses in order to appeal to listeners holding varying, even incongruous or contradictory, ideological views of environmental issues. This approach utilizes the complexity and indistinctness of persons' ideological or epistemological positions as opportunities (as opposed to constraints) for innovative environmental thinking. Herndl and Brown (1996) indicate that divergent positions may "seem incommensurate, not only because they represent opposing interests, but because of differences in institutional, disciplinary, and social discourses" (p. 19). Yet, they posit that this is not an insurmountable barrier, suggesting that rhetoric "intended for a large public audience is most successful when it combines the rhetorical resources of more than one kind of discourse" (p. 19).

In the purposeful construction of a hybrid, seemingly oppositional discourses such as those that position the natural world as a setting for ecocentric, aesthetic appreciation or ethnocentric, outdoor recreation at the same time that they overlook personal complicity in human influence or impact, may demonstrate strategic utility for motivating dialogue. For example, the respondent who defines the natural world as "forests, canyons, mountains, some lakes and rivers, oceans, etc.," and who spends "most of [her] daily life . . . hiking, swimming, boating, and absorbing all that the natural world has to offer" also sees the natural world as "always [having] been here" or "made natural without outside influences" (WP.a24). In several other instances, respondents characterize their relationship with the natural world in terms of recreational activities; nature is a place "to camp, hike, water ski, snow ski, climb, etc. . . ." (B.a2). Such discursive formations, high-lighted through the lenses of both eco- and ethnocentrism include overt objecti-fications of the natural world as an entity for human use or enjoyment.

Attention to multiplicity within an utterance, and analysis of the heteroglossic nature of discourse and the hybridity of individual voices, allows environmental speakers to respond constructively and creatively with complex, even ambivalent, perspectives. Within the framework of the strategic hybrid, the "natural" quality of (and therefore the inherent right to engage in) human activities, such as the hiking or swimming mentioned by the respondents, might be highlighted in order to call attention to the negative impact of air and water pollution on the natural world as well as on human communities. By appealing to internal discourses to express com-plex views of the natural world, environmentalists may begin to interrupt detach-ment from the consequences of humans' personal impacts, and advance awareness of mutual interrelation or interdependence between individual persons and the other-than-human world (Capra, 1996; Rogers, 1998). As opposed to the use of purely anthropocentric language which may run counter to environmentalist prin-ciples, such goals may also be accomplished, for instance, by framing references to impacts within ecocentric discourse, such as the idea of *loving the land to death*; a notion that is common within environmental circles, but is likely a novel—and po-tentially thought-provoking—concept for public audiences.

A specific goal of strategic hybridity is to nurture environmentally aware nego-tiations of the organic play between differing voices within persons' internally persuasive discourses (Bakhtin, 1975/1996), thereby influencing the environmen-tal consciousness of those who hear and respond to the messages. "When some-one else's ideological discourse is internally persuasive for us and acknowledged by us, entirely different possibilities are opened up. Such discourse is of decisive significance in the evolution of an individual consciousness" (Bakhtin, 1975/1996, p. 345). In this way, hybrids—conscious and organic—provide the underpinnings of a strategy—strategic hybridity—for crafting effective environmental appeals that may reflect the dialogic, socioideologic tensions inherent in many views of, and discourses about, the natural world.

Conclusions

As noted in the chapter's epigraph, environmental perspectives may reveal "any number of conventional or idiosyncratic constructions of the person–planet rela-tion" (Killingsworth & Palmer, 1992, p. 4), even within discursive communities. Hybridity accounts for the complexity of many forms of discourse and helps us to move beyond the limitations of monologic views and expressions of environmen-tal issues. We argue that narrowly focused goals and messages in public, political, or other dominant discourses, the taking of either/or positions, and reinforcing of exclusive categories of thinking about the world may result in consequences that are not in the interests of environmentalism. First, such messages reify polarized views of, and ideological positions in relation to, the natural world. They may, for instance, reinforce characterizations of environmental advocates as "radicals" who are out of touch with the public interest or who reject progress (Plec & Marafiote, 2004) alienating large segments of the population that both embrace technology and desire a healthy environment.[7] Consequently, narrowly directed messages can create obstacles for identifying positions and policies that may be viewed with some interest (or less skepticism) by those with ambivalent or some-what unsympathetic perspectives toward the natural world. Second, messages tar-geting those who explicitly identify as environmentalists may not speak to those members of the general public who hold environment-friendly views, but reject the *environmentalist* label and do not recognize themselves in the rhetoric used by many environmental groups. Hybridity offers a way to mitigate these potential consequences and instead both identify and appeal to those who already hold multiple and complex views of the natural world, and who have already learned to internalize and negotiate seemingly divergent perspectives. Therefore, in crafting rhetorical messages, environmental advocates and organizations can potentially benefit by emulating the writer of prose who, according to Bakhtin,

> does not purge words of intentions and tones that are alien to him [*sic*], he does not destroy the seeds of social heteroglossia embedded in words, he does not eliminate those language characterizations and speech mannerisms . . . glimmering behind words and forms, each at a different distance from the ultimate semantic nucleus of his work, that is, the center of his own personal intentions. (Bakhtin, 1975/1996, p. 298)

Rather, advocates for the natural world can further their goals through strategic environmental hybridity, the purposeful inclusion of diverse environmentally re-lated discourses that may appeal to a broader public audience. With such goals in mind, we believe that the concepts of *organic hybridity* and *strategic hybridity*, grounded in theories of dialogism, will help environmentalists interpret, under-stand, and appeal to audiences who have internalized multiple and potentially contrasting ideological understandings of the person–planet relation. One signifi-cant limitation may be the possible appropriation of such a strategy by those wish-ing to obstruct environmentally friendly goals, and appeal to consumerist and hu-man-centered, ethnocentric attitudes. Although we recognize that strategies such as strategic hybridity can be used by actors on different sides of environmental is-

---

[7]For example, the January/February, 2005, issue of *Sierra*, the magazine of the Sierra Club, reports that while 61% of Sierra Club members regularly participate in outdoor recreational activities (and 100% are members of an environmental organization), almost 90% own computers, and 27% drive sport utility vehicles (SUVs) (Hattam, 2005).

sues, our intention is that environmental theory be utilized in an ethical manner, with the goal of maintaining a healthy ecosystem.

By developing both organic and strategic hybridity as theoretical frameworks that expand the explanatory power of environmental communication theories, we advance an environmental communication approach that can both be used to interpret incongruencies in environmental discourses, and purposefully invoked to appeal to diverse audiences. Additionally, the examination of organic hybrid utterances possibly allows us to anticipate emerging formations, and intervene— through strategic hybridity—in the reproduction of repressive, centripetal, monoglossic representations such as nature/culture dualisms. Given the predominance of anthropocentric discourse and nature/culture dualisms, two primary possibilities for environmentalists are to try to change these views and/or to acknowledge and utilize them in effective and ethical ways. Although we strongly support the former, we focus here on the possibilities of the latter (which we also believe can contribute to positively changing these views). Hybridity potentially mitigates problems posed by all-or-nothing perspectives on environmentalism. Moreover, hybridity offers a subject position that resides in the overlapping spaces within the individual's consciousness and colloquial discourses about the natural world.

In this chapter, we position organic hybridity and strategic hybridity in relation to the larger aims of dialogism and democratic praxis. We should not strive merely to bring many different voices to the discursive decision-making table, although this is a crucial step toward both dialogism and democracy; we should also seek critical theories that enable us to recognize the diverse languages and speaking styles inherent in the multivoiced utterances of any one participant.

ACKNOWLEDGMENTS

The authors would like to thank Christine Oravec and Marianne Neuwirth for their contributions during the initial phases of the research for this chapter. Earlier versions of the chapter were presented at the 2002 National Communication Association (NCA) convention and the 2003 Conference on Communication and Environment.

REFERENCES

Ashcroft, B., Griffiths, G., & Tiffin, H. (2000). *Post-colonial studies: The key concepts.* New York: Routledge.
Bakhtin, M. M. (1996). *The dialogic imagination: Four essays by M. M. Bakhtin.* In M. Holquist (Ed.) (C. Emerson & M. Holquist, Trans.). Austin: University of Texas Press. (Original work published 1975)
Bhabha, H. (1994). *The location of culture.* New York: Routledge.

Bramwell, A. (1989). *Ecology in the 20th century: A history.* New Haven, CT: Yale University Press.

Cantrill, J. G. (1996). Perceiving environmental discourse: The cognitive playground. In J. G. Cantrill & C. L. Oravec (Eds.), *The symbolic earth: Discourse and our creation of the environment* (pp. 76–94). Lexington: University Press of Kentucky.

Capra, F. (1996). *The web of life: A new scientific understanding of living systems.* New York: Doubleday.

Corbett, J. B. (2002). A faint green sell: Advertising and the natural world. In M. Meister & P. J. Japp (Eds.), *Enviropop: Studies in environmental rhetoric and popular culture* (pp. 141–160). Westport, CT: Praeger.

Crowley, T. (1989). Bakhtin and the history of the language. In K. Hirschkop & D. Shepherd (Eds.), *Bakhtin and cultural theory* (pp. 68–90). Manchester, England: Manchester University Press.

Curtin, D. (1997). Women's knowledge as expert knowledge: Indian women and ecodevelopment. In K. J. Warren (Ed.), *Ecofeminism: Women culture, nature* (pp. 82–98). Bloomington: Indiana University Press.

Danow, D. K. (1991). *The thought of Mikhail Bakhtin: From word to culture.* New York: St. Martin's Press.

DeLuca, K. M. (1999). *Image politics: The new rhetoric of environmental activism.* New York: Guilford Press.

Evernden, N. (1992). *The social creation of nature.* Baltimore: Johns Hopkins University Press.

Foertsch, J. (1998). The impact of electronic networks on scholarly communication: Avenues for research. *Discourse Processes, 19,* 301–328.

Glover, K. S. (2000). Environmental discourse and Bakhtinian dialogue: Toward a dialogic rhetoric of diversity. In N. W. Coppola & B. Karis (Eds.), *Technical communication, deliberative rhetoric, and environmental discourse: Connections and directions* (pp. 37–54). Norwood, NJ: Ablex.

Hattam, J. (2005, January/February). *Sierra* readers, by the numbers. *Sierra, 90,* 47–48.

Hauser, G. (1995). Vernacular dialogue and the rhetoricality of public opinion. *Communication Monographs, 65,* 83–107.

Hauser, G. (1999). *Vernacular voices: The rhetoric of publics and public spheres.* Columbia, SC: University of South Carolina Press.

Herndl, C. G., & Brown, S. C. (Eds.). (1996). Introduction. *Green culture: Environment rhetoric in contemporary America* (pp. 3–20). Madison: University of Wisconsin Press.

Hirschkop, K. (1989). Introduction: Bakhtin and cultural theory. In K. Hirschkop & D. Shepherd (Eds.), *Bakhtin and cultural theory* (pp. 1–38). Manchester, England: Manchester University Press.

Holquist, M. (Ed.). (1996). Introduction; Glossary. In *The dialogic imagination: Four essays by M. M. Bakhtin* (pp. xv–xxxiii). Austin: University of Texas Press.

Killingsworth, M. J., & Palmer, J. S. (1992). *Ecospeak: Rhetoric and environmental politics in America.* Carbondale: University of Southern Illinois Press.

McGee, M. C. (1990). Text, context, and the fragmentation of contemporary culture. *Western Journal of Speech Communication, 54,* 274–289.

Morson, G. S., & Emerson, C. (1990). *Mikhail Bakhtin: Creation of a prosaics.* Stanford, CA: Stanford University Press.

Nash, R. (1967). *Wilderness and the American mind* (3rd ed.). New Haven, CT: Yale University Press.

Oelschlaeger, M. (1991). *The idea of wilderness: From prehistory to the age of ecology.* New Haven: Yale University Press.

Ono, K. A., & Sloop, J. M. (1995). The critique of vernacular discourse. *Communication Monographs, 62,* 19–46.

Ono, K. A., & Sloop, J. M. (1999). Critical rhetorics of controversy. *Western Journal of Communication, 63,* 526–538.

Ono, K. A., & Sloop, J. M. (2002). *Shifting borders: Rhetoric, immigration, and California's Proposition 187.* Philadelphia: Temple University Press.

Opie, J., & Elliot, N. (1996). Tracking the elusive jeremiad: The rhetorical character of American environmental discourse. In J. G. Cantrill & C. L. Oravec (Eds.), *The symbolic earth: Discourse and our creation of the environment* (pp. 9–37). Lexington, KY: The University Press of Kentucky.

Pechey, G. (1989). On the borders of Bakhtin: Dialogisation, decolonisation. In K. Hirschkop & D. Shepherd (Eds.), *Bakhtin and cultural theory* (pp. 39–67). Manchester, England: Manchester University Press.

Peterson, T. P., & Horton, C. C. (1998). Rooted in the soil: How understanding the perspectives of landowners can enhance the management of environmental disputes. In C. Waddell (Ed.), *Landmark essays on rhetoric and the environment* (pp. 165–194). Mahwah, NJ: Hermagoras Press.

Plec, E., & Marafiote, T. (2004). Characterizations and articulations in vernacular environmental discourse. *Journal of the Northwest Communication Association, 33,* 1–19.

Rogers, R. A. (1998). Overcoming the objectification of nature in constitutive theories: Toward a transhuman, materialist theory of communication. *Western Journal of Communication, 62,* 244–272.

Spivak, G. C. (1988). *In other worlds: Essays in cultural politics.* London: Routledge.

Ulman, H. L. (1996). "Thinking like a mountain": Persona, ethos, and judgement in American nature writing. In C. G. Herndl & S. C. Brown (Eds.), *Green culture: Environment rhetoric in contemporary America* (pp. 46–81). Madison: University of Wisconsin Press.

Wells, B., & Wirth, D. (1997). Remediating development through an ecofeminist lens. In K. J. Warren (Ed.), *Ecofeminism: Women culture, nature* (pp. 300–314). Bloomington: Indiana University Press.

Young, R. J. C. (1995). *Colonial desire: Hybridity in theory, culture, and race.* New York: Routledge.

CHAPTER FOUR

Influences on the Recycling Behavior of Young Adults: Avenues for Social Marketing Campaigns[1]

Olaf Werder
University of New Mexico

As a result of a heightened interest in the environmental movement, environmental issues emerged on the agenda of U.S. research in sociology, psychology, and economics in the late 1960s and early 1970s. Researchers began exploring factors that contributed to environmental quality as social problems (Dunlap, 2002). In comparison, communication scholars have turned to the environment much later. Early studies have not particularly focused on what people can do to improve their abilities to advocate a particular approach to the environment (Cantrill, 1993). However, due to the efforts in allied fields, advocacy-oriented inquiries quickly became a main focus of communication research. Initial efforts were concentrated on broad-based environmental concern and awareness among residents across demographic segments. Consequently, early findings indicated that educated, young, liberal adults appeared to be the most concerned segment. Gradually, social and attitudinal correlates of concern were added (Dunlap, 2002).

Twenty-eight years ago, Henion and Wilson (1976) correctly predicted that the uniformity of the proenvironmental segment would dissipate alongside continuous growth in size of the group. Rather than simply encouraging everybody in the segment to engage in some proecological activity, they suggest that the future challenge for marketers would be to identify the specific attitudes associated with a consumer's willingness to engage in a specific action (e.g., recycling). In 1993, Cantrill argued that it would be

---

[1] A previous version of this chapter was presented at the 2003 National Communication Association (NCA) convention in Miami, FL.

unreasonable to expect that those who are exposed to environmental advocacy and listen to the arguments for or against [pro-environmental activities] will do so with an unbiased frame of reference. In particular, if we want to assist those who advocate environmental policy, we must attend to the myriad of ways in which people make sense out of environmental discourse itself. (p. 68)

Moreover, the ultimate goal of environmental advocacy is usually behavior change or adjustment. Sand (1999) reported that despite strong environmental concerns, self-reported conservation behaviors had not changed since 1993. Given the strong environmental attitudes and widespread agreement that conservation is positive, it is surprising that there is still little "conservation behavior" (Vining & Ebreo, 2002). For example, many people agree that recycling is useful in principle but do not recycle for various reasons. This impression gives rise to Geller's (1987) suggestion about a need for greater collaboration between the fields of environmental psychology and applied behavioral analysis.

Waste management issues, as a matter of fact, have become a key concern of the government, the private sector, and the general public (Taylor & Todd, 1995). People appear sensitive to conservation issues, and many seem to hold positive attitudes toward recycling programs. Despite these positive attitudes, participation in different voluntary waste management programs varies widely (McCarty & Shrum, 1994). Notwithstanding a growing literature on the behavioral research on recycling (Ebreo, 1999; Shrum, Lowrey, & McCarty, 1994; Stern & Oskamp, 1987), little is known about the factors that influence individual waste-management behavior or how beliefs and attitudes relate to behavior. According to Shrum et al. (1994), most studies examine only a small number of variables and create models that lack integrative power. "Rather than depend on simplistic linear models that tell us that Conservation Attitude A predicts Conservation Behavior B, conservation psychologists should embrace two common and interrelated marketing research strategies: market segmentation and product positioning" (Bixler, 2003, p. 154). The challenge for social-marketing campaigns lies in finding the crucial elements that unify a specific target segment and match advocacy messages to characteristics of this group.

As most researchers of environmental psychology and behavior bring with them the theories and methodologies of their disciplines, a variety of models of conservation behavior have emerged within the last two decades. Most recycling models have analyzed the cognitive (attitudinal) antecedents or dispositions believed to guide the behavior (Kok & Siero, 1985; Vining & Ebreo, 1992). Some of the more popular models in attitude research on recycling behavior have been based on the Theory of Planned Behavior (Ajzen, 1991) as this theory has shown to be a useful template for (a) analyzing the complex psychological origins of recycling among a specific target segment and (b) identifying the most relevant predictors as a guide for advocacy campaigns directed at that segment. The current study selected young educated adults as the target group because this segment ranks among the more undecided populations when it comes to recycling (Goldenhar & Connell, 1992–1993). In other words, because an undecided person does not hold personal values toward recycling, an attitude toward a recycling act is formed outside convictions. As a result, this person might occasionally support a prorecycling message while it is in the market, but ceases to engage in the activity once the educational prompt is removed.

REVIEW OF CONCEPTUAL MODELS
OF CONSERVATION BEHAVIOR

In the late 1980s, research turned toward a better understanding of environmentally significant behavior. Balderjahn (1988), for instance, used causal modeling to investigate the relationship between a series of predictors and ecological consumption patterns. Others, such as Granzin and Olsen (1991), started examining the relationship between environmental protection activities (donating items, recycling newspapers, walking for conservation causes) and different demographic and psychosocial variables. Many of the early studies focused on demographic and sociocultural differences between conservation and nonconservation groups.

Sociocultural Explanations

Gender, education, and socioeconomic status are among the early variables in research that specifically focused on demographic characteristics to predict ecological concern and involvement in environmental protection. Later, the role of family, organized religion, and community were added as equally important in explaining willingness to change or maintain a new conservation behavior (Monroe, 2003).

The findings, however, are inconsistent. For example, younger persons seem to hold greater ecological concern (Van Liere & Dunlap, 1980), whereas older persons are more likely to recycle (Vining & Ebreo, 1990). Although findings tend to describe the ecological concerned consumer as young and better educated (Balderjahn, 1988), politically liberal (Dunlap, 1975; Mohai, 1985), and higher in socioeconomic status (Buttel & Flinn, 1978; Kinnear, Taylor, & Ahmed, 1974; Vining & Ebreo, 1990), the variables are limited to describing generic ecological concern rather than specific proenvironmental behaviors. Finally, many studies corroborate no reliable relationship between demographics and proenvironmental behavior (Antil, 1984; Granzin & Olson, 1991). As a result, conservation research began analyzing the link between a particular proenvironmental behavior and various independent predictors such as a person's beliefs about the self as a part of nature and society, external influence factors, the role of education in the modification of a behavior, and feelings of empowerment regarding one's ability to contribute to conservation activities.

Personal Beliefs

The first construct refers to individuals' ways of conceptualizing themselves as participants in social and environmental settings. It is composed of two variables: environmental activism and environmental sensitivity. Each touches on a different aspect of the individual's psychological condition.

Environmental activism describes the state of active participation in environmental initiatives. It is born out of a deep belief in the individual that personal conservation action is the responsibility of every human being (Stern, 2000). According to altruism models (Geller, 1995; Schwartz, 1977), beliefs of social responsibility appear to be an important predictor of philanthropic behavior. For instance, Webster (1975) found that people high in social responsibility (the display of altruistic behavior patterns) are also highly involved in their community. Furthermore, the theories assert that people who hold personal values that lead them to actively care about performing altruistic tasks also engage in conservation activities. Evidence for the significance of these theories has been provided by Bratt (1999), Kaiser and Shimoda (1999), and Vining, Linn, and Burdge (1992).

Environmental sensitivity is largely defined as the emotional bond someone has with nature and compassion toward its protection (Sanders, 2003). Although studies of emotional involvement and other feeling-oriented variables toward nature have been vastly overlooked in the ". . . pursuit of cognitive structures that predict conservation behavior" (Vining & Ebreo, 2002), these approaches show great promise. Geller and his colleagues' (e.g., Geller & Clarke, 1999; Geller, Winett, & Everett, 1982) research on behavior-based self-perception offers some advice and directions for future endeavors.

Behavior Modification

The information people use to direct a specific behavior encompasses their knowledge derived from education as well as ongoing interpersonal and mass-mediated influences. How people learn is the subject of applied behavioral analyses and learning theories (De Young, 1990; Geller, 1987) often directed at younger age groups. These approaches have been popular in encouraging behavior change in recycling.

Initial efforts were confined to documenting levels of public learning and, as a result, concern for the environment across differing sectors of society. Research has shown that citizens with more environmental knowledge are more environmentally concerned and engaged (De Young, 1990). Attention is primarily paid to documenting the correlates of environmental concern (Dunlap, 2002) as well as the impact of concern on behavior. Hines, Hungerford, and Tomera (1987) proposed that greater education about the interaction between people and nature would manifest itself in a greater commitment toward proenvironmental activity. Most studies, however, have not demonstrated a direct correlation of concern with conservation behavior based on sociodemographic cluster and have shed little light on the nature of environmental concern via learning.

In general, the idea of having obtained the proper knowledge and skills to act competently and responsibly has become known as *environmental literacy* (Hungerford & Volk, 1990). However, the relationship between ecological literacy and specific conservation activities remains largely unexplored (Pickett, Kangun, & Grove, 1993).

In a literature review of 31 experiments studying direct connections between learned information and behavioral outcome (Porter, Leeming, & Dwyer, 1995), two thirds have been found to study manipulations of behavioral antecedents (e.g., prompts, commitments to perform the behavior), and one third have analyzed manipulations of behavioral outcomes (e.g., rewards, penalties, feedbacks). Whereas these learning approaches to environmental behaviors and consequences have resulted in more tangible, environmentally appropriate behaviors (Jordan, Hungerford, & Tomera, 1986), they have appeared to be more effective in promoting short-term change rather than permanent change (Deci & Ryan, 1985; Geller, 2002). Thus, motivating people solely through the use of educational information does not appear to be a viable permanent solution to conservation problems (Vining & Ebreo, 2002).

Given the reduced knowledge and experience of young adults, it is reasonable to expect that conservation behavior is initiated as a result of interpersonal influences. The social acceptability of a conservation behavior has been found to form a critical barrier to the performance of the behavior among this group, and must be addressed in intervention campaigns (Monroe, 2003). Vining and Ebreo (1990), for instance, have found that young adults who start to recycle relied on environmental information from friends. As a matter of fact, the influence of others on specific behaviors has attained a key role in determining behavioral intentions in the Theory of Reasoned Action (Ajzen & Fishbein, 1980; Fishbein & Ajzen, 1975). Goldenhar and Connell (1992–1993) and Jones (1990) provided empirical examples for recycling behavior. According to Deci and Ryan's (1985) self-determination theory, some people's behavior may be entirely the result of personal influences. Research on recycling (e.g., Schultz, 1998) suggested that the provision of feedback about individual and group-level norms can result in beneficial behavioral changes.

Perceived Empowerment

Finally, a number of researchers have focused on the question of volitional control. Many of these approaches have derived from the concept of *internal and external locus of control* (Rotter, 1966). This concept refers to the subjective probability that one is capable of executing a certain course of action. As a general rule, the greater the feeling of personal empowerment, the stronger is the intention to perform the behavior under consideration.

Within the field of conservation psychology, personal empowerment conveys the awareness in a person that he or she is able to take proenvironmental action (McKenzie-Mohr, 2000). It is defined as a belief that individual effort can make a difference in the solution to an environmental problem. The degree to which a person feels little control over the performance of the behavior inhibits this behavior even if attitudinal beliefs and external influences are conducive toward the action. Research on recycling has shown that individuals recycle to the degree to which they feel empowered enough that their behavior influences a desired outcome (Goldenhar & Connell, 1992–1993).[2]

HYPOTHESES

Unfortunately, much of the past behavioral science research has studied general environmental concern rather than more restricted topics (Oskamp et al., 1991). To generate a controlled study in line with earlier statements about specific conservation behaviors, we needed to identify a particular activity that was simple, not driven by rules, and widely available for participation. Recycling is used in this project as it is a good example for an action that typically offers little direct benefit to the individual but that often involves substantial personal cost with respect to time and effort (Smith, Haugtvedt, & Petty, 1994). As more landfills are filled to their capacity, community recycling is regarded by many as the primary solution to waste management (Hershkowitz, 1998). For some years now, various municipalities in the United States have engaged in serious efforts to promote waste avoidance in the form of reducing, reusing, or recycling waste with varying acceptance rates within the citizenry in particular young adults.

The review of the literature provides the background and structure for the key premise of this study that personal beliefs, external behavior modification attempts, and self-empowerment thoughts predict recycling behavior. More specifically, we hypothesize the following relationships regarding self-reported recycling activity:

> *H1:* Young adults who hold positive beliefs about community activism and emotional connections toward nature are more likely to recycle than those who do not.

---

[2]This view originates from concepts of *social exchange theory* (Thibaut & Kelley, 1959). The key tenet of this theory is that human behavior is in essence an exchange, particularly of rewards or resources of primarily material character (wealth) and secondarily of symbolic attributes. Presumably, such exchange transactions permeate all social phenomena, including group processes and intergroup relations, which are conceived as sets of joint outcomes of voluntary individual actions induced by rewards. In this view, exchange transactions constitute the foundation of social life, and of group processes and relations particularly.

H2: Young adults who are likely to be influenced by others' opinions and environmental education are more likely to recycle than those who do not.

H3: Young adults who have a strong sense of self, that is, feel self-empowered, are more likely to recycle than those who do not.

Because previous research has shown that sociocultural and demographic variables are related to self-reported recycling behavior, we add the following research question:

RQ1: What roles do demographic and sociocultural variables play in explaining the recycling behavior of young adults?

METHODOLOGY

This study uses survey methodology to assess the relationship between young adults' personal beliefs, feedback from others, self-image, and their recycling behavior. First, a correlation analysis explores the relationships between the variables, followed by a regression analysis to test the significance of effects of the hypothesized predictors.

Subjects and Procedures

The survey was administered to a convenience sample of 250 students in various communications courses at two large universities: one in the southeastern United States and one in the southwestern United States. Students received course credit for their participation. Twenty-six surveys were eliminated because they had substantial missing data reducing the final sample size to 224 respondents. Of these, 53% were female, 75% were White, 19% were Hispanic, 3% were African American, and 8% were Asian, and their mean age was 23 (range: 19–53). Median income of their parents was $62,000, 41% considered themselves fairly to strongly religious, and they split evenly between a liberal, moderate, and conservative political orientation.

Measures and Validation

Five psychological constructs (environmental activism, environmental sensitivity, interpersonal influences, environmental literacy, and perceived empowerment) were chosen for our conceptual framework. In addition to these key measures, data were collected on a number of control variables. Specifically, we assessed respondents' age, gender, socioeconomic status, religious devoutness, and political orientation.

In most cases, established measures relevant to our constructs could be taken directly from the literature. In a few instances, measures were incompatible with our goals, and they had to be adapted to the study. All of the measures were examined for internal consistency, validity, and dimensionality. The intercorrelations, reliabilities, and descriptive statistics of the key measures are provided in Table 4.1.

All multiitem scales were assessed using 5-point Likert scales. The final instrument contained multiitem measures of environmental activism, environmental sensitivity, interpersonal influences, environmental literacy, and perceived empowerment. A self-reported recycling scale was created for the dependent variable.

Environmental Activism. The idea of social responsibility as a predictor of proenvironmental activism is based on Geller's (1995) notion of altruism. That is, people who are active in their community out of a sense of duty will incorporate conservation behavior patterns more readily than those who are less active. This measure of choosing to be involved was assessed with Miller's (1977) Community Involvement Scale. The 5-item scale displays strong reliability ($\alpha = .83$).

Environmental Sensitivity. Many approaches to explaining recycling are based on the assumption that humans are rational animals that systematically use or process the information available to them and that the information is used in a reasonable way to arrive at a behavioral decision. However, many so-called rational decisions have been found to be the result of affective relationships to the decision object. In order to tap the emotional sensitivity construct, we combined items from Davis's (1983) Sympathy Scale and Pickett et al.'s (1993) Commitment to the Environment Scale. The measure displays satisfactory reliability ($\alpha = .77$).

Interpersonal Influences. The influence of others on individuals' behavior has been demonstrated first with the Theory of Reasoned Action. Research on the motivating factors of young people's behavior has also repeatedly shown an influence of role models. Our measure includes items from the original Subjective Norm Scale (Ajzen & Fishbein, 1980) and the Susceptibility to Interpersonal Influence Scale (Bearden, Netemeyer, & Teel, 1989). The resultant scale shows strong reliability ($\alpha = .86$).

Environmental Literacy. As noted earlier, research has shown that citizens with more environmental knowledge are more environmentally concerned and engaged. Specifically, measures were selected to assess both general environmental knowledge and specific knowledge about local recycling. The scale is a combined construct from Allen and Ferrand's (1999) Educating-the-Self Scale and Pickett et al.'s (1993) Ecological Knowledge Scale. This measure displays good reliability ($\alpha = .79$).

TABLE 4.1

Correlations for Key Measures

Variable	Mean	SD	Correlation										
			A	B	C	D	E	F	G	H	I	J	K
A. Behavior	1.54	.50	(.91)										
B. Activism	1.50	.56	.27	(.83)									
C. Sensitivity	1.80	.66	.14	-.10	(.77)								
D. Influences	2.35	.82	.19	.03	.27	(.86)							
E. Literacy	2.61	.88	.12	.31	-.18	-.21	(.79)						
F. Empowerment	2.76	.58	.41	-.12	.13	-.05	.12	(.89)					
G. Age	24	5	-.32	.11	-.27	.11	-.01	-.26	(NA)				
H. Gender	NA	NA	-.14	.15	.09	-.02	.13	.01	-.04	(NA)			
I. Parent income	3.81	1.38	.12	-.05	.06	.16	-.20	-.09	-.16	.03	(NA)		
J. Religiosity	3.43	1.21	-.06	-.14	.09	.09	-.12	-.02	-.08	-.01	.23	(NA)	
K. Political view	2.01	.85	-.14	.02	.37	.05	-.30	-.19	.08	-.09	.07	.10	(NA)

Note. Correlations > .14 or < −.14 are significant at $p < .05$; correlations > .20 or < −.20 are significant at $p < .01$; reliability coefficients are listed on the diagonal.

85

Perceived Empowerment.   This construct is an expression of locus of control, that is, the individual's perception that he or she is able to carry out a specific action. The measure was assessed using Ellen, Wiener, and Cobb-Walgren's (1991) Perceived Consumer Effectiveness (PCE) Scale and Smith-Sebasto's (1992) Revised Perceived Environmental Control Measure (RPECM). This scale displays strong reliability (α = .89).

Dependent Variable: Recycling Behavior.   This scale measures the extent to which college students self-report to engage in recycling of household trash and show dispositions to reuse containers, purchase recyclable products, and avoid wasteful packages. The construct is a combination of the recycling behavior subscale of Allen and Ferrand's (1999) Environmentally Friendly Behavior Scale and Schwepker and Cornwell's (1991) Purchase Intention (PI) Scale. The new measure shows high degree of convergent validity (interindicant r = .74) and overall strong reliability (α = .91).

Using LISREL 8 (Jöreskog & Sörbom, 1996), each scale was subjected to a confirmatory factor analysis to assess their dimensional structure and discriminant validity. This step followed Churchill (1992) who suggested that the examination of borrowed measures for reliability and validity is particularly needed when the borrowed scales were developed in a different context. The suggested model demonstrated a satisfactory degree of fit ($\chi^2(147)$ = 126.7, CFI = .88, RMSEA = .08) indicating that the data match the latent structure of each construct. As a test of discriminant validity, the baseline model (in which correlations between the latent constructs were freely estimated) was compared against a series of alternative models in which the correlations between related pairs of constructs (activism and sensitivity, influence and literacy) were constrained to unity (Burroughs & Rindfleisch, 1992). In each case, the constrained model showed a statistically significant increase in chi-square. As a result, sufficient discriminant validity between the construct measures seems to exist. The final question items are shown in Table 4.2.

RESULTS

The hypotheses are tested using forced-entry multiple regression. The separate regression models examine the extent to which the four hypothesized constructs (demographics, personal beliefs, behavior modification, and empowerment) explain variations in the continuous recycling behavior variable. A multicollinearity diagnosis of the independent variables was performed prior to the regression analysis. Only moderate correlations (less than .36) emerged between the variables indicating that the method is appropriate for proper interpretations. Findings are presented in Table 4.3.

TABLE 4.2

Survey Items

Recycling Behavior

I will change my lifestyle (e.g. separate out recyclables) to help the environment.
I purchase products in recyclable packages over similar products in nonrecyclable packages.
I recycle items such as glass bottles, aluminum cans, and paper regularly.
I don't recycle often because it inconveniences me.[a]
I often purchase some products in larger packages with less frequency to avoid waste.

Environmental Activism

I contribute money to organizations promoting conservation efforts.
I visit community organizations to get information about recycling.
I hardly ever discuss conservation issues with more than one person.[a]
I would assume leadership of action programs if it would help the environment.
I belong or am interested to join community/campus organizations that take a stand on issues.

Environmental Sensitivity

I would be willing to pay an additional tax if it would decrease the waste problem.
I am willing to stop buying products from companies that pollute even if it is inconvenient.
I often have tender, concerned feelings for the environment and wonder what I can do to help.
I am not willing to go out of my way to recycle since it's the government's job to take care of waste pollution issues.[a]
I would vote for a law that fines people who don't recycle.

Personal Influences

I would only recycle if other people in my neighborhood/complex do so as well.
I often consult others to help choose the best alternative to recycle.
I know as well as anybody where recycling sites are and when recycling occurs.[a]
It is important to me that others like the brands I buy and use.
I recycle about as much as my family does.

Environmental Literacy

There is a difference between recyclable and biodegradable packaging.
I educate myself regularly to learn what I can do to help solve environmental problems.
Environmental laws are not as strict in the United States as in other countries.
Trees are an example of a renewable resource.
The city in which I live is running out of places to dispose of its trash.

Perceived Empowerment

Recycling has become so complex today that I have lost trust in doing it right.[a]
The recycling practices of a person like myself will influence the quality of the environment.
People like me have little influence over local recycling conditions.[a]
If I were inclined, I am sure I could convince others to take action on recycling.
The local recycling situation could be affected if I would voice my concern.

---

[a]Indicates a reverse-coded statement.

TABLE 4.3

Regression Results of Recycling Behavior

Variables	Sociocultural B/SE	Sociocultural Partial	Personal Beliefs B/SE	Personal Beliefs Partial	Behavior Modification B/SE	Behavior Modification Partial	Perceived Empowerment B/SE	Perceived Empowerment Partial	Combined Effect B/SE	Combined Effect Partial
Gender	.17*/.08	.18							.23**/.07	.29
Age	-.03**/.01	-.31							-.02**/.00	-.28
Parent income	.04/.03	.11							.05/.03	.16
Religiosity	-.04/.03	-.10							-.02/.03	-.08
Political view	-.08/.05	-.14							-.06/.05	-.12
Activism			.26*/.07	.29					.33**/.07	.41
Sensitivity			.13*/.06	.18					.06/.06	.08
Influences					.14*/.05	.22			.12*/.05	.23
Literacy					.09/.05	.16			.01/.04	.03
Empowerment							.35**/.07	.41	.33**/.06	.43
Constant	2.30**/.26		.92**/.17		.98*/.20		.57*/.19		.22/.38	
	$F = 4.88**$		$F = 7.58**$		$F = 4.33*$		$F = 27.49**$		$F = 10.27**$	
	$R^2 = .16$		$R^2 = .10$		$R^2 = .06$		$R^2 = .18$		$R^2 = .45$	
	Adj. $R^2 = .13$		Adj. $R^2 = .09$		Adj. $R^2 = .05$		Adj. $R^2 = .16$		Adj. $R^2 = .41$	

Note. $* < .05$ significance, $** < .01$ significance, B = regression coefficient, SE = standard error.

The first hypothesis asserts that personal beliefs about social activism and emotional sensitivity toward nature significantly predict recycling behavior among young adults. In isolation, the regression is significant ($F = 7.58$, $p < .01$), yet the R^2 of .10 indicates that little variation in the dependent variable is explained. In the combined function, only activism ($B = .33$, $p < .01$) remains a significant predictor. Therefore, Hypothesis 1 is only partially supported.

The second hypothesis proposes that recycling behavior of young adults is determined by behavior modification via learning (literacy) or the influences of important others. This hypothesis is only partially supported as well. Although the function is significant ($F = 4.33$, $p < .05$) in isolation, only the interpersonal influences variable reaches significance ($B = .12$, $p < .05$) in the combined function.

Finally, the third hypothesis states that perceived self-empowerment predicts recycling behavior among young adults. By itself, this single-item regression is strongly significant ($F = 27.49$, $p < .01$) and explains 18% of the variation in recycling behavior ($R^2 = .18$). Moreover, in the combined function the empowerment scale remains significant ($B = .33$, $p < .01$), supporting the hypothesis.

The research question asks about the significance of demographic and sociocultural variables in explaining the recycling behavior of young adults. Among all sociocultural variables, only gender ($B = .23$, $p < .01$) and age ($B = -.02$, $p < .01$) emerge as significant predictors for this group of respondents.

Some additional insight is gained by a deeper analysis of the combined effects in comparison to the isolated effects. Although many variables drop in significance when inputted in the combined function, the self-empowerment variable maintains its significance, and the community activism and gender variables even increase their significance as a predictor. Furthermore, the explanatory power of the equation (adjusted $R^2 = .41$) is significantly improved in the combination of all variables.

In general, the combined effects analysis reveals that the perception of self-empowerment, a personal belief in the value of community activism, and gender emerge as key predictors of young adults' recycling behavior.

DISCUSSION

The purpose of this exploratory study was to discover which variables can best determine whether college students will engage in recycling behavior. Similar to other research that identified attitudes and personality factors as key influences on conservation behavior (Ellen et al., 1990; Newhouse, 1990), the current study found an altruistic drive to be socially active in the community and faith in one's own abilities as the main predictors of distinguishing between young adults who recycle and those who do not.

Consistent with previous findings (Balderjahn, 1988; Van Liere & Dunlap, 1980), the results show only a marginal influence of demographic variables (with the exception of gender) in distinguishing the degree of participation in recycling. The modest effect of age supports the assumption that younger persons hold greater ecological concern (Van Liere & Dunlap, 1980), but it contradicts the argument that proenvironmental involvement increases with age (Vining & Ebreo, 1990). The effect of gender supports findings indicating that women are more likely to participate in voluntary activities (Hill, Rugin, Peplau, & Willard, 1979).

Monroe (2003) argued that the planning, research, and execution of social marketing campaigns is not at all that different from commercial campaigns. However, conservation behavior is usually more inconvenient and requires more effort as many other types of behavior without any obvious returns on investment. As a result, direct persuasion attempts in the context of environmental conservation activities have not yielded the most promising results, especially among young adults.

In general, this study supports the claim of some scholars (e.g., Thøgersen, 1996) that recycling should be treated as an instance of prosocial behavior because of its benefits to society and the environment. They assert that attitudes regarding this type of behavior are not based on thorough calculation, but they are a function of the person's beliefs in what is the right or wrong thing to do. Maybe this also explains why personal influences and education only had a reduced influence on recycling in this study.

One of the primary goals of conservation education has long been to stimulate environmentally responsible behavior via an increase in knowledge. With this in mind, education has been a key component of advocacy campaigns. However, although efforts directed at older publics have been met with success, young adults today have grown up with a greater awareness of conservation topics. As a result, educational campaigns will have fewer effects on the behavior for this target group. Therefore, it should not be surprising that environmental literacy and sensitivity, albeit present, do not significantly affect behavior among our respondents. The mean score of environmental literacy in this study ($X = 2.6$) shows satisfactory familiarity with ecological issues. The failure to link this knowledge with personal involvement shows that young adults know about recycling and related waste management problems but do not see how this has anything to do with them.

Another finding of the study echoes Bem's (1972) classic theory of self-perception, which contends that "[I]ndividuals come to know their own attitudes, emotions and other internal states by inferring them from observations of their own overt behavior and/or the circumstances in which behavior occurs" (p. 2). More specifically, many people have been brought up to value thriftiness in some form. As a result, they tend to feel a pang of guilt when acting wastefully. Opportunities to redeem their wastefulness consequently become important for them to feel good about themselves. For some, recycling becomes a token gesture; for others it's a perpetually important concern. In either case, their overt action and experiences with recycling influence their subsequent perceptions and feelings.

Hereby, the perception of empowerment, or the lack thereof, appears to be driven by the knowledge of how recycling influences the environmental equilibrium and how a single individual can make a difference in that respect. This knowledge base is an acquired faculty that guides individuals in interpreting and evaluating behavior patterns.

The aforementioned arguments support the strong connection that was found between empowerment and recycling behavior of this group. The present study shows that the more a student feels strongly about being able to contribute to helping with the waste problem, the more likely the recycling behavior will occur.

LIMITATIONS

Several caveats associated with the study should be mentioned. First, the sample has been designated from two communities of university students studying communication. As such, the sample does not provide enough randomness to be entirely representative of the population to which the study attempts to generalize. A larger, more scientific sample of young adults would provide the desired generalizable information.

The assessment of recycling activity based on self-reports poses a problem of accuracy and validity. It is possible that respondents who have no opinion or attitude on recycling, either because they have little knowledge of the issue or have not thought about it may behave randomly when responding to a survey question. In that case they might feel pressure to offer an opinion in response to a question when in fact they have none. Because respondents usually wish to project a positive image and self-image, their random choices from the offered alternatives will increase the amount of random variation in the variable. Because recycling is perceived as socially desirable, respondents may systematically overestimate the extent to which they perform this behavior (Vining & Ebreo, 2002). As a measure to prevent this problem, the questionnaire continuously offered the respondents the option to choose a "don't know" answer.

Furthermore, the selection and quantity of items for each scale can attenuate the correlation between measures and harm scale reliability. Although the relationships that were found appear to be significant, future research should be careful about how recycling behavior measures are worded (Schwepker & Cornwell, 1991). It could be argued that some of the items (e.g., the item in the dependent variable, "I will change my lifestyle to help the environment") suggest more of an attitudinal component than a behavior component. Clear word choices that delineate attitude from behavior should be used to avoid confounding effects.

The low significance levels of certain variables can potentially be explained with the choices made to construct the instrument. The recurring issue in survey methodology of validity (do respondents give accurate answers that reflect true measurements of the construct) versus reliability (are there sufficient probes to arrive at a conclusive answer repetitively) is at the heart of this issue. Another concern regarding this issue is the aforementioned generic nature of the questions. An implied episodic version (a threat scenario connected to recycling) might have led to a different outcome.

Although there is no assurance that any measure achieves complete reliability of these constructs, the study attempted to achieve statistically acceptable reliability of the group of items, hypothesized to measure separate aspects of the same concept, by checking for their internal consistency via the Cronbach's alpha test.

Finally, the current study cannot effectively conclude that a strong relationship of any parameter to self-reported recycling behavior will automatically lead to actual recycling behavior. There is some theoretical evidence that positive intentions lead to positive behavior. There is disagreement, however, about what respondent variable one should choose when discussing implications for communication efforts (based on the argument that any communication effort can at most influence an intention to do something but not the actual act of doing). The ultimate proof of having measured the complete model would be to actually measure recycling behavior (e.g., counting the amount of recycled garbage near the residences of the target population).

CONCLUSION

Considering the social and political implications for a successful recycling program, it is important that communities convince their residents to engage in a more proenvironmental behavior pattern. Given the fact that this is one among many social and environmental activities that a person engages in voluntarily without any repercussion if one chooses not to participate, it is essential that recycling programs reach people at a level that they understand and that motivates them to comply.

As a result, social marketing theory has made significant contributions into the area of conservation behavior more recently (McKenzie-Mohr, 2000). It basically suggests that positioning the idea of an ideal in a way that acceptance of this description by potential consumers of the issue will result in a favorable action toward the ideal solution for the issue. To do so, the first step is the careful identification of the behavior and understanding of benefits and barriers to this behavior for the various diverse audiences of a conservation message. These key factors are then incorporated into media campaigns or other tools directed at those exclusive segments. Attention needs to be paid to provide information and reminders about the consequences of the behavior, social acceptability, the ease with which action can be done, and effectiveness of the behavior to solve the problem (Monroe, 2003). Ultimately, recycling promoters need to achieve a situation in which individuals are internally motivated to engage in prorecycling behavior and attitude patterns.

Although studies of recycling behavior have been plentiful in the literature since 1987, very few studies have focused on particular population segments. With environmental campaigns played out in the presence of different and multiple publics, it is a mistake to assume that a single format is appropriate for everybody. Long-term residents of a community may have a different perception and a greater stake in community-based issues, such as recycling, than a more transient and emotionally less connected university student segment. This life situation combined with age may render this group more undecided on the issue.

The results provide the basis for future research endeavors that will examine more closely an impact of targeted social marketing campaigns on recycling. Experimental studies could apply the findings of this study as well by introducing stimuli in promotional messages to a population of young adults incorporating cues that underscore a specific variable connected to recycling. With a basic understanding of what motivates today's young adults to participate in specific conservation activities, one can tailor more effective policies and associated communication. Identifying a target group's beliefs and motivations is germane to finding the right persuasive language in advocacy campaigns.

Communicators ought to welcome ideas from marketing related to market segmentation and product positioning. For instance, marketers have been known to use a method called "cool hunting" for some time now to attract a younger market segment to a brand. Hereby, marketers locate and query trendsetters among a cohort to (a) incorporate their opinions into brand messages and (b) use these individuals as brand ambassadors. The successful youth smoking prevention campaign, "Truth," has already demonstrated that this technique is applicable for cause-related campaigns (Sly, Hopkins, Trapido, & Rey, 2001).

Therefore, people who are undecided about recycling should be communicated to in a way that appeals to their sets of values and standards. Effective communication could contribute in motivating a student segment toward the activity of recycling. Media advertising's alleged ineffectiveness in reaching this group might actually be the victim of misperception about the role of message and medium. While the nonoccurrence of a presupposed change in attitude or behavior is often attributed to the communication channel, the real reason might have been a misinterpretation of the real determinants of recycling.

REFERENCES

Ajzen, I. (1991). The theory of planned behavior. *Organizational Behavior and Human Decision Processes, 50*, 179–211.

Ajzen, I., & Fishbein, M. (1980). *Understanding attitudes and predicting social behavior.* Englewood Cliffs, NJ: Prentice-Hall.

Allen, J. B., & Ferrand, J. I. (1999). Environmental locus of control, sympathy, and proenvironmental behavior: A test of Geller's actively caring hypothesis. *Environment & Behavior, 31*(3), 338–353.

94   WERDER

Antil, J. H. (1984). Socially responsible consumers: Profile and implications for public policy. *Journal of Macromarketing, 4*, 18–32.

Balderjahn, I. (1988). Personality variables and environmental attitudes as predictors of ecologically responsible consumption patterns. *Journal of Business Research, 17*, 51–56.

Bearden, W. O., Netemeyer, R. G., & Teel, J. E. (1989). Measurement of consumer susceptibility to interpersonal influence. *Journal of Consumer Research, 15*(4), 473–481.

Bem, D. J. (1972). Self-perception theory. In L. Berkowitz (Ed.), *Advances in experimental social psychology* (Vol. 6, pp. 1–60). New York: Academic Press.

Bixler, R. (2003). Segmenting audiences and positioning conservation interventions. *Human Ecology Review, 10*(2), 154–155.

Bratt, C. (1999). The impact of norms and assumed consequences on recycling behavior. *Environment & Behavior, 31*(5), 630–656.

Burroughs, J. E., & Rindfleisch, A. (1992). Materialism and well-being: A conflicting values perspective. *Journal of Consumer Research, 29*(3), 349–370.

Buttel, F. H., & Flinn, W. (1978). Social class and mass environmental beliefs: A reconsideration. *Environment & Behavior, 10*(3), 433–450.

Cantrill, J. G. (1993). Communication and our environment: Categorizing research in environmental advocacy. *Journal of Applied Communication Research, 21*, 66–95.

Churchill, G. (1992). Better measurement practices are critical to better understanding of sales management issues. *Journal of Personal Selling & Sales Management, 17*, 73–80.

Davis, M. H. (1983). Measuring individual differences in empathy: Evidence for a multidimensional approach. *Journal of Personality and Social Psychology, 44*(1), 113–126.

Deci, E. L., & Ryan, R. M. (1985). *Intrinsic motivation and self-determination in human behavior.* New York: Plenum Press.

De Young, R. (1990). Recycling as appropriate behavior: A review of survey data from selected recycling education programs in Michigan. *Resources, Conservation and Recycling, 3*, 1–13.

Dunlap, R. E. (1975). The impact of political orientation on environmental attitudes and actions. *Environment & Behavior, 7*(3), 428–454.

Dunlap, R. E. (2002). Environmental sociology. In R. Bechtel & A. Churchman (Eds.), *Handbook of environmental psychology* (Vol. 2, pp. 160–171). New York: Wiley.

Ebreo, A. (1999). Reducing solid waste. *Environment & Behavior, 31*(1), 107–134.

Ellen, P. M., Wiener, J. L., & Cobb-Walgren, C. (1991). The role of perceived consumer effectiveness in motivating environmentally conscious behaviors. *Journal of Public Policy & Marketing, 10*, 102–117.

Fishbein, M., & Ajzen, I. (1975). *Belief, attitude, intention and behavior: An introduction to theory and research.* Reading, MA: Addison-Wesley.

Geller, E. S. (1987). Applied behavior analysis and environmental psychology: From strange bedfellows to a productive marriage. In D. Stokols & I. Altman (Eds.), *Handbook of environmental psychology* (Vol. 1, pp. 361–388). New York: Wiley.

Geller, E. S. (1995). Actively caring for the environment: An integration of behaviorism and humanism. *Environment and Behavior, 27*(2), 184–195.

Geller, E. S. (2002). The challenge of increasing proenvironment behavior. In R. Bechtel & A. Churchman (Eds.), *Handbook of environmental psychology* (Vol. 2, pp. 525–540). New York: Wiley.

Geller, E. S., & Clarke, S. W. (1999). Safety self-management: A key behavior-based process for injury prevention. *Professional Safety, 44*, 29–33.

Geller, E. S., Winett, R. A., & Everett, P. B. (1982). *Environmental preservation: New strategies for behavior change.* New York: Pergamon.

Goldenhar, L. M., & Connell, C. M. (1992–1993). Understanding and predicting recycling behavior: An application of the theory of reasoned action. *Journal of Environmental Systems, 22*, 91–103.

Granzin, K. L., & Olsen, J. E. (1991). Characterizing participants in activities protecting the environment: A focus on donating, recycling, and conservation behaviors. *Journal of Public Policy & Marketing, 10*(2), 1–27.

Henion, K., & Wilson, W. (1976). The ecologically concerned consumer and locus of control. In K. Henion & T. Kinnear (Eds.), *Ecological marketing*. Chicago: American Marketing Association.

Hershkowitz, A. (1998). In defense of recycling. *Social Research, 65*(1), 148–218.

Hill, C., Rugin, Z., Peplau, L., & Willard, S. (1979). The volunteer couple: Sex differences, couple commitment, and participation in research on interpersonal relationships. *Social Psychology Quarterly, 42*, 415–420.

Hines, J., Hungerford, H. R., & Tomera, A. (1987). Analysis and synthesis of research on responsible environmental behavior: A meta-analysis. *Journal of Environmental Education, 18*, 1–8.

Hungerford, H. R., & Volk, T. L. (1990). Changing learner behavior through environmental education. *Journal of Environmental Education, 16*(3), 8–22.

Jones, R. E. (1990). Understanding paper recycling in an institutionally supportive setting: An application of the theory of reasoned action. *Journal of Environmental Systems, 19*, 307–321.

Jordan, J. R., Hungerford, H. R., & Tomera, A. (1986). Effects of two residential workshops on high school students. *Journal of Environmental Education, 18*(1), 15–22.

Jöreskog, K., & Sörbom, D. (1996). *LISREL 8 User's Reference Guide*. Lincolnwood, IL: Scientific Software, Inc.

Kaiser, F. G., & Shimoda, T. A. (1999). Responsibility as a predictor of ecological behavior. *Journal of Environmental Psychology, 19*, 243–253.

Kinnear, T. C., Taylor, J. R., & Ahmed, S. A. (1974). Ecologically concerned consumers: Who are they? *Journal of Marketing, 38*(2), 20–24.

Kok, G., & Siero, S. (1985). Tin recycling: Awareness, comprehension, attitude, intention and behavior. *Journal of Economic Psychology, 6*(2), 157–173.

McCarty, J. A., & Shrum, L. J. (1994). The recycling of solid wastes: Personal values, value orientations, and attitudes about recycling as antecedents of recycling behavior. *Journal of Business Research, 30*(1), 53–62.

McKenzie-Mohr, D. (2000). Fostering sustainable behavior through community-based social marketing. *American Psychologist, 55*(5), 531–537.

Miller, D. C. (1977). *Handbook of research design and social measurement* (3rd ed.). New York: David McKay Co.

Mohai, P. (1985). Public concern and elite involvement in environmentally-conservation issues. *Social Science Quarterly, 66*(4), 820–838.

Monroe, M. C. (2003). Two avenues for encouraging conservation *behavior. Human Ecology Review, 10*(2), 113–125.

Newhouse, N. (1990). Implications of attitude and behavior research for environmental conservation. *Journal of Environmental Education, 22*, 26–32.

Oskamp, S., Harrington, M. J., Edwards, T. C., Sherwood, D. L., Okuda, S. M., & Swanson, D. C. (1991). Factors influencing household recycling behavior. *Environment and Behavior, 23*(4), 494–519.

Pickett, G. M., Kangun, N., & Grove, S. J. (1993). Is there a general conserving consumer? A public policy concern. *Journal of Public Policy & Marketing, 12*(2), 234–243.

Porter, B. E., Leeming, F. C., & Dwyer, W. O. (1995). Solid waste recovery: A review of behavioral programs to increase recycling. *Environment & Behavior, 27*(2), 122–152.

Rotter, J. B. (1966). Generalized expectancies for internal versus external control of reinforcement. *Psychological Monographs, 80*.

Sand, L. K. (1999, April). 1999 Earth Day poll: Environmental concern wanes. *The Gallup Poll Monthly*, 38–44.

Sanders, C. D. (2003). The emerging field of conservation psychology. *Human Ecological Review, 10*(2), 137–149.

Schultz, P. W. (1998). Changing behavior with normative feedback interventions: A field experiment on curbside recycling. *Basic and Applied Social Psychology, 21*, 25–36.

Schwartz, S. H. (1977). Normative influences on altruism. In L. Berkowitz (Ed.), *Advances in experimental social psychology* (Vol. 10, pp. 221–279). New York: Academic Press.

Schwepker, C. H., & Cornwell, T. B. (1991). An examination of ecologically concerned consumers and their intention to purchase ecologically packaged products. *Journal of Public Policy & Marketing, 10*(2), 77–101.

Shrum, L. J., Lowrey, T. M., & McCarty, J. A. (1994). Recycling as a marketing problem: A framework for strategy development. *Psychology & Marketing, 11*, 393–416.

Sly, D., Hopkins, R., Trapido, E., & Ray, S. (2001). Influence of a counteradvertising media campaign on initiation of smoking: The Florida "truth" campaign. *American Journal of Public Health, 91*(2), 233–238.

Smith, S. M., Haugtvedt, C. P., & Petty, R. E. (1994). Attitudes and recycling: Does the measurement of affect enhance behavioral prediction? *Psychology & Marketing, 11*(4), 359–374.

Smith-Sebasto, N. J. (1992). The revised perceived environmental control measure: A review and analysis. *Journal of Environmental Education, 23*(2), 24–33.

Stern, P. C. (2000). Toward a coherent theory of environmentally significant behavior. *Journal of Social Issues, 56*(3), 407–423.

Stern, P. C., & Oskamp, S. (1987). Managing scarce environmental resources. In I. Altman & D. Stokols (Eds.), *Handbook of environmental psychology* (pp. 1043–1088). New York: Wiley.

Taylor, S., & Todd, P. (1995). An integrated model of waste management behavior—A test of household recycling and composting intentions. *Environment and Behavior, 27*(5), 603–630.

Thibaut, J., & Kelley, H. (1959). *The social psychology of groups.* New York: Wiley.

Thøgersen, J. (1996). Recycling and morality: A critical review of the literature. *Environment and Behavior, 28*(4), 536–558.

Van Liere, K. D., & Dunlap, R. E. (1980). The social bases of environmental concern: A review of hypotheses, explanations and empirical evidence. *Public Opinion Quarterly, 44*(2), 181–197.

Vining, J., & Ebreo, A. (1990). What makes a recycler? A comparison of recyclers and nonrecyclers. *Environment and Behavior, 22*(1), 55–73.

Vining, J., & Ebreo, A. (1992). Predicting recycling behavior from global and specific environmental attitudes and changes in recycling opportunities. *Journal of Applied Social Psychology, 22*, 1580–1607.

Vining, J., & Ebreo, A. (2002). Emerging theoretical and methodological perspectives on conservation behavior. In R. Bechtel & A. Churchman (Eds.), *Handbook of environmental psychology* (Vol. 2, pp. 541–558). New York: Wiley.

Vining, J., Linn, N., & Burdge, R. J. (1992). Why recycle? A comparison of recycling motivations in four communities. *Environmental Management, 16*(6), 785–797.

Webster, F. (1975). Determining the characteristics of the socially conscious consumer. *Journal of Consumer Research, 2*, 188–196.

CHAPTER FIVE

Rejuvenating Nature in Commercial Culture and the Implications of the Green Commodity Form

Mark Meister
Kristen Chamberlain
Amanda Brown
North Dakota State University

Commercial culture often relies on popular trends, opinions, and movements in manufacturing popularity. Premised on commercialization, defined by Mosco (1996) as a "process that specifically refers to the creation of a relationship between an audience and an advertiser" (p. 144), commercial culture constructs popularity rhetorically. Since the 1970s, advertisers and marketers have incorporated "green" themes in attempts to manufacture the popularity of certain products and services (Hoch & Franz, 1994). Most of the claims in these advertisements embellished or falsified a relationship to nature, leading to a widespread critique of green marketing (Hendy, 1996; Messaris, 1997).

Regardless of the critique of such marketing practices, many green marketers hold steadfast to their emerging and burgeoning craft: Nature remains, and will always be conceived in American culture as a commodity form, precisely because it operates in a capitalist system. Although this claim is contentious among environmental ethicists and scholars, Fitchett and Prothero (1999) argued that any stance—anthropocentric or ecocentric—that condemns capitalism and commodity relations is doomed to failure.[1] Fitchett and Prothero (1999), pointed out that just as the politicians, activists, policymakers, and bureaucrats have a part to play in shaping environmental policy, so do consumers. The green movement can either continue to battle against the institutions and structures that sustain and distribute commodity relations, or it can embrace them to further the cause. As

---

[1] An *anthropocentric* vision of nature refers to a "human centered" manifestation of nature while an *ecocentric* manifestation is a "nature centered" vision of nature.

Fitchett and Prothero (1999) pointed out, battling against the commodity form invites failure because the ideology of capitalism will always sustain itself despite "whatever revolutionary changes are made against it" (p. 274). Because the commodity form operates in the realm of popular culture and the media, which is, according to Prothero (2000) a "productive, persuasive, and communicative medium [that] can be used just as successfully by those seeking to achieve environmental enlightenment as it can for those who aspire to ecological martyrdom" (p. 46), the success of the ecological movement seems to be largely dependent on the use and mobilization of commodity culture rather than its rejection (Reisner, 1998).

Ultimately, the green commodity form does not challenge the broader ethical concern of consumption. As we discuss in this chapter, the consumption of nature in green commodities, such as nature-based beauty products, may increase public awareness about environmental issues, but it also increases public demand and for nature "as a product," whereby nature becomes a reflection of consumer desires. Nature is certainly commodified in green marketing practices. But as we argue in the case of AVEDA™, nature is part of a conservative discourse that redefines environmental ethics as a consumer rather than a citizen activity. In this discourse, advocacy for the environment simply involves being a "conscious" consumer who buys environmentally "friendly" (themed) products and does not address the embedded ethical questions of redefining environmental advocacy as consumerism.

One company that redefines environmental advocacy as a consumer practice is AVEDA. The AVEDA Corporation and its founder, Horst Rechelbacher, successfully manufacture popularity by commercializing nature-based beauty products. As an alternative to chemical-based beauty products, Horst Rechelbacher, created AVEDA in 1978, after being inspired by a trip to India (Tannen, 2002). While in India, Rechelbacher discovered Ayurvedic Medicine, "an ancient Indian practice that focuses on the seven centers of energy or 'chakras' that govern physical, mental, emotional, and spiritual functioning" (Product News, 1999). According to Rechelbacher, AVEDA describes itself as "both a company and a philosophy" (AVEDA, 1997, para. 1), and as an environmentally aware manufacturer of cosmetics (Tannen, 2002). The popularity of AVEDA's holistic philosophy is not only manufactured in its "natural products" but also in its advertising budget and distribution (AVEDA, 1997, para. 3).

AVEDA's popularity and commercial success are a reflection of Rechelbacher's professed philosophy and commitment to environmental causes (Tannen, 2002). AVEDA does not publish annual sales figures, but since 1997 when Rechelbacher sold AVEDA to Estee Lauder Corporation (Cardona, 1999), Estee Lauder's stock price has steadily grown (Davis, 2001; Tannen, 2002). The Estee Lauder Corporation reported earnings of $83 million for the third fiscal quarter of 2003 (Tannen, 2002). Rechelbacher (1999) broadly promotes holism and rejuvenation, noting in his book, *AVEDA™ Rituals: A Daily Guide to Natural Health and Beauty*, "We must understand, as a global community, that achieving happiness and wellness is a holistic process. For most of us it is one that requires changes in every part of our lifestyle, changes that include a daily practice of mind-, body-, and spirit-nurturing rituals" (p. 1). Prothero (1996) noted that nature-based marketing should demonstrate a holistic research design and consistency with the environmental policies of the sponsoring company.

This chapter investigates green commodity discourse as evident in AVEDA's commercial rhetoric and green marketing. Specifically, we focus in our analysis on Rechelbacher's books, *Rejuvenation: A Wellness Guide for Women and Men*, and *AVEDA™ Rituals: A Daily Guide to Natural Health and Beauty*, as well as the 2002 AVEDA product catalogue. Although Rechelbacher's books are not intended to be overt advertisements for AVEDA products, but as introductions to the "ancient wisdom of meditation, yoga, massage, and spiritual development with modern scientific research" (Rechelbacher, 1999, p. 7), both books promote AVEDA products throughout. One critic of the books pointed out that each book contains "thinly veiled commercials for AVEDA products" (Hagloch, 1996, p. 116). Thus, we do not distinguish between Rechelbacher's books and AVEDA product guides because both facilitate the green commodity form in the "manufacturing" of popularity in commercial culture. By including in our analysis Rechelbacher's books and the 2002 AVEDA product guide, we illustrate how the AVEDA commercial campaign is consistent with Fitchett and Prothero's (1999) suggestions for embracing the green commodity for increasing public awareness. Ultimately, the green commodity form (as depicted by AVEDA) fails to change ecological consciousness while reinforcing consumer consciousness.

We support our contention by illustrating the conservative jeremiadic features of AVEDA discourse. Herein, AVEDA seemingly redefines the traditional jeremiadic rhetorical form so that it is palatable for consumer culture. As a result, environmental awareness may increase, but so too does consumerism. AVEDA illustrates how commercial industries often embrace social issues for bolstering market shares and profits. As our analysis shows, the promotion of social responsibility within commercial culture does not invite consumers to question the ethical implications about the green commodity form.

We begin by outlining the relationship between advertising and nature in a capitalistic society and discuss the possibility of how green commodities can facilitate public awareness about environmental issues and sustainability. Our focus here is first on green marketing in general. We then discuss green marketing in relation to the cosmetic and personal-care industry. Next, we further profile AVEDA's green-commodity discourse and point out in our analysis that AVEDA's commercial rhetoric operates as a commercial jeremiad, intent on rejuvenating public awareness about environmental issues and not on challenging consumption. This study is not concerned with illustrating the direct effects of AVEDA advertising on consumer choices. Rather, we use rhetorical theory and analysis to explicate AVEDA's commercial discourse in shaping ideological and social attitudes that call for audiences to reconsider sustainability and nature preservation. We conclude by outlining the implications of a commercial jeremiad as represented in AVEDA's green marketing.

GREEN ADVERTISING, MARKETING, AND THE COSMETICS INDUSTRY

According to Williams (cited in During, 1999, p. 420) advertising is "the official art of modern capitalist society" that characterizes commercial culture. Because advertising refers to paid-for messages that attempt "to transfer symbols onto commodities to increase the likelihood that the commodities will be found appealing and purchased" (Fowles, 1996, p. 13), advertising, noted Williams, "has become involved in the teaching of social and personal values" (cited in During, 1999, p. 421). Advertising rhetorically aligns products and services to a larger human system of socially constructed knowledge that are lived through lifestyle choices. "Advertisements," noted Williamson (1978), ". . . must take into account not only the inherent qualities and attributes of the products they are trying to sell, but also the way in which they can make the properties mean something to us" (p. 12). In all, advertising is not only a creative discourse, as Messaris (1997) suggested, but more significantly, a political discourse. Advertising is political in that it facilitates meaning creation. Goldman and Papson (1996) concluded: "[t]he dilemma faced by corporate advertisers today is how to cut through the clutter and get viewers to notice their message. Advertisers often respond with even more spectacular executions" (p. 27). Williamson (1978) explained the significance of images in advertising: "[t]hings 'mean' to us, and we give this meaning to the product, on the basis of an irrational mental leap invited by the form of the advertisement" (p. 43). With reference to AVEDA discourse, we agree with Messaris and Williams's contention that advertising constitutes political claims. AVEDA provides a clear illustration of the political nature of advertising because, as our analysis illustrates, AVEDA advocates lifestyle changes by commercializing the traditional jeremiadic rhetorical construct.

Motivated primarily by the need to be distinguished from competing visuals and messages, advertising appropriates rhetoric from wherever it wants: artful invention, cliche, and sensational picturing are equally mannered devices use to "hail" viewer's attention (Fowles, 1996). Because advertising shapes, reinforces, reflects, creates, and recreates social values, it potentially presents itself as a powerful social and ideological system that encourages lifestyle choices. The phrases *green washing*, or *green* advertising, for example, describes how products and services rhetorically identify with environmental issues (Campbell, 1999; Goldman & Papson, 1997).

Beginning with Fisk's (1974) *Marketing and the Ecological Crisis*, critical attention to "green" advertising and marketing surfaced. The focus during the 1970s and 1980s was to advertise and market products and services in association with environmental issues (such as recycling programs) to facilitate social responsibility. This emphasis enhanced the image of sponsoring companies as socially responsible by associating its products and services with environmental issues. Social critics, Kilbourne, Banerjee, Gulas, and Iyer (1995) attempted to clarify the nature of green advertising, thus demonstrating that the concept is far more complex than the extant marketing literature suggests. The complexity of understanding how nature is commercialized in green advertising and marketing relates to the ambiguity and abstractness of nature as a symbol. Nature is a powerful symbol in advertising because it is ". . . captive of our language community; the environment, beyond its physical presence, *is* a social creation" (Cantrill & Oravec, 1996, p. 2, emphasis in original). Peterson (1997) agreed, noting "[h]umans add language to nature, thus enabling them to manipulate and transform their origins" (p. 3). Nature functions rhetorically: It encompasses ideologies and philosophies as well as profit-driven incentives and motives. One example of this is the concept of *sustainability*. Peterson (1997) pointed out that the concept of sustainability rhetorically encompasses and simultaneously implies respect and concern for nature with economic growth. Meister and Japp (1998) pointed out that the concept of sustainability "hides" the paramount motive of economic development by implying associations with nature.

Personal-care products, such as shampoos, soaps, moisturizers, hair-care products, and cosmetics, have been marketed products with misleading claims about their environmental benefit or their "natural" or "organic" ingredients. Each year, consumers, primarily women, spend over $28 billion on cosmetics alone (Erickson, 1997). Rechelbacher (1999) pointed out that through the simple act of using cosmetics and personal-care products, people are exposing themselves to harmful chemicals. One of the most harmful chemicals, phthalates, are found in 72% of beauty products (Women's voices for the earth, 2002). Phthalates are linked to, "reproductive birth defects, infertility and other illnesses" (Women's voices for the earth, 2002, para. 2). Researchers for the Centers for Disease Control argue that phthalates found in the urine of young women are linked to cosmetics.

In addition to phthalates, beauty products, such as hair dye, are potentially linked to cancer (Erickson, 2002). Many products contain chemicals that have not been tested for toxicity. Erickson (2002) explained, "Although most cosmetic companies voluntarily test their products for common sensitivities, researchers from the National Research Council found that, of the tens of thousands of commercially important chemicals, only a few have been subjected to extensive toxicity testing and most have scarcely been tested at all" (para. 3). While many beauty products contain harmful chemicals, the consumer is, quite often, not aware of the chemicals contained in the products they use. Because beauty products, especially fragrances, contain hundreds of ingredients, it is impossible to list all of them on a product's package (Pickrell, 2002). Moreover, through legal loopholes, such as "trade secrets," companies can keep harmful chemicals off ingredient lists.

Specifically, Pickrell (2002) contended that the cosmetics industry insulates itself from criticism for using phthalates because of "trade secrets." "By claiming the chemicals are fragrance or that they're trade secrets," notes Pickrell (2002, para. 7), "companies can legally keep phthalates off ingredient lists."

As an alternative to harmful products, consumers are increasingly purchasing products that contain natural or organic ingredients (Bender, 2002; Fost, 1996). AVEDA is the industry leader in promoting all-natural beauty products (Bender, 2002; Environmentally friendly, 1991). However, an article in *Drug & Cosmetic Industry* (1991), a leading trade magazine in the industry, criticizes AVEDA for exploiting consumers' awareness of the environmental issues (Environmentally friendly, 1991). Seemingly, exploitation occurs because it is easy for consumers to become confused by beauty products that claim to be natural. Erickson (2002) explained, "With all the new 'natural' products on the market, it's easy to be fooled. Just because a product boasts oatmeal, aloe vera or other plant-based ingredients, doesn't mean it's chemical free" (para. 7). Furthermore, there are no accepted standards as to what constitutes a "natural" product (Bender, 2002; Erickson, 2002).

Given the prominence of nature as a powerful symbol in commercial culture for marketing beauty products (and advertising's general potential for shaping popular support for environmental issues), an examination of AVEDA's green marketing illustrates how popularity about environmental issues is commodified. In what follows, we discuss AVEDA's discourse as a commercial jeremiad to illustrate how popularity is manufactured through powerful rhetorical strategies that promote rejuvenation in nature.

THE POSSIBILITIES OF A GREEN COMMODITY: AVEDA AS COMMERCIAL JEREMIAD

Fitchett and Prothero (1999) claimed that green marketing needs to embrace the commodity form because it encourages environmentally responsible behavior in both consumers and organizations. Although the term *commodity* can be broadly defined as "materialist," as a physical product or service that has a value in terms of utility and exchange, many green marketers argue that commodities can also increase public awareness. Rather than vilifying the commodity form, green marketing uses it for promoting responsible ecological behavior. Prothero (2000) clarified: "A green commodity discourse needs to be developed that employs positive, persuasive, and communicative characteristics in such a manner that enables ecological objectives to be prioritized and achieved" (p. 51). In this vein, green marketers embrace capitalism as an ideology that can further ecological causes. Accordingly, green marketers can work within the capitalist system, re-forming the system so that it is not forever aligned with environmental exploitation. Capitalism provides a valuable and culturally accepted discourse that can be aligned with the merits of ecological responsibility to popularize and communicate the message to a wider public.

In essence, green marketing posits that the commodity form needs piety. Piety is defined by Burke (1953) as "loyalty to the sources of our being" (p. 71). Green commodities should be aligned with social responsibility, concern, and piety. For too long, the argument continues, the notion of piety and industry have been at odds; distinct rhetorical forms whereby corporate piety functions as a rhetorical window-dressing for capitalistic motives. Yet, Prothero's (2000) challenge of creating a green commodity, inculcated with piety, is rhetorically intriguing. We continue our discussion of the green commodity form by focusing on the rhetorical construct of the jeremiad (a construct grounded on piety). We argue that the traditional jeremiad is modified for purposes of employing a green commodity form that "rejuvenates" ecological perspectives. The commercial jeremiad incorporates the commodity form with a sense of piety. Rather than evoking fear, the commercial jeremiad differs from the traditional jeremiadic form by evoking piety and concern about social issues.

The Commercial Jeremiad

In the tradition of public-address scholarship, the jeremiad calls for the return to key values and beliefs (Bercovitch, 1978; Buehler, 1998). The notion of piety is central to these claims. The *jeremiad* is widely defined as a "political sermon" because it furthers a political or social agenda (Jasinski, 2001). As such, the jeremiad is a powerful form of oratory because the audience experiences an integrated combination of appeals (core values, deviation from core values, redemptive possibilities, and a vision of the future). Over its long evolution and study, the jeremiad has been used to further conservation (Buehler, 1998), to eulogize (Murphy, 1990), and to comment on the state of American moral decay (Jendrysik, 2002; Johannesen, 1985). Whether religious or secular, the morality, or piety, of the audience is in question.

The jeremiad identifies a covenant with a chosen people by making a public proclamation of decline (Jendrysik, 2002; Owen, 2002). This lamentation accuses the chosen of becoming impious, often corrupt, and in many cases, prophesizes about the decline of social values. Through immoral action or inaction, the chosen people are in physical or spiritual jeopardy. Still another rhetorical function of the jeremiad is to predict redemption (Owen, 2002). Redemption allows for the moral covenant to be reestablished for the chosen. For critics of contemporary America, the corruption can only be stopped with a return to the teaching of traditional Christian values (Jendrysik, 2002). Another important feature of the jeremiad is that it acts as a bridge connecting the present to the past (Owen, 2002). This is an important aspect of the third function of the jeremiad, the promise of redemption. Reclaiming piety for the chosen people can be achieved by remembering an ideal moral past.

Finally, the jeremiad performs in two ways: it mobilizes the audience by creating fear and desperation and it builds optimism by prophesizing redemption. Jeremiads combine "lamentation with a firm optimism about the eventual fate of the community. That vision of the 'shining city on the hill' worked to unite the audience in pursuit of the goal and to reaffirm the values of the community" (Murphy, 1990, p. 403).

Because the jeremiad is a political statement, it faces special rhetorical challenges. In colonial times, the Puritans shared a common belief system that makes the appeal applicable to a mass audience based on common religious values. Today, because of diversity, it is more difficult to appeal to a belief system that is shared by all members of an audience (Jendrysik, 2002). To overcome the difficulty, the jeremiad must be vague and ambiguous so as to appeal to a wider audience while at the same time not alienating potential allies. Such is the case within the contemporary environmental movement. The traditional jeremiadic form is often the rhetorical device of choice by many environmentalists opposed to the commodity form. However, laminations about the plight of society's overconsumption, the commodification of nature, and the absence of an environmental ethic, are at the heart of many of the criticisms (Benton, 1995; Dadd & Carothers, 1991; Glickman, 1999; Irvine, 1991; Tokar, 1997). Yet, a discourse of sustainability—a discourse promoting holism, piety, concern for future generations, social concern, humanism, and profits—exhibits how the traditional rhetorical form becomes ambiguous for commercial uses.

As a major "player" in the commercial culture of beauty products, AVEDA prophesizes about the possibility of the nature commodity for ecological motives. Like the traditional jeremiad, the commercial jeremiad evokes fear, but not to the extent of its traditional form. In the case of the commercial jeremiad, fear becomes concern, the consumer and consumption is not lamented as entirely evil, and the chosen people are described as only partially sinful. Moreover, in the commercial jeremiad, consumers are the chosen people, obliged to "rejuvenate" themselves by considering the possibilities of sustainability—holism, piety, and social responsibility—all the while participating in the exchange of goods and services. This is particularly the case of AVEDA. In the analysis that follows, we point out the commercial jeremiadic characteristics of AVEDA's promotional materials. Specifically, we outline how the commercial jeremiadic form facilitates the more market-friendly response of concern rather than fear, whereby the consumer and consumption are not evil, but simply in need of rejuvenation. We conclude by addressing how the green marketing of AVEDA succeeds for purposes of rejuvenating consumer desires, but ultimately fails in transforming consumers' ethical perceptions about nature.

AVEDA: Prophesizing Rejuvenation Through Nature

Rechelbacher's two books and the AVEDA product catalogue set forth the idea of "environmental piety" that functions as a commercial jeremiad. Specifically, AVEDA's commercial jeremiad laments the decline of beauty both in humanity and nature, promotes sustainability as the key to rejuvenating beauty in both humanity and nature, and promotes holism by calling consumers to action in defense of beauty and the environment. Moreover, emphasis on the natural as spiritual connects contemporary consumers to a past where people were more in touch with nature, an idyllic past where some of today's chemical-laden, mass-produced products were not available. As such, AVEDA promotes a rhetorical vision very similar to that of sustainability, a commercialized jeremiad that embraces the commodity form: "Our mission at AVEDA is to care for the world we live in, from the products we make to the ways in which we give back to society. At AVEDA, we strive to set an example for environmental leadership and responsibility, not just in the world of beauty, but around the world (AVEDA, 2002, p. 2).

Lamenting the Decline of Beauty.   To understand why AVEDA's commercial jeremiad laments the decline of beauty, it is necessary to profile Rechelbacher's background. In 1978, Rechelbacher created AVEDA, a company with an environmental agenda (Rechelbacher, 1999). As a hairdresser in the 1960s and 1970s, Rechelbacher experimented with his own line of cosmetics and hair-care supplies, while also becoming interested in Eastern philosophy and religion (The rake secrets of the city, 2002). In the mid-1970s, Rechelbacher changed his fast-paced lifestyle. "About this time, a light went on. He saw with clarity new connections between his spiritual interests, his business ventures, and his personal interests" (The rake secrets of the city, 2002, para. 3). Rechelbacher went on to receive a degree in Ayurvedic medicine, the study of essential oils derived from plants and flowers that takes a holistic approach to health (Rechelbacher, 1987). The Ayurvedic philosophy takes into account the balance of one's mind, body, and spirit. Furthermore, the philosophy seeks a balance between the self and the environment (Rechelbacher, 1999).

In his study of Ayurvedic medicine, Rechelbacher became convinced that beauty products should be plant-based, as opposed to products based on petrochemicals (Rechelbacher, 1999). Since its inception, AVEDA has grown to include more than 650 plant-based products, in addition to destination and day spas (Rechelbacher, 1999). Rechelbacher (1999) argued that AVEDA makes intelligent use of earth's resources. Rechelbacher clarifies his desire for creating AVEDA while condemning current cosmetic manufacturing:

> Petrochemicals are chemicals derived from petroleum. Used as raw materials in the manufacture of many household products, petrochemicals are nonrenewable and

can be toxic to people and other living organisms. They pollute our soil, water, air, and bodies. They are widely used because they cost significantly less than pure ingredients extracted from plants. But increased consumption of plant-derived materials would lower their cost, eventually ensuring the ability of the natural environment to sustain itself. For this reason, AVEDA is committed to promoting the use of plant-derived materials. (Rechelbacher, 1999, p. x)

As a hairstylist, Rechelbacher observed the negative health impacts of beauty products that included petrochemicals such as polyvinyl chloride, or PVC. "In fact, a lot of my friends are dead, all died of lung cancer—in the 1950s, 60s, hairsprays killed a lot of people, customers and stylists. Nobody knew. Now we know, one molecule of polyvinyl chloride—PVC—can cause cancer" (The rake secrets of the city, 2002, para. 57). Rechelbacher (1987, 1999) also contended that household and beauty products should come from natural sources, not petrochemicals.

In Rechelbacher's (1987) first book, *Rejuvenation: A Wellness Guide for Women and Men*, he emphasizes the jeremiadic concept that humans are destroying the natural environment—a lamentation about the decline of human beauty as a reflection on the decline of nature's beauty: "Ecological collapse is occurring in large and small ways throughout the world because humankind no longer practices environmental balancing" (p. 8). In the product catalogue, AVEDA substantiates this prophecy. One section is entitled "Vanishing Voices, Lost Knowledge," which reinforces AVEDA's sensitivity toward environmental losses and improper social practices. The caption on an empty page reads "Extinct: Sexton Mountain Mariposa Lily: The Sexton Mountain Mariposa Lily has vanished from North America—an asphalt highway was built over the lily's single remaining natural habitat in Oregon" (2002, p. 12). The caption goes on to offer three Web sites where concerned individuals can go to learn more about vanishing species and what they can do to help. The remainder of "Vanishing Voices, Lost Knowledge" discusses the importance of biological and linguistic extinctions, pointing out that languages are culturally significant and their loss deprives us all of important multicultural wisdom. For example, the article discusses the vital link between language tradition in the Penean culture of Brothero and indigenous environmental wisdom on how "modern influences" (2002, p. 13) are negatively contributing to the demise of the Penean traditional language.

Connecting the Past and the Present.   Like the traditional jeremiad, the commercial jeremiad, exhibited by AVEDA, connects the present with the past. Rechelbacher appeals to a simpler time, linking the past to a more natural way of life, and in doing so, provides a critique of Western culture, and how it contributes to displacing indigenous knowledge and the wisdom of nature. "Today, there's a return to some of the world's oldest medical practices, global processes that rely heavily on natural plant and herb remedies . . . all of which utilize phytochemicals and herbs to balance the body" (Rechelbacher, 1999, p. xiv). For Rechelbacher, the nostalgic past was a time of indigenous wisdom and knowledge of nature. "When most of the Western world turned its attention to this relatively new symptom-relieving modern medicine, a few wise practitioners . . . spent their time and energy rediscovering the ancient knowledge of nature" (Rechelbacher, 1999, p. xiii). Like the traditional jeremiadic form, Rechelbacher harkens a more spiritual time whereby nature was perceived as nurturing.

The rhetorical significance of recalling a simpler and more pious past with nature is central to Rechelbacher's attack on petrochemical-based beauty products. Consistent with the jeremiadic form, Rechelbacher prophesizes against an inherent evil, an evil "brought about by the mass production of petrochemical additives in skin and hair products" (Rechelbacher, 1999, p. xv). For Rechelbacher, indigenous wisdom about nature is vital to recalling a glorious past without petrochemicals.

Promoting Sustainability.   In order to live in balance with the earth, Rechelbacher (1987, 1999) argued that we must acknowledge the importance of plants in sustaining life. To sustain life, one must sustain the earth. Rechelbacher (1999) explained:

> We can't separate ourselves from the environment anymore because we rely on it so strongly, especially since our population keeps growing. If we keep depleting and polluting the environment, we end up destroying our own systems. To reinstate a healthy balance we all must become activists by supporting only those activities and companies that put back into the environment as much as they take out. (p. 10)

Rechelbacher (1999) argued that better living is a choice, one that requires changes in one's lifestyle: "While we are definitely moving in the right direction toward a more holistic view of our world and health, there's still a long way to go. The pursuit of happiness—not to mention beauty, health, and the ever-elusive fountain of youth—still leads many people to silver-bullet, quick-fix solutions" (p. xv).

AVEDA's piety regarding environmentalism and nature is very apparent. In fact, all of Rechelbacher's literature, including the product catalogue, is designed to showcase piety. This piety of business (profit) appears to be secondary and is very subtly interwoven into the language of the catalogue and Rechelbacher's books. Redemption for restoring beauty in humanity and in nature requires, according to AVEDA, a return to nature: "Ours may be the planet's most exquisite, interesting, memorable aromas—each individual, subtly and delightfully varying from bottle to bottle, season to season—as unique as the plants from which they come. Let your body choose what rare gift it needs today" (AVEDA, 2002, p. 14). This passage is an example of the pairing of the pieties of consumerism and environmentalism. The beginning of the skin-care section of the catalogue states "AVEDA skin care is an extraordinary blend of art and science that works in harmony with your system to nourish and restore balance" (p. 30). Rechelbacher introduces products in the process of introducing a lifestyle change that is possible (at least at this point) through the consumption of AVEDA products. Rechelbacher (1999) noted: "Youthful, healthy radiant skin. Radiant beauty. These are things we all wish for. We can achieve them—by nurturing our skin properly, by promoting a clean environment, by being as peaceful as we can be, and by getting in touch with our own individual beauty and style" (p. 108).

Significantly, Rechelbacher rhetorically recalls a glorious past and human redemption in the present to bolster his attack on the cosmetics industry's use of chemicals. In what follows, Rechelbacher's commercial jeremiad comes full circle. Rechelbacher's prophecy calls on "beautiful people" to engage in consumerism in order to reinvest in the glorious past of indigenous wisdom and knowledge about nature.

Holism and the Beautiful People. In the commercial jeremiad professed by AVEDA, the chosen people are consumers not only concerned with beauty, but with holism (a prominent theme in sustainability). "Imagine a butterfly wing that flutters the air that stirs the wind that creates the storm that roils the sea that crashes the coastline of a distant shore. We are part of everything and everything is part of us. We are all one" (AVEDA, 2002, p. 10). This statement and others like it hold an important place at the beginning of the 2002 product catalogue. The statement about the butterfly wing heads up a two-page spread entitled, "The Web of All Things," which discusses the ideology that all parts of the world are interconnected. Preceding "The Web of All Things" are sections, "Our Mission," which states AVEDA's mission in ten different languages and "Awaken," which is a short treatise on AVEDA's connection to nature and its commitment to environmental practices. Moreover, such sublime statements target those people who have the resources to be concerned with beauty both in its human form and in nature.

In his dedication to preserving the natural environment, Rechelbacher acts as a ". . . living billboard for his healthcare products" (Peiken, 2003). In an effort to retain his mental and physical health, Rechelbacher sold AVEDA to Estee Lauder. "Frustrated by the constant pressure of running a business that had outgrown him, he decided to focus on the things that mattered most to him: meditation and activism" (The rake secrets of the city, 2002, para. 6). Not only is Rechelbacher dedicated to the production of environmentally friendly beauty products, he is also devoted to a healthy and balanced lifestyle outside the realm of beauty products. Rechelbacher (1999) warned that our current environmental practices place humans on a downward trajectory:

> Lately, so many of us have forgotten how our actions can and often do interfere with the collective urge for the preservation of life that we have begun to put our surroundings and the life they hold in danger. If we think only about filling our bellies

and don't consider whether our food is treated with pesticides that harm the environment, then we may not have a healthy environment for much longer. And it won't matter, because as part of the life that this Earth sustains, we as people are endangered ourselves. In the end, if we don't make a change, we won't be here anymore, either. (p. 9)

The use of nature in green advertising, as documented earlier, is a powerful strategy, particularly in the cosmetics industry. The piety of environmentalism in the commercial jeremiad does not, however, evoke fear, but more passively, concern for the environment. Even though more and more Americans identify themselves as environmentally conscious, many are unwilling to sacrifice or change their lifestyle in defense of the environment. AVEDA understands that the environment is important to its consumers, but for marketing purposes, it promotes environmental concern rather than fear. Simply concern in protecting the environment is conceived as piety in the consumption of AVEDA products. When promoting "Love," one of AVEDA's "Pure-fumes" the catalogue claims "Our sandalwood is carefully collected from sustainable managed forests in Australia because of concern over the worldwide poaching and deforestation of sandalwood trees. So you can enjoy AVEDA Love with a clear conscience, knowing you made a difference" (2002, p. 18). Thus, concern for the environment is transformed into a rationale for purchasing products.

On page 50 of the 2002 product catalogue and on the AVEDA Web site, under the heading of "Protect the Planet," AVEDA highlights concern rather than fear. "How do you make a difference? Just by making smart choices. Choose products and aromas like ours, that care for you and the Earth." AVEDA is saying that environmentalists do not have to be activists, do not have to recycle, and do not have to do anything except shop AVEDA. AVEDA does not shun activism or recycling, indeed, they promote these activities. Yet, the activism facilitated by AVEDA is a consumer activity and not one grounded in facilitating an environmental ethic.

CONCLUSION AND IMPLICATIONS

The connection between consumerism and environmentalism fosters AVEDA's successful use of the commercial jeremiad. This is evident in the increasing quantity of beauty and personal-care products marketed as green or containing natural ingredients.

The commercial jeremiad, as we have profiled here, exhibits many of the same characteristics of the traditional jeremiadic form. AVEDA's commercial jeremiad laments the decline of beauty, draws connections between the past and the present, promotes sustainability, and promotes holism. Similarly, the traditional and commercial jeremiad remain conservative political positions. Murphy (1990) argued that the traditional jeremiad serves as a rhetoric of social control that limits "the kinds of choices they [Americans] can make about the future" (p. 412). Traditionally, the jeremiad is a conservative rhetorical form, because it is grounded in piety; a conformist moral notion.

Effectively, the commercial jeremiad, like the traditional jeremiad, prophesizes; it targets a chosen people in need of rejuvenation. But as demonstrated in the discourse of Rechelbacher and AVEDA, the tone of the prophecy and the identification of consumers as chosen for rejuvenation rather than redemption mark the commercial jeremiad as ultimately appealing not for social or cultural change, but for reinforcing the status quo (Murphy, 1990). Potentially, consumers become interested in environmental issues by accepting the commodity form because they see how such issues relate to consumption interests. In semiotic terms, the visual sign of nature in green marketing becomes arbitrarily "connected, disconnected, and reconnected to commodities [and] needs become insatiable" (Goldman & Papson, 1997). The irony of the commercial jeremiad, in the case of AVEDA and Rechelbacher, is that as its prophecy about nature and sustainability potentially increases cultural awareness about nature and advocacy for preserving it, the fact remains that consumer-based behaviors do not encourage a personal shift toward a conservation or preservation ethic. The commercial jeremiad posits the piety of nature and environmentalism (or human rights, or peace and justice), but its call for rejuvenation fails to bring about redemptive cultural change precisely because it is involved in consumptive-based practices. Both the jeremiad and consumerism are conservative and conformist activities that may generate public awareness about environmental issues, but fail to productively critique and reject the piety of both the market and the green commodity in promoting an environmental ethic that challenges consumption.

For example, AVEDA's discourse about nature and beauty may make the environmental ethic of justice more difficult because class distinctions rise as AVEDA's products are less affordable than alternative beauty and personal-care products. AVEDA's concern for nature becomes an economic issue reinforcing anthropocentricism. Rechelbacher (1999) acknowledged that the pure and natural ingredients used in AVEDA products cost more than the petroleum-based products in most beauty products. AVEDA also emphasizes the culture of beauty. Helping the environment is connected to enhancing one's outer physical beauty. Ultimately, AVEDA's marketing discourse advocates consumption by engaging only those who are concerned with beauty and who can afford its "natural" beauty products.

Thus, the aesthetic and sublime vision of nature found in green marketing provides a compelling and popular rhetorical vision, but does not bring about the ultimate change needed for embracing an environmental ethic. The regularity of the symbolic representations of nature creates among consumers an unwarranted sense of knowledge about, or familiarity with, nature (McKibben, 1992). Because images of nature are a constant part of the cultural landscape—not only through the images of nature in the commercial jeremiad of AVEDA, but also through the poster and calendar art, cartoons and picture postcards, that are woven into our daily lives—we may come to feel we know nature. DeLuca and Demo (2000) noted that visual images of nature "cultivate and propagate an image of sublime nature, but, to be precise, a spectacularly sublime nature reduced to domestic spectacle, a nature both sublime and a source of sustenance for the civilized tourist" (p. 254). Because nature is domesticated, thereby catering to human desires, Tokar (1997) suggested that an environmental backlash is taking shape. Tokar (1997) pointed out that three related phenomena—the absorption of the mainstream environmental movement by the political status quo, the emergence of corporate environmentalism, and the proliferation of "ecological" products in the marketplace—have helped fuel the perception of a declining popular commitment to environmental protection.

In this way, green marketing, like that of AVEDA, will continue propagating the nature/culture dichotomy; defining, shaping, and promoting "sustainable" lifestyles that seemingly encourage environmental awareness while necessitating and rejuvenating consumptive practices. What is disturbing, from an ethical communication standpoint, is how themes related to consumption and technology, are redefined into a pseudo-environmental ethic. Leopold (1966) foreshadowed this shift, noting:

> Perhaps the most serious obstacle impeding the evolution of a land ethic is the fact that our educational and economic system is headed away from, rather than toward, an intense consciousness of land. Your true modern is separated from the land by many middlemen, and by innumerable physical gadgets. He has not [a] vital relation to it; to him it is the space between cities where crops grow. Turn him loose for a day in the land, and if the spot does not happen to be a golf link or a "scenic" area, he is bored stiff. Synthetic substitutes for wood, leather, wool, and other natural land products suit him better than the originals. In short, land is something he has "outgrown." (pp. 261–262)

Green marketing, and its vision of sustainability, facilitates a consumption ethic, rather than an environmental ethic predicated on interconnectedness with the diversity of nature. Leopold's (1966) land ethic, for example, poignantly points out the importance of diversity in facilitating a conservation ethic. The prominence of green marketing, like that of AVEDA, requires consumers to be conscious of the blurred ethical boundaries. If the concealment surfaces in an analysis of print advertisements, there certainly exist broader levels of abstraction on televised green marketing. The motion, sound, music, and context of green advertising and marketing on television provides further "synthetic substitutes" that threaten our connectedness to nature—a connectedness more complex and more significant than the commodity form. Just as AVEDA markets itself as environmentally friendly, so are other companies hoping to capitalize from today's emphasis on environmentalism. Critical examination of these products and their marketing strategies is vital, because, as noted by Prothero (1996, 2000), the commodity form can popularize environmental awareness. Specifically, research that focuses on how the commercial jeremiad invites spiritual understanding of the commodity form creates an interesting dynamic between nature and science that raises interesting scholarly questions. Finally, as many of these products are marketed largely toward women, a feminist perspective could be particularly insightful.

REFERENCES

AVEDA™: Creating a healthy business. (1997). [Electronic Version]. *Drug & Cosmetic Industry, 161,* 16–19.

AVEDA™: The art and science of pure flower and plant essences. (2002). [Product catalogue]. New York:

Bender, M. (2002). Makeup goes organic. [Electronic Version]. *Heath, 16,* 118–126.

Benton, L. M. (1995). Selling the natural or selling out? Exploring environmental merchandising. *Environmental Ethics, 17,* 3–22.

Bercovitch, S. (1978). *The American jeremiad.* Madison, WI: University of Wisconsin Press.

Buehler, D. O. (1998). Permanence and change in Theodore Roosevelt's conservation jeremiad. *Western Journal of Communication, 62,* 439–458.

Burke, K. (1953). *Permanence and change: An anatomy of purpose.* Berkeley: University of California Press.

Campbell, C. (1999). Consuming goods and the good of consuming. In L. B. Glickman (Ed.), *Consumer society in American history: A reader* (pp. 19–32). Ithaca:, NY Cornell University Press.

Cardona, M. M. (1999, May 3). AVEDA™ gets bigger budget, new ad look under Lauder. [Electronic Version]. *Advertising Age,* p. 10.

Dadd, D. L., & Carothers, A. (1991). A bill of goods? Green consuming in perspective. In C. Plant & J. Plant (Eds.), *Green business: Hope or hoax?* (pp. 11–20). Philadelphia: New Society Press.

Davis, D. (2001). Lauder signals significant change. [Electronic Version]. *Global Cosmetic Industry, 168,* n.a.

DeLuca, K. M., & Demo, A. T. (2000). Imaging nature: Watkins, Yosemite, and the birth of environmentalism. *Critical Studies in Media Communication, 17,* 241–261.

During, S. (1999). *The cultural studies reader, 2nd edition.* London: Routledge.

Environmentally friendly: The bandwagon gathers speed. (1991). *Drug and Cosmetic Industry, 149,* 31–35.

Erickson, K. (1997). Make-up call: Beauty doesn't have to be a chemical stew. [Electronic Version]. *E, 8,* 42–43.

Erickson, K. (2002). *Drop dead gorgeous: Protecting yourself from the hidden dangers of cosmetics.* Chicago: Contemporary Books.

Fisk, G. (1974). *Marketing and the ecological crisis.* New York: Harper & Row.

Fitchett, J. A., & Prothero, A. (1999). Contradictions and opportunities for a green commodity. *Advances in Consumer Research, 26,* 272–275.

Fost, D. (1996). Improving on nature: Enhancing functionality in the search for the ideal ingredient. [Electronic Version]. *Drug & Cosmetic Industry, 159,* 46–49.

Fowles, J. (1996). *Advertising and popular culture.* Thousand Oaks, CA: Sage.

Glickman, L. B. (1999). Born to shop? Consumer history and American history. In L. B. Glickman (Ed.), *Consumer society in American history: A reader* (pp. 1–14). Ithaca, NY: Cornell University Press.

Goldman, R., & Papson, S. (1996). *Sign wars: The cluttered landscape of advertising.* New York: Guilford.

Hagloch, S. B. (1996). [Review of the book, *Aveda™ rituals: A daily guide to natural health and beauty*]. *Library Journal, 124*, 96.

Hendy, D. (1996). The green theory: Facing the challenge of environmentally-friendly packaging in the 'natural products' market. [Electronic Version]. *Drug & Cosmetic Industry, 158*, 26–29.

Hoch, D., & Franz, R. (1994). Eco-porn versus the constitution: Commercial speech and the regulation of environmental advertising. *Albany Law Review, 58*, 441–466.

Irvine, S. (1991). Beyond green consumerism. In C. Plant & J. Plant (Eds.), *Green business: Hope or hoax?* (pp. 21–29). Philadelphia: New Society Press.

Jasinski, J. (2001). *Sourcebook on rhetoric: Key concepts in contemporary rhetorical studies.* Thousand Oaks, CA: Sage.

Jendrysik, M. S. (2002). The modern jeremiad: Bloom, Bennett, and Bork on American decline. *Journal of Popular Culture, 36*, 361–382.

Johannesen, R. L. (1985). The jeremiad and Jenkin Lloyd Jones. *Communication Monographs, 52*, 156–171.

Kilbourne, W. E., Banerjee, S., Gulas, C. S., & Iyer, E. (1995). Green advertising: Salvation or oxymoron? [Electronic Version]. *Journal of Advertising, 24*, 7–19.

Leopold, A. (1966). *A Sand County almanac.* London: Oxford University Press.

McKibben, B. (1992). *The age of missing information.* New York: Plume.

Meister, M., & Japp, P. M. (1998). 'Sustainable development' and the 'global economy': Rhetorical implications for improving the 'quality of life.' *Communication Research, 45*, 223–234.

Messaris, P. (1997). *Visual persuasion: The role of images in advertising.* Thousand Oaks, CA: Sage.

Mosco, V. (1996). *The political economy of communication: Rethinking and renewal.* Thousand Oaks, CA: Sage.

Murphy, J. M. (1990). "A time of shame and sorrow": Robert F. Kennedy and the American jeremiad. *Quarterly Journal of Speech, 76*, 401–414.

Owen, S. A. (2002). Memory, war, and American identity: Saving Private Ryan as cinematic jeremiad. *Critical Studies in Media Communication, 19*, 249–282.

Peiken, M. (2003, March 27). Visual arts: Horst offering focus on the art of being whole. Retrieved May 30, 2003, from www.twincities.com/mld/twincities/entertainment/5486785.htm?template= contentModules/printstory

Peterson, T. R. (1997). *Sharing the earth: The rhetoric of sustainable development.* Columbia: University of South Carolina Press.

Pickrell, J. (2002). Beauty products may damage fetal development. [Electronic Version]. *Science News, 162*, 36.

Product News. (1999). [Electronic Version]. *Global Cosmetic Industry, 164*, 70.

Prothero, A. (1996). Environmental decision making: Research issues in the cosmetic and toiletries industry. *Marketing Intelligence and Planning, 14*, 19–25.

Prothero, A. (2000). Greening capitalism: Opportunities for a green commodity. *Journal of Macromarketing, 20*, 46–56.

The rake secrets of the city. (2002, December). Horst: The Rakish interview. The fine art of living well. Retrieved May 30, 2003, from www.rakemag.com/printable.asp?catID=46&itemID=861&pg=all

Rechelbacher, H. (1987). *Rejuvenation: A wellness guide for women and men.* Rochester, VT: Thorsons Publishers, Inc.

Rechelbacher, H. (1999). *AVEDA™ rituals: A daily guide to natural health and beauty.* New York: Henry Holt & Co.

Reisner, M. (1998). Green expectations: Making money the environmental way. [Electronic Version]. *The Amicus Journal, 20*, 19–24.

Tannen, M. (2002, September 15). Beauty and the feast: A new age cosmetics guru entertains the flock. [Electronic Version]. *The New York Times Magazine*, p. 106.

Tokar, B. (1997). Marketing the environment. In C. Plant & J. Plant (Eds.), *Green business: Hope or hoax?* (pp. 32–37). Philadelphia: New Society Press.

Williamson, J. (1978). *Decoding advertisements: Ideology and meaning in advertising.* London: Marion Boyars.

Women's voices for the earth: Should beauty products contain chemicals linked to birth defects and infertility? (2002, November 18). *US Newswire,* p. 1.

CHAPTER SIX

Rhetoric of the Perpetual Potential: A Case Study of the Environmentalist Movement to Protect Orangutans

Stacey K. Sowards
University of Texas at El Paso

Orangutans, the most endangered great ape species (Nadler, Galdikas, Sheeran, & Rosen, 1995), have received substantial international attention due to their rapidly declining numbers and deforestation of their habitats. Ninety percent of all orangutans live in Indonesia; the other 10% are located in Malaysia. There are an estimated 15,000 to 24,000 orangutans on Borneo (an island shared by Indonesia, Brunei, and Malaysia) and Sumatra (Knott, 2003). Since the beginning of the 20th century, the number of orangutans has dwindled so that only 7% of Bornean and 14% of Sumatran orangutans remain (Rijksen & Meijaard, 1999). In the 1990s alone, Rijksen and Meijaard (1999) estimate that thousands of orangutans have died.

Orangutans are considered a keystone, umbrella, or flagship species. Rijksen and Meijaard (1999) define an umbrella species as "one whose home range is large enough, and its habitat requirements wide enough, that when it becomes the focus of protective management the entire structure of the original biological diversity of its range is automatically protected as well" (pp. 78–79, citing Stork, 1995). Because orangutans need undeveloped territory and a variety of foods, possess eating habits that disperse seeds, and indicate the general health of the rain forest ecosystem, they are an important representative species for forest protection (Mittermeier & Konstant, 2000; Rijksen & Meijaard, 1999; Smits, 1992).

However, habitat destruction, the pet and souvenir trade, population pressures, lack of focus on conservation and preservation, and lack of protected areas and law enforcement have critically endangered orangutans and threatened the health of their rain forest habitat (Nadler et al., 1995; Rijksen, 1995; Rijksen &

Meijaard, 1999; Sugardjito, 1995).[1] Indonesia's rain forest is the fastest disappear-ing forest in Asia, and the second fastest in the world after Brazil's rain forest. In fact, more than 12,000 km^2 of forested area disappear every year in Indonesia due to logging and unsustainable agriculture (Potter, 1995), a problem compounded by the 1997 and 1998 forest fires that ravaged Indonesia (Kessler, 1997; Rijksen & Meijaard, 1999). Rijksen and Meijaard (1999) report that the immediate impact of logging can cause a 50% to 100% decline of orangutan density, and after a 5-year period, orangutan density in logged areas is only 40% of the orangutan density in unlogged areas.

Illegal logging ventures and the pet trade are in part a result of dire economic situations in Indonesia. Djiwandono (1998) reports that the Indonesian economic crises in 1997 and 1998 were among the worst of the century. These economic cri-ses combined with Indonesia's growing population of over 200 million have left many people in poverty. Furthermore, the Indonesian government has pursued aggressive transmigration policies, moving Indonesians from more densely popu-lated areas, such as Java and Madura, to Kalimantan (the Indonesian side of Bor-neo) and Sumatra (Dove, 1998; Sunderlin, 2002). This practice has increased pop-ulation pressures in areas that used to be sparsely populated. Logging, diamond, coal, and gold mining, and the pet and souvenir trade have become lucrative ven-tures. Rijksen and Meijaard (1999) report that only .5% of the orangutan market is detected in Indonesia and Malaysia. In fact, "more orangutans were sold ille-gally in Taiwan between 1990 and 1993 than are housed in all the world's zoos" (Nadler et al., 1995, p. v).

Several environmental, nongovernmental organizations (ENGOs) and prima-tologists are committed to protecting orangutans and their rain forest habitats. In this study, I examine how primatologists and ENGOs create rhetorical appeals to prevent orangutan extinction. Specifically, I examine the rhetoric (in English and Indonesian) of primary international orangutan protection organizations, such as the Borneo Orangutan Survival Foundation (BOSF), the Balikpapan Orangutan Society–USA (BOS–USA), the Orangutan Foundation International (OFI), the Gunung Palung Orangutan Conservation Program, and The Nature Conser-vancy. These organizations are the most prominent orangutan organizations in the world, operating in Indonesia, Asia, Australia, Europe, and North America. I

---

[1]The official status of endangered or threatened species depends on various organizational defini-tions. For example, the International Union for Conservation of Nature and Natural Resources (IUCN) classifies species into nine categories, including *extinct, extinct in the wild, critically endan-gered, endangered, vulnerable, near threatened, least concern, data deficient,* and *not evaluated* (Baillie et al., 2004). Orangutans in Borneo (*pongo pygmaeus*) are listed as endangered and orangutans in Suma-tra (*pongo abelii*) are listed as critically endangered (The IUCN Red List™, 2004). The U.S. Fish and Wildlife Service, responsible for administering the Endangered Species Act of 1973, lists two catego-ries, endangered and threatened (U.S. Fish & Wildlife Service, 2004). Orangutans are listed as endan-gered, although both Borneo and Sumatran orangutans are classified under *pongo pygmaeus* (U.S. Fish & Wildlife Service, n.d.). The Convention on the International Trade in Endangered Species (CITES) has rated the orangutan as highly endangered (Appendix 1 status).

also analyze popular press items by internationally renowned orangutan prima-tologists such as Biruté Galdikas, Carel van Schaik, and Anne Russon.[2] In this chapter, I argue that these organizations and primatologists have crafted a rhetoric of the perpetual potential through their discourse on the endangered status of orangutans and their emphasis on the perpetual ability to engage in environmen-tal activism to save orangutans. That is, there is always potential to save orang-utans, despite their endangered status and dwindling numbers. In what follows, I provide a framework for understanding the perpetual potential, using J. Robert Cox's (1982) discussion of uniqueness, precariousness, and timeliness to illustrate how environmental discourse relating to orangutans functions in Western[3] and Indonesian contexts and languages. These elements of Cox's study of the irrepara-ble illustrate the rhetorical function of the perpetual potential, demonstrating that orangutans have been *perpetually* on the brink of extinction in the past, present, and future, yet the *potential* to save them from extinction continues. Finally, I conclude with some implications for environmental activism for Indonesian and Western audiences.

APOCALYPTIC RHETORIC AS ENVIRONMENTAL APPEAL

Apocalyptic rhetoric has been used historically and contemporarily by various groups for different reasons. In the latter half of the 20th century, rhetorical prac-tices outlining the apocalypse of a nuclear holocaust were widespread. Several re-ligious groups have garnered media attention for their predictions of an impend-ing apocalypse that have ended in mass suicides and killings. Stephen O'Leary (1994) explains that religious apocalyptic rhetoric is "discourse that reveals or makes manifest a vision of ultimate destiny, rendering immediate to human audi-ences the ultimate End of the cosmos in the Last Judgment" (p. 6). According to O'Leary (1994), religious groups have used three primary topoi: evilness, author-

---

[2]Biruté Galdikas is considered the first person to study wild orangutans long term, starting in 1971 (Galdikas, 1995), although Herman Rijksen, a Dutch primatologist, also began his studies of wild orangutans in Sumatra around the same time (Rijksen & Meijaard, 1999). Carel van Schaik is another primatologist who has engaged in long-term studies of orangutans in Sumatra, and most recently, Kalimantan. Anne Russon is well known for her studies of reintroduced orangutans. Cheryl Knott is also well known for her studies of wild orangutans in West Kalimantan, documented in several *National Geographic* essays. I also have included several popular press items, such as Anne Russon's (2000) book, *Orangutans: Wizards of the Forest*; van Schaik's (2004) book, *Among Orangutans: Red Apes and the Rise of Human Culture*; Galdikas' (1995) autobiography, *Reflections of Eden: My Years with the Orangutans of Borneo*; and Galdikas and Briggs' (1999) book, *Orangutan Odyssey*.

[3]The terms *third world, first world, developing nation, developed nation, West, East, North*, and *South* are problematic in their dichotomization of two distinct epistemologies and ontologies. However, for lack of a better term, I refer to non-Indonesian audiences as Western audiences, meaning European, North American, and/or Australian audiences.

ity, and temporality. These topoi allow rhetors to construct a sense of urgency in a "dramatic contest of good and evil" (p. 6). Revelation and order also play an important role in apocalyptic rhetoric, as Brummett (1991) explains: "[the apocalyptic] is a mode of thought and discourse that empowers its audience to live in a time of disorientation and disorder by revealing to them a fundamental plan within the cosmos" (p. 9). Brummett (1988) further observes that the "Apocalyptic advises its audiences on what to do *now* in the secular world to prepare for *eternity*" (pp. 60–61, emphasis in original).

In a similar fashion, many environmentalists have called attention to environmental problems, claiming an environmental apocalypse is at hand (e.g., see Brummett, 1991; Cox, 1982; Killingsworth & Palmer, 1992, 1995, 1996; O'Leary, 1994). Apocalypse has become a "standard feature of environmentalist polemic" (Killingsworth & Palmer, 1996, p. 21). Many environmentalists, such as Rachel Carson, Paul Ehrlich, and Murray Bookchin, have been accused of apocalyptic or hysterical rhetoric that creates undue worry, emotionally charged reactions, and the politicization of science (Killingsworth & Palmer, 1992, 1995, 1996). As Killingsworth and Palmer (1996) explain, "The hyperbole with which the impending doom is presented—the image of total ruin and destruction—implies a need for ideological shift" (p. 41).

Unlike religious apocalyptic rhetoric, where followers believe they will find a better life in the Last Judgment, audiences of environmental rhetoric may despair in their inability to change impending environmental disasters. Shabecoff (1996) argues "[environmentalist] arguments are often weakened and their successes limited by the Chicken Little syndrome—too frequently, too shrilly, and much too inaccurately warning that the Apocalypse is near" (p. 76). For example, Cantrill (1996) maintains that too much information can increase passivity, creating what Cantrill calls environmental default mechanisms, or rhetorical obstacles that prevent audience action. Cantrill (1993) argues that people act if they perceive an immediate and social self-interest in the action. Furthermore, public participation is often necessary in creating effective environmental policies and activism (Belsten, 1996; Katz & Miller, 1996; Waddell, 1996). If the public is not interested in an environmental issue, public activism is significantly less likely.

J. Robert Cox's (1982) analysis of the irreparable in environmental discourse is a particularly useful framework for understanding the nature of the perpetual potential. Similar to apocalyptic rhetoric, the *irreparable* represents an action or inaction that may cause permanent loss or total destruction. However, the irreparable can also refer to loss of a specific and valued object or person that does not result in or is not caused by total devastation (Perelman & Olbrechts-Tyteca, 1969). Cox (1982) contends that rhetorical strategies employing the "locus of the irreparable" are rooted in audience choice. Such discourse emphasizes *uniqueness;* failure to act will cause the loss of the irreplaceable: "Loss of the unique is even more poignant when juxtaposed against the usual, the ordinary, the vulgar, that which is fungible or interchangeable" (p. 229, citing Perelman & Olbrechts-Tyteca, 1971). A second feature of this rhetoric is the focus on *precariousness*. The unique is either fragile, or "secure, but threatened by radical intrusion" (Cox, 1982, p. 230). Here again, with audience action, precariousness can be diminished because the audience has the choice to take action to avoid loss. Finally, a third feature of the irreparable is *timeliness*, the need for urgent action to avoid risk of the irreparable. The extraordinary becomes further threatened with time, yet present action can postpone or prevent the irreparable. Environmental discourse in many Western contexts urges audiences to act to save the irreparable for the future.

In the case of orangutan protection, environmental organizations employ appeals both to apocalypse and to the irreparable through their emphasis on total destruction of orangutan habitats and the rapidly declining number of remaining orangutans. However, a better lens to understand such discourse is the *perpetual potential*, because it offers a rhetorical alternative to the hopelessness of doomsday and apocalyptic rhetoric. The first feature of perpetual potential rhetoric is the emphasis on the irreparable or a pending apocalypse that is always and already occurring in perpetuity (past, present, and future). References to the past allow environmental organizations to create the probability for a future apocalypse, whereas the present brings the possibility for change and activism. As long as orangutans exist, ENGOs and primatologists will continue to importune audiences to save orangutans. In essence, urgency and the apocalypse are arbitrary, yet continual. Stephen O'Leary (1994) observes that the apocalypse "has already occurred; it is always about to occur; it is here now and always has been" (p. 220). It is always the present where a choice to avoid an apocalypse or the irreparable can be made, even though the present exists in perpetuity.

Such *arguments of perpetuity* focus on both the possibilities of environmental disaster and for environmental solutions. It is this balance between apocalyptic predictions and hopefulness that illustrate the second function of this rhetoric of the perpetual potential. Time threatens the unique, but is also full of possibility and potentiality. As Richard Rorty (1999) argues, hope is a key aspect of our humanity. Even if we are "limited to such fuzzy and unhelpful answers because what [we] hope is not that the future will conform to a plan, will [fulfill] an immanent teleology, but rather that the future will astonish and exhilarate" (p. 28). For example, Tarla Rai Peterson (1997) contends that the rhetoric of sustainable development "offers hope that there will be a tomorrow toward which we should direct our energies" (p. 171). Environmental organizations and activists see hope in the measures they take. This practice of working for a cause that seems doomed to fail might best be described as acting in Burke's comic frame. As Burke (1984) explains:

> A comic frame of motives, as here conceived, would not only avoid the sentimental denial of materialistic factors in human acts. It would also avoid the cynical brutality that comes when such sensitivity is outraged, as it must be outraged by the acts of others or by the needs that practical exigencies place upon us. . . . In sum, the comic frame should enable people *to be observers of themselves, while acting*. Its ultimate

would not be *passiveness*, but *maximum consciousness*. (pp. 170–171, emphasis in original)

The comic frame provides environmental leaders and primatologists the ability to believe in the perpetual potential, that orangutans will eventually be saved despite the seemingly environmental apocalyptic circumstances playing out in Indonesia. The alternative, Burke's tragic frame, would allow audience members to envision a world carried to its entelechial end, passiveness, and hopelessness, as does religious apocalyptic rhetoric (O'Leary, 1994) and the rhetoric of the irreparable (Cox, 1982). The hopeful nature of the perpetual potential, enacted in the comic frame, provides inspiration for both the leaders and the supporters of the orangutan cause.

THE PERPETUAL POTENTIAL IN ORANGUTAN PROTECTION

In creating a rhetoric of the perpetual potential, ENGOs importune audiences to recognize the general necessity of rain forest protection and the pernicious effects of environmental degradation. E. O. Wilson, the famed Harvard biologist, demonstrates how orangutan organizations and primatologists use apocalyptic rhetoric:

> The worst thing that can happen—will happen—is not energy depletion, economic collapse, limited nuclear war, or conquest by a totalitarian government. As terrible as these catastrophes would be for us, they can be repaired within a few generations. The ongoing process that will take millions of years to correct is the loss of genetic and species diversity by destruction of natural habitats. This is the folly our descendants are least likely to forgive us. (E. O. Wilson, quoted on BOS–USA fundraiser invitation, 2000)

Yet, their rhetoric is also marked by hopefulness, or the perpetual potentiality that these situations can be reversed. Such rhetoric is prevalent in Indonesian and Western contexts, both in English and Indonesian,[4] through appeals to uniqueness, precariousness, and timeliness.

Orangutans and Rain Forest Habitat: Uniqueness

Environmental organizations and primatologists employ rhetoric that establishes the uniqueness of orangutans and their rain forest habitat in several ways. Although the extinction of orangutans would mean the loss of biodiversity and a

---

[4]These rhetorical appeals also occur in German, Dutch, French, and Japanese through the branch organizations of Indonesian nongovernmental organizations (NGOs); however, in this study, I have not included appeals of the apocalypse in these languages.

single unique species, environmental organizations attempt to prove that orang-utans are unique above and beyond other endangered species, such as the panda bear, the Sumatran tiger, or the Javanese rhinoceros. Because so many environ-mental organizations seek to prove that their particular cause is unique, the uniqueness factor may have diminished persuasive appeal. The unique species has become the ordinary rather than the extraordinary because of the many species that are threatened with extinction. However, environmental organizations and primatologists working to protect orangutans engage in several rhetorical prac-tices that make orangutans more unique and extraordinary, at least to their audi-ences, than other endangered species. The perpetual potential is defined by the idea that environmental organizations have to prove perpetually that their cause is increasingly unique in comparison to other causes.

Much of the environmental literature regarding orangutans focuses on the di-versity of their rain forest habitats. Russon (2000) demonstrates the complexity of the forest by its sheer size and number of species alone:

> For a hint of that diversity, consider this: Borneo and Sumatra represent only 1.3 percent of Indonesia's land mass but they support 10 percent of its known plant spe-cies, 12.5 percent of its mammals, and 17 percent of its other vertebrates. Borneo alone has 10,000 to 15,000 species of flowering plants. That is as rich as the whole of Africa, which is forty times as large, and 10 times as rich as the British Isles. In addi-tion, Borneo has 3,000 species of trees, 2,000 orchids, and 1,000 ferns . . . Borneo's animal life is no less diverse. It is known to support about 222 mammals, 420 birds, 166 serpents, 100 amphibians, and 394 freshwater fish, not to mention the inverte-brates, by far the most numerous animal species in tropical rainforests. Many of these life forms are endemic, or unique to the island. (pp. 116–118)

The possibilities for new discoveries of unknown species demonstrate the perpet-ual potential. Even in the face of deforestation, the possibilities become perpetual. Because humans can never know what was lost, environmental organizations fo-cus on what we can know in the perpetual potential of biodiversity. A poster in In-donesian also emphasizes the uniqueness of orangutans and their habitat:

> Say goodbye . . . Wide scale forest conversion, illegal logging, and forest fires have caused significant forest damage. This loss of habitat is the largest threat for the orangutan and tarsius, the largest and smallest arboreal primates in the world . . . Help us protect forests in East Kalimantan, one of the most special places in the world. (The Nature Conservancy, 2002, translated from Indonesian)

In these examples, orangutans and their habitats are special, unique, and endemic. Continued deforestation will result in the loss of these unique species and habi-tats, and possibly species that are not yet known to humankind.

However, environmental rhetoric that emphasizes the vast number of "unique" species also may have the effect of rendering those species ordinary because there are so many "special" places, animals, and plants. Emphasizing that orangutans are a keystone species is one way to make orangutans unique in comparison to other species. For example, the BOSF brochure contends:

> When the situation for orangutans is good, it is a sign that the situation for tropical rain forests is good. . . . orangutans are known as an "indicator species," because the existence of an orangutan population in a certain region indicates there is a healthy ecosystem in that area. (Balikpapan Orangutan Survival Foundation, n.d., translated from Indonesian)

Many environmental organizations employ the keystone species argument as a reason to protect other species. This rhetorical practice makes orangutans and other keystone species more unique, while providing a rationale for protecting other species, such as the rain forest habitat, that may also be unique.

Even though orangutans have become unique through their keystone species status, environmental organizations strive to make the orangutan more extraordinary. The proliferation of keystone species that are endangered throughout the world means that even keystone species may become the usual or ordinary in environmental rhetoric. Thus, environmental groups attempt to establish that orangutans are unique above all other animals by privileging their Great Ape status and connecting them to humans:

> The word Orangutan literally means "forest person" in the Malay and Indonesian languages. It's an apt description, as orangutans share 97% of the genetic makeup of humans, and are intelligent, cognizant beings. They are completely unique in the great ape world, as they are the only Asian great ape. Their mainly solitary lifestyle, fruit-eating diet, and arboreality makes them like no other primate in the world. (Balikpapan Orangutan Society–USA, n.d.)

This close genetic makeup means that orangutans are a close relative to humans in the primate order, a fact that ENGOs and primatologists capitalize to bridge orangutan and human worlds. Their literature refers to orangutans as "our beautiful red-haired brothers and sisters," "among our cousins," and "one of our closest living relatives" (Balikpapan Orangutan Survival Foundation & Balikpapan Orangutan Society–USA brochures, n.d.; Galdikas & Briggs, 1999; Russon, 2000).

In addition to discussing orangutans' unique genetic composition and behaviors, environmental organizations and primatologists also anthropomorphize and personify orangutans. Every orangutan in a reintroduction center or in the wild has a story that personifies them. For example, the *1992 Orangutan Year Report* contains the stories and at least one photograph for every orangutan at the Wanariset reintroduction center, as the story of Hero illustrates:

> Hero is not afraid of anybody. He likes to disturb anybody who dares to near his the [sic] cage. He does however not really bite people or other orangutans. When he is

being hold [sic] he points out his hands in the direction where he wants you to go. He eats well but does not like to drink. He likes to play with the chains in his cage and is close friends with Alto especially when it concerns teasing Bento . . . Hero is still a very naughty boy. Alto keeps clinging to him all the time and Hero is always the initiator of their sex games. Whenever someone passes by his cage he will attempt to tore [sic] their shirts and make silly faces and sounds to those people . . . He is just as before, naughty, playful and not afraid of any person or other orangutan. He still invites the other orangutans to wrestle with him. (Smits, 1992, pp. 151–152)

Hero's story demonstrates his personification through his naughtiness, playfulness, and fearlessness. These behaviors create an endearing visualization of an orangutan who is much like a human baby or child. Another example in the BOS–USA newsletter demonstrates how orangutans are distinguished from other types of animals: "One day in March they brought in a small orangutan. She had been found under someone's house, probably being fed scraps like a dog. She was in a pathetic state, down to skin and bones, her hair in dreadlocks, but worst of all . . . there was little life left in her eyes" (Nielsen, 1999, p. 7). These stories personify and separate orangutans from other animals, rendering them uniquely connected to humans, and thus worthy of attention and environmental protection.

The effects of anthropomorphism through photographs, documentaries, and narratives establish the uniqueness of orangutans, but also create the perpetual potential through the images etched into the minds of the audiences. As Donna Haraway (1989) suggests, "To make an exact image is to insure against disappearance, to cannibalize life until it is safely and permanently a spectacular image, a ghost. The image arrested decay. That is why nature photography is so beautiful and so religious—and such a powerful hint of an apocalyptic future" (p. 45). For example, BOS–USA and BOSF's Web sites feature an orangutan named Theo. He is pictured in a plastic basket with a blanket. What the photographs do not tell the viewer is that Theo died a few days after the pictures were taken. Once the viewer has seen the orangutan in a photograph or read an orangutan's story, that image is fixed forever. Orangutans may diminish in number, but their stories and images will always remain. Even though Theo has died, he lives on through the rhetoric of perpetual potential. This rhetorical practice indicates that the possibility to save orangutans like Theo will always exist.

The perpetual potential in the case of orangutans is marked by advocacy that anthropomorphizes and elevates their uniqueness. Anthropomorphism can be an effective way to connect with audiences because it establishes common ground or characteristics between the human and nonhuman worlds in what Kenneth Burke (1969) calls consubstantiality: "both joined and separate, at once a distinct substance and consubstantial with another" (p. 21). Audiences see shared substance or likeness in orangutans through anthropomorphism. Primatologists argue that we can learn about humans through the study of nonhuman primates, because of their similarities in genetic composition, cognition, and behaviors. The perpetual potential emerges from what we do not know about orangutans that might teach us something about the human condition. In the case of orangutans, environmental organizations and primatologists continue to find ways to make orangutans more unique to their audiences, searching for the perpetual potential in audience identification with their cause.

At the same time, the perpetual potential appears in discourse relating to the "not yet known" species, waiting to be discovered in orangutan habitat. The possibilities for learning about biology and medicine through rain forest habitats also illustrate both the perpetuity and the potentiality of such rhetoric. New species could be discovered regularly and continually, offering the possibilities for curing diseases and discoveries about the world that we do not yet know: "Every time an acre of rainforest is burned or chopped down, we might lose a cure for cancer or AIDS" (Dedicated to the conservation, n.d.). Yet this potentiality only exists if environmental organizations are successful in protecting orangutans and their habitat. Finally, the perpetual potential exists through the fixity and finality of documentaries, photographs, and narratives that tell an orangutan's story at one particular moment in time, even though preserved forever. The rhetoric of the perpetual potential develops the possibilities and hopefulness for environmental successes. Unlike rhetoric of apocalypse and the irreparable, the perpetual potential establishes possibilities for discovering the unknown through appeals of uniqueness.

Hanging in the Balance: Precariousness

Precariousness is a fundamental element of the perpetual potential. To demonstrate that orangutans face a precarious situation, environmental organizations use language that indicates orangutans are on the brink or verge of extinction. Several events have made arguments of precariousness effective, such as the 1997–1998 forest fires that had a devastating effect on habitats and orangutan populations. The illegal pet trade also causes a larger number of females of reproductive age to be killed, thus reducing chances for replenishing orangutan populations. Because it is difficult to estimate how many orangutans remain and how many would be needed to sustain a wild population, orangutans are and will remain perpetually on the brink of extinction. Environmental organizations indicate that their actions and audience support can secure orangutan populations, yet they also simultaneously emphasize their precarious situation.

For instance, Anne Russon (2000) argues that humans have been a threat to orangutans throughout history:

> This tiny community, like the whole species, is on the verge of collapse, and this damage has been suffered at human hands. The threat is not new—it can be traced back 40,000 years to when humans first reached the isle of Borneo. But it is now on the point of sending orangutans into extinction, as it has sent so many other species in the last century. (p. 194)

Humans have been responsible for the decline of orangutans for 40,000 years, but it is now that orangutans face extinction. Similarly, an Indonesian newspaper article discussing the Gunung Palung Orangutan Conservation Program illustrates the threats orangutans face:

> In the middle of illegal logging at TNGP (Gunung Palung National Park) there is still a bright spot stored away; the natural and priceless riches [of the forest]. One of those is the orangutan, whose genetic relationship is very closely related to humans. These natural riches are threatened with extinction, because of the measures taken by those who destroy TNGP. Frankly, those people act blindly in response to the animals that are protected. (*Orangutan di TNGP terancam punah* [Orangutans in TNGP threatened with extinction], 2002, p. 18, translated from Indonesian)

This article implies that this dire situation is caused by illegal logging and lack of law enforcement, which threatens the forest and orangutans. Yet, a "bright spot" still exists, creating the potentiality that environmental organizations might be able to save orangutans.

Organizations and primatologists also emphasize the potential cataclysmic effects of the illegal pet trade. Although the organizations and primatologists recognize that deforestation is the greatest threat to orangutans, the pet trade factor is pervasive throughout their literature. Galdikas (1995) describes the pet trade in vivid detail:

> For every orangutan put up for sale, at least eight or nine die. To capture an infant orangutan, one must first kill the mother. But captured infants rarely survive for long. A single survivor like Sugito represents three or four other infants who died in captivity, as well as four or five mother orangutans who were murdered during their capture . . . All of this adds up to terrible slaughter. (pp. 134–135)

This passage represents the extensive coverage of the pet trade in newsletters, documentaries, news programs, books, and brochures. The Gunung Palung Orangutan Conservation Program outlines a similar scenario written in Indonesian:

> Do you know that for every one orangutan that becomes a pet, it is estimated that six other orangutans are killed? Orangutans are animals that reproduce very slowly. Female orangutans give birth approximately every eight years and only produce one infant for each birth. The result is that the orangutan population has a very slow growth rate. (*Ancaman kepunahan orangutan* [Orangutans threatened with extinction], n.d., p. 1, translated from Indonesian)

These activists clearly express the devastation that the pet trade causes to orangutan populations. The imbalanced number of female orangutans killed compared to male orangutans reflects a crisis for replenishing orangutan populations.

Furthermore, primatologists and ENGOs contend that the orangutan situation mirrors the future for humans. Orangutans reflect a troubled worldwide ecosystem, one for which humans are responsible. In the documentary, "Orangutans: Just Hangin' On," the narrator summarizes this approach in the conclusion to the film:

> So as the experts say, it's important for humanity to realize what's happening to the orangutans is actually happening to us. The orangutan situation is a microcosm of what is happening to humanity. Because of human greed, the rain forest, the world's natural air conditioner, is declining. Orangutans are a mirror of our own situation . . . If they are just hanging on, then so are we. (Searle, n.d.)

Such passages urge audience members to recognize the immediacy of environmental problems and take responsibility because these same problems threaten the existence of (human) life. The focus on human responsibility in many of these examples illustrates the precarious situation for orangutans but also provides the potentiality for reversing this situation.

Precariousness is ubiquitous in environmental rhetoric, continually reflecting the brink of environmental disaster year after year, decade after decade. In part, this emphasis on the perpetual brink of disaster occurs because environmental advocates need to maintain supporters and attract new followers. As Brummett (1988) suggests, apocalyptic religious rhetoric appeals to the already converted. The perpetual potential enables environmental organizations to reach out to new supporters, while maintaining the activity of current members. A hopeful note may move existing members to act repeatedly. However, to avoid member discouragement, environmental organizations use new disasters and more critical brinks, urging audience action that can reduce the precariousness of the disaster. The perpetual potential emphasizes the continual nature of environmental disasters, but also provides the potential to address those problems.

Threats and Opportunities: Timeliness

In the case of orangutans and other such endangered species, timeliness functions to encourage audiences to take immediate action, to prevent a worsening scenario. As with precariousness, timeliness as a rhetorical practice emphasizes urgency to save the unique. However, this sense of urgency is arbitrary and continues over time. Harré, Brockmeier, and Mülhäusler (1999) contend that temporal dimensions of environmental issues mean that humans do not know how much time will lead to predicted environmental disasters. Species disappear at an alarming rate, yet no one knows how many disappear or how many species are needed to maintain viable ecosystems. The potential to save orangutans is always now, rather than in the future. The anticipation of the future extinction of orangutans or rain forests requires that audiences act to save them, even if their efforts are futile. Environmental organizations often employ temporal appeals that provide activist frameworks to prevent the extinction of orangutans.

Environmental organizations and primatologists often outline the past, present, and future within a single exhortation. For example, the Balikpapan Orangutan Survival Foundation creates a sense of impending crisis in the present, yet also explains how serious the problem has become based on what existed in the past:

> Suitable orangutan habitat in Indonesia and Malaysia has declined by more than 80% in the last 20 years. The impact on the species goes beyond the removal of trees. The number of trees taken, the techniques for their removal, the accessibility of the site, enforcement against illegal hunting and poaching, and encroachment of slash and burn cultivation will determine how much habitat is lost and for how long. (Orangutans at risk, 1997)

ENGOs use words and phrases such as urgent, immediate, crisis, "orangutans face extinction," and "they will vanish from the wild within the next decade unless immediate action is taken!" (Balikpapan Orangutan Society–USA, n.d.). The past becomes very important in these rhetorical practices because environmental organizations have to demonstrate great loss to indicate that the crisis has become the present, and extinction will become the future. The urgent tone of this discourse creates a sense of the perpetual potential in that the crisis is the same every year, with slightly different numbers for habitat size and remaining orangutans. Yet, audiences can always take immediate action to prevent the extinction of orangutans.

The future also plays a role in the construction of the precarious situation facing orangutans. As the Gunung Palung Orangutan Conservation Program explains, "Experts estimate that within the next 20 years, orangutans will become extinct, unless there are dramatic changes in forest management and conservation. This situation has caused orangutans to be classified in the group of critically endangered species" (Mulyadi, 2002, translated from Indonesian). Similarly, Cheryl Knott (2003) gives a shorter time frame for orangutans: "As illegal loggers close in, the race is on to safeguard the home of Borneo's imperiled orangutans. At the current rate of habitat destruction, orangutans could be extinct in the wild in 10 to 20 years. We must stop this trend—the alternative is unthinkable." Audience action in the present can change the outcome of the future, and it is always now that is the critical time to enact changes. Here again, the present is always the critical time for action to avoid the unthinkable future of extinction and massive habitat loss.

The urgent and numerous environmental threats facing orangutans may leave audience members overwhelmed by the enormity and intractability of these threats, even as they feel compelled to participate and act to protect orangutans and habitat. ENGOs establish a theme of successful action to encourage membership and donations. For example, the Balikpapan Orangutan Survival Foundation offers several ways in which members can donate various sums of money and how the individual can help orangutans:

> Become a member of the Balikpapan Orangutan Society; sponsor or co-sponsor an orangutan; donate money to the Balikpapan Orangutan Society; start a campaign at work or in your school; call your local newspaper or radio station drawing their attention upon the disastrous situation in Kalimantan; draw the attention of friends and relatives upon this situation and make them support [it]. (How to help, 2000)

BOS–USA also encourages similar actions, but expands their list to include a letter writing campaign to U.S. Senators and Representatives to support the Great Ape Conservation Act and a list of items to avoid including rayon, tropical plywood, and palm oil. Their brochure encourages people to join, learn, teach, and fundraise to help BOS–USA become more effective. Orangutan Foundation International (OFI) also includes a "How to help" section on their Web site. They encourage people to become members of OFI, purchase OFI items (such as T-shirts, Galdikas' books, and posters), donate money, and promote outreach and education (How to help, 2000). These options contribute to the audience's ability to act, because they are all actions an individual could enact in their everyday lives.

In order to create rhetorical success, ENGOs also dedicate substantial rhetorical space to developing their successes and creating a sense of hope for the future. If the situation orangutans face is hopeless, action is significantly less likely. Hence, organizations and primatologists develop rhetoric that is built around human responsibility for orangutans and the success of acting on that responsibility. The organizational Web sites, brochures, and newsletters discuss successes in rehabilitation, reintroduction, law enforcement, and legislation. Successes encourage audiences to continue or begin activist measures and give hope for saving orangutans. Russon (2000) concludes her book with a plea for audience action:

> There are no hopeless causes—only hopeless people and expensive causes. Orangutan survival may be an expensive cause, but only hopeless people would abandon it. If this seems a sad end to my tale, remember that the end of my tale is not the end of orangutans. . . . The real end of the tale is, after all, up to us. (p. 203)

This passage summarizes the rhetorical practices emphasizing audience choice. Russon observes that there is hope in saving orangutans even though the situation is dire. Thus, the potential to choose activism is perpetual, spanning the past, present, and future.

CONCLUSION: IMPLICATIONS FOR INTERNATIONAL ENVIRONMENTAL ACTIVISM

Environmental organizations and primatologists working to protect orangutans have created international and Indonesian awareness about the threats facing the imminent extinction of orangutans with a modicum of success, especially for

Western audiences. Their rhetoric establishes orangutans as a unique phenomenon, one that is emblematic of the diverse environmental problems in Indonesia. Stressing the urgent and immediate dangers for orangutans produces a perpetual potential rhetoric that emphasizes the precariousness and timeliness for environmental action. The perpetual potential offers an alternative rhetorical practice to the doomsday appeals of apocalypse and the irreparable. Although the perpetual potential may employ such rhetoric, appeals of hopefulness and environmental activism are also important rhetorical aspects of the perpetual potential. The perpetual potential provides a different framework for understanding the rhetorical practices of environmental organizations, moving beyond apocalypse, hysteria, and the irreparable. However, across cultural contexts, the rhetoric of the perpetual potential has different effects. As Cantrill (1993) suggests, environmental organizations should come to understand how these appeals succeed or fail, depending on their audiences.

Indonesian Audiences

Extant literature on environmental rhetoric outside of the United States is very limited in scope, although many scholars have recognized this absence of literature and called for more studies (Bennett & Chaloupka, 1993; Carbaugh, 1996; Depoe, 1997; Herndl, 1997; Jagtenberg & McKie, 1997; C. L. Jennings & B. H. Jennings, 1993; Muir & Veenendall, 1996; Myerson & Rydin, 1997; Ross, 1996). Environmental movements in the Southern hemisphere may focus more on equity issues and environmental degradation that directly and immediately affects local communities (Guha, 1989). As James Cantrill (1993) suggests, cultural factors play a powerful role in how people respond to environmental rhetorical appeals. He notes that people act out of self-interest; thus, environmental appeals must address those who have the greatest interest in a particular cause, while emphasizing the self-interests and values of those communities.

However, Indonesians do not value orangutans in the same way that many Western primatologists and environmental activists do. Westerners in Europe and North America essentially created the field of primatology and the importance of understanding primate behavior. Galdikas (1995) explains how many of her Indonesian assistants already knew much of what she studied in orangutan behavior: "Much of what we scientists 'discover' is known by local people. . . . Having decided in advance what is important, we are ecstatic when we find it" (p. 154). Indigenous communities have little understanding of or value for primatology and academic research. Sellato (1994) explains that many indigenous communities in Borneo are based on traditional lifestyles that rely or used to rely on hunting and gathering, with either little or no monetary income. Many local people have little concept of those who have enough money and resources to travel to Indonesia and study orangutans as a full-time job. Consequently, studying and working to protect orangutans as a livelihood might seem absurd to some local communities.

Furthermore, claims that orangutans are on the brink of disaster may seem counterintuitive to local people. Orangutans are primarily solitary, arboreal animals, yet interaction with humans is inevitable as population pressures encroach into orangutan habitat. As Galdikas (1995) notes, "Wherever the forests have been cut down, virtually every town and village houses captive orangutans. Many other people tell of orangutan 'pets' who died" (p. 135). Loggers encounter orangutans that are often captured and sold (if not killed). Orangutan pets can be a status symbol for local leaders, military officials, and other prominent figures in these communities. For these Indonesians, orangutans do not seem on the brink of extinction. Local people in Sumatra and Borneo are probably the most important audience, since their activities have more direct effect on forests and orangutan populations. However, local people may see orangutans as pests or threats to their gardens and fields when food scarcity or forest fires force orangutans into neighboring villages. Translating rhetorical appeals of the precarious into Indonesian does not work because Indonesian audiences do not see orangutan populations disappearing in the same way that Western audiences come to understand claims of the precarious.

Finally, rhetorical appeals relating to the perpetual potential of orangutans do not motivate many Indonesian audiences, largely because their attention is directed elsewhere. Since 1998, Indonesians have experienced major economic and political turmoil, such as the transition from former president Suharto's authoritarian regime to a democracy that has elected four presidents in 6 years. Decentralization from federal power to provincial and district-level power has also created chaotic governance throughout much of Indonesia. Furthermore, widespread poverty and increasing food prices have directed the attention of Indonesians toward meeting their basic needs. In 1999, the number of Indonesian households below the official poverty line (approximately $10 per month in U.S. currency) in Kalimantan was 2.2 million (19.87%), while in Sumatra, it was 8.6 million (19.81%; Perdana & Maxwell, 2004).[5] Other estimates put the Indonesian poverty rate at 27%, unemployment at 10.5%, and the per capita income at $3,200 in U.S. currency (Indonesia economy, 2004). Timeliness is certainly an issue for addressing poverty, unemployment, health care, education, and political and economic concerns. Environmental concerns fall to the bottom of this list, rendering environmental rhetorical appeal of perpetual potential largely ineffective.

---

[5]In 1997–1998, Indonesia experienced severe economic problems, which led to a significant currency devaluation and disproportionately increased economic pressures for the poor (Sowards, 1998). In 1999, the official poverty line was listed as Rp 89,845 (Rupiah, Indonesia's currency) for urban communities and Rp 69,420 for rural communities per month (Perdana & Maxwell, 2004); a conservative estimate for earnings during this period would be less than $10 per month in U.S. currency. In 2001, the poverty line was at Rp 100,011 for urban communities and Rp 80,832 for rural communities per month, the equivalent to just over $10 in U.S. currency for urban communities and just less than $10 in U.S. currency for rural communities. Indonesians currently face rising costs for food and gasoline, as the Indonesian government attempts to cover the costs of heavily subsidized products, such as rice, tea, sugar, and gasoline.

Although much more could be said about the rhetorical strategies and effects of environmental campaigns in Indonesia, the rhetorical appeal of the perpetual potential seems largely ineffective for local communities near and in orangutan habitat. In part, this lack of concern for orangutans can be attributed to widespread poverty and social unrest in Borneo and Sumatra. However, the connection between poverty and environmental issues is complicated because many of the poor or indigenous communities have a vested interest in maintaining their ecosystems for fresh water, food, and livelihoods (Bill, 2005). Indonesian environmental groups, primatologists, and some indigenous groups have been very active in pursuing forest protection and law enforcement (Dove, 1998), but two major factors in deforestation, rampant corruption and lack of law enforcement, contribute to what Garrett Hardin (1968) calls the "tragedy of the commons" (p. 1243). Forests and marine ecosystems in Indonesia are exploited in part because these are common areas, and whoever exploits them first, profits financially in the short term. The tragedy of the commons effect is further exacerbated by the controversial and unclear aspects of property rights and people's forests (*hutan rakyat*), conflicts between transmigrants and indigenous people, and an emphasis on maximizing short-term profit, often by outsiders (Atje & Roesad, 2004).

In short, the political, economic, and cultural aspects of environmental issues in Indonesia are complex. These factors indicate, however, that long-term environmental protection of orangutans is not at the forefront of most Indonesians' minds. It is not clear to many Indonesian audiences that protecting orangutans offers any long-term benefit. The orangutan does not work as an effective rhetorical appeal, because the survival of the orangutan has very little meaning or implication for the immediate survival of humans in Indonesia. Environmental organizations may want to carefully consider this lack of connection to the orangutan, and focus instead on establishing direct connections to the effects of deforestation and the short- and long-term benefits of the rain forest habitat.

Western Audiences

Although much of this chapter has centered on the rhetorical function of the perpetual potential, environmental organizations must take caution in delineating between what moves an audience and what does not, as Cantrill (1993) suggests. The continual apocalyptic threats outlined repeatedly in environmental rhetoric can have undesired effects. O'Leary (1994) explains:

> When appeals to the irreparable are overused, they become ineffective; fear may turn from an incentive for action into fatalism, or threats may appear no longer credible. Rhetors who seek to ensure the survival and effectiveness of their movements must take care that their discourse does not contain the seeds of apocalyptic despair. (p. 216)

Organizational members may come to see environmental issues as hopeless, even though there might be some solutions or activist measures taking place. Potential members may be interested in a particular cause, but find the doomsday rhetoric overwhelming. Several prominent U.S. Americans have called attention to the problems associated with environmentalism's appeal to the apocalyptic or irreparable (Shellenberger & Nordhaus, 2005; Werbach, 2005). They contend that environmentalism is dead, in part, because U.S. American environmentalism has lacked vision for the future (Werbach, 2005). For orangutans, the future seems bleak, primarily because of illegal logging and lack of law enforcement. Despite the numerous activities of environmental organizations over the last 15 years, the plight of the orangutan is simply not improving. Audiences may only see a desperate situation that cannot be remedied.

However, rhetorical appeals to the perpetual potential offer an alternative to apocalyptic or irreparable rhetoric. The continual emphasis on the possibilities for success creates a rhetoric that offers a hopeful vision for the future. The rhetoric of the perpetual potential emphasizes uniqueness, precariousness, and timeliness, but functions in Burke's (1984) comic frame, offering hopeful solutions to its audiences. Furthermore, the implied values of intrinsic worth, advancement of science and medicine, and development of human knowledge found in perpetual potential rhetoric speak to the values of many Western audiences. Rhetoric of potentiality attempts to avoid passivity, pessimism, and cynicism, while creating motive for meaningful commitment to a cause. Because Western audiences may be overwhelmed by rhetorical appeals to the irreparable and apocalypse in environmental discourse, the perpetual potential may offer a viable alternative to understanding such rhetorical practices. The perpetual potential, with its emphasis on hopeful solutions and a vision for a future of possibilities, allows audiences to avoid a world of apocalypse and the irreparable through recognition of successful environmental organizations' endeavors and individual actions.

ACKNOWLEDGMENTS

This research project was funded by a J. William Fulbright research grant (2000–2001), and two grants from the International Institute at California State University, San Bernardino (2002, 2003). The author thanks Sonja Foss, William Waters, Richard Pineda, Larry Erbert, Tom Ruggiero, Kenneth Yang, and three anonymous reviewers for their helpful suggestions to improve this chapter.

REFERENCES

Ancaman kepunahan orangutan [Orangutans threatened with extinction]. (n.d.). Gunung Palung Orangutan Conservation Program [information sheet], 1–3. Ketapang, West Kalimantan, Indonesia: Gunung Palung Orangutan Conservation Program.

Atje, R., & Roesad, K. (2004, February). Who should own Indonesia's forests?: Exploring the links between economic incentives, property rights and sustainable forest management (WPE 076). Jakarta, Indonesia: Centre for Strategic and International Studies. Available at http://www.csis.or.id/papers/ wpe076

Baillie, J. E. M., Bennun, L. A., Brooks, T. M., Butchart, S. H. M., Chanson, J. S., Cokeliss, Z., Hilton-Taylor, C., Hoffmann, M., Mace, G. M., Mainka, S. A., Pollock, C. M., Rodrigues, A. S. L., Stattersfield, A. J., & Stuart, S. N. (2004). *2004 IUCN red list of threatened species™: A global assessment*. Gland, Switzerland: International Union for Conservation of Nature and Natural Resources.

Balikpapan Orangutan Society–USA. (n.d.). *BOS–USA: Balikpapan Orangutan Society–USA: Dedicated to the conservation of the orangutan and their rainforest habitat* [brochure]. Aptos, CA: BOS–USA.

Balikpapan Orangutan Survival Foundation. (1997). *Orangutans at risk* [brochure]. Balikpapan, East Kalimantan, Indonesia.

Balikpapan Orangutan Survival Foundation. (n.d.). *Orangutan terancam bahaya* [Orangutans dangerously threatened] [brochure]. Balikpapan, East Kalimantan, Indonesia.

Belsten, L. A. (1996). Environmental risk communication and community collaboration. In S. A. Muir & T. L. Veenendall (Eds.), *Earthtalk: Communication empowerment for environmental action* (pp. 27–42). Westport, CT: Praeger.

Bennett, J., & Chaloupka, W. (1993). Introduction: TV dinners and the organic brunch. In J. Bennett & W. Chaloupka (Eds.), *In the nature of things: Language, politics, and the environment* (pp. vii–xvi). Minneapolis, MN: University of Minnesota Press.

Bill, R. (2005, March 24). No more living dangerously: Environment at a crossroads. *The Jakarta Post* [online]. Retrieved on March 23, 2005, from http://www.thejakartapost.com/outlook/pol10b.asp

Brummett, B. (1988). Using apocalyptic discourse to exploit audience commitments through "transfer." *The Southern Communication Journal, 54,* 58–73.

Brummett, B. (1991). *Contemporary apocalyptic rhetoric.* Westport, CT: Praeger.

Burke, K. (1969). *A rhetoric of motives.* Berkeley, CA: University of California Press.

Burke, K. (1984). *Attitudes toward history* (3rd ed.). Berkeley, CA: University of California Press.

Cantrill, J. G. (1993). Communication and our environment: Categorizing research in environmental advocacy. *Journal of Applied Communication Research, 21*(1), 66–95.

Cantrill, J. G. (1996). Perceiving environmental discourse: The cognitive playground. In J. G. Cantrill & C. L. Oravec (Eds.), *The symbolic earth: Discourse and our creation of the environment* (pp. 76–94). Lexington, KY: The University Press of Kentucky.

Carbaugh, D. (1996). Naturalizing communication and culture. In J. G. Cantrill & C. L. Oravec (Eds.), *The symbolic earth: Discourse and our creation of the environment* (pp. 38–57). Lexington, KY: The University Press of Kentucky.

Cox, J. R. (1982). The die is cast: Topical and ontological dimensions of the locus of the irreparable. *Quarterly Journal of Speech, 68,* 227–239.

Dedicated to the conservation of orangutans and their rain forest home. (n.d.). Retrieved May 16, 2005, from the Balikpapan Orangutan Society–USA Web site at http://www.orangutan.com

Depoe, S. (1997). Environmental studies in mass communication. *Critical Studies in Mass Communication, 14*(4), 368–372.

Djiwandono, J. S. (1998). The rupiah—One year after the float. *The Indonesian Quarterly, 26*(3), 170–177.

Dove, M. R. (1998). Local dimensions of 'global' environmental debates: Six case studies. In A. Kalland & G. Persoon (Eds.), *Environmental movements in Asia* (pp. 44–64). Richmond, Surrey, England: Curzon Press.

Galdikas, B. M. F. (1995). *Reflections of eden: My years with the orangutans of Borneo.* New York: Little, Brown & Co.

Galdikas, B. M. F., & Briggs, N. (1999). *Orangutan odyssey.* New York: Harry N. Abrams.

Guha, R. (1989). Radical American environmentalism and wilderness preservation: A Third World critique. *Environmental Ethics, 11,* 71–83.

Haraway, D. (1989). *Primate visions: Gender, race, and nature in the world of modern science.* New York: Routledge.

Hardin, G. (1968, December 13). The tragedy of the commons. *Science, 162*, 1243–1248.

Harré, R., Brockmeier, J., & Mühlhäusler, P. (1999). *Greenspeak: A study of environmental discourse.* Thousand Oaks, CA: Sage.

Herndl, C. G. (1997). Response to Professor Depoe. *Critical Studies in Mass Communication, 14*(4), 373–374.

How to help. (2000). Retrieved January 30, 2000, from the Balikpapan Orangutan Survival Foundation Web site at http://www.redcube.nl/help.htm

How to help. (2000). Retrieved April 5, 2000, from the Orangutan Foundation International Web site at http://orangutan.org/how2help/index.php

Indonesia economy—2004. (2004). *2004 CIA world factbook.* Retrieved March 23, 2005, from http://www.immigration-usa.com/wfb2004/indonesia/indonesia_economy.html

The IUCN red list of threatened species.™ (2004). Retrieved March 23, 2005, from www.iucnredlist.org/search/details.php?species=17975

Jagtenberg, T., & McKie, D. (1997). *Eco-impacts and the greening of postmodernity.* Thousand Oaks, CA: Sage.

Jennings, C. L., & Jennings, B. H. (1993). Green fields/brown skin: Posting as a sign of recognition. In J. Bennett & W. Chaloupka (Eds.), *In the nature of things: Language, politics, and the environment* (pp. 173–194). Minneapolis: University of Minnesota Press.

Katz, S. B., & Miller, C. R. (1996). The low-level radioactive waste siting controversy in North Carolina: Toward a rhetorical model of risk communication. In C. G. Herndl & S. C. Brown (Eds.), *Green culture: Environmental rhetoric in contemporary America* (pp. 111–140). Madison, WI: The University of Wisconsin Press.

Kessler, P. (1997). Forest fires in Kalimantan and their impact on food. *Long Call, 5*, 6–7.

Killingsworth, M. J., & Palmer, J. S. (1992). *Ecospeak: Rhetoric and environmental politics in America.* Carbondale, IL: Southern Illinois University Press.

Killingsworth, M. J., & Palmer, J. S. (1995). The discourse of "environmentalist hysteria." *Quarterly Journal of Speech, 81*, 1–19.

Killingsworth, M. J., & Palmer, J. S. (1996). *Millennial ecology*: The apocalyptic narrative from *Silent spring* to *global warming*. In C. G. Herndl & S. C. Brown (Eds.), *Green culture: Environmental rhetoric in contemporary America* (pp. 21–45). Madison, WI: University of Wisconsin Press.

Knott, C. (2003). Code red. *National Geographic, 204*(4), 76–81. [Online version]

Mittermeier, R. A., & Konstant, W. R. (2000, January 25). World's vanishing primates. *The Christian Science Monitor.* Retrieved from http://www.csmonitor.com

Muir, S. A., & Veenendall, T. L. (1996). Introduction. In S. A. Muir & T. L. Veenendall (Eds.), *Earthtalk: Communication empowerment for environmental action* (pp. xiii–xviii). Westport, CT: Praeger Publishers.

Mulyadi, A. (2002). *Species we may lose in the next twenty years* [brochure]. Ketapang, West Kalimantan, Indonesia: Program Konservasi Orangutan Gunung Palung.

Myerson, G., & Rydin, Y. (1997). The future of environmental rhetoric. *Critical Studies in Mass Communication, 14*(4), 376–379.

Nadler, R., Galdikas, B., Sheeran, L., & Rosen, N. (1995). Preface. In R. Nadler, B. Galdikas, L. Sheeran, & N. Rosen (Eds.), *The neglected ape* (pp. v–vii). New York: Plenum Press.

The Nature Conservancy. (2002). *Ucapkan selamat tinggal . . .* [Say goodbye . . .] [poster]. Samarinda, East Kalimantan, Indonesia.

Nielsen, L. D. (1999). Nyaru Menteng—a new orangutan rescue center in Central Kalimantan. *Voices from the Wilderness, 2*(1), 7.

O'Leary, S. D. (1994). *Arguing the apocalypse: A theory of millennial rhetoric.* New York: Oxford University Press.

Orangutans at risk. (1997). Retrieved January 30, 2000, from the Balikpapan Orangutan Survival Foundation Web site at http://www.redcube.nl/bos/risk.htm

Orangutan di TNGP terancam punah [Orangutans in TNGP threatened with extinction]. (2002, May 26). *Pontianak Post,* p. 18.

Perdana, A. A., & Maxwell, J. (2004). *Poverty targeting in Indonesia: Programs, problems and lessons learned* (WPE 083). Jakarta, Indonesia: Centre for Strategic and International Studies. Available at http://www.csis.or.id/papers/WPE083

Perelman, C., & Olbrechts-Tyteca, L. (1969). *The new rhetoric: A treatise on argumentation.* Notre Dame, IN: University of Notre Dame Press.

Peterson, T. R. (1997). *Sharing the earth: The rhetoric of sustainable development.* Columbia, SC: University of South Carolina Press.

Potter, D. (1995). Environmental problems in their political context. In P. Glasbergen & A. Blowers (Eds.), *Environmental policy in an international context* (pp. 85–110). London: Open University of the Netherlands.

Rijksen, H. D. (1995). The neglected ape?: NATO and the imminent extinction of our close relative. In R. D. Nadler, B. Galdikas, L. K. Sheeran, & N. Rosen (Eds.), *The neglected ape* (pp. 13–21). New York: Plenum Press.

Rijksen, H. D., & Meijaard, E. (1999). *Our vanishing relative?: The status of wild orangutans at the close of the twentieth century.* Dordrecht, The Netherlands: Kluwer Academic.

Rorty, R. (1999). *Philosophy and social hope.* New York: Viking/Penguin Press.

Ross, S. M. (1996). Two rivers, two vessels: Environmental problem solving in an intercultural context. In S. A. Muir & T. L. Veenendall (Eds.), *Earthtalk: Communication empowerment for environmental action* (pp. 171–190). Westport, CT: Praeger.

Russon, A. E. (2000). *Orangutans: Wizards of the rain forest.* Buffalo, NY: Firefly Books.

Searle, M. (Director and Producer). (n.d.). *Orangutans: Just hangin' on* [film]. Storyteller, Prime Time.

Sellato, B. (1994). *Nomads of the Borneo rain forest: The economics, politics, and ideology of settling down* (S. Morgan, Trans.). Honolulu: University of Hawaii Press.

Shabecoff, P. (1996). *A new name for peace: International environmentalism, sustainable development, and democracy.* Hanover, NH: University Press of New England.

Shellenberger, M., & Nordhaus, T. (2005, January 13). The death of environmentalism: Global warming politics in a post-environmental world. *Grist Magazine.* Retrieved February 16, 2005, from http://www.grist.org/news/maindish/2005/01/13/doe-reprint/

Smits, W. (1992). *1992 Year report: Orangutan reintroduction at the Wanariset I Station, Samboja, East Kalimantan, Indonesia.* Available from Wanariset Station, East Kalimantan, Indonesia.

Sowards, S. K. (1998). The International Monetary Fund and implications of the 1997–1998 negotiations with Indonesia. *The Indonesian Quarterly, 26*(3), 235–254.

Sugardjito, J. (1995). Conservation of orangutans: Threats and prospects. In R. D. Nadler, B. Galdikas, L. K. Sheeran, & N. Rosen (Eds.), *The neglected ape* (pp. 45–49). New York: Plenum Press.

Sunderlin, W. D. (2002). Effects of crisis and political change, 1997–1999. In C. J. Pierce Colfer & I. A. Pradnja Resosudarmo (Eds.), *Which way forward?: People, forests, and policymaking in Indonesia* (pp. 246–276). Washington, DC: Resources for the Future Press.

U.S. Fish & Wildlife Service. (n.d.). *Endangered species and threatened wildlife and plants.* Retrieved March 23, 2005, from http://endangered.fws.gov/50cfr_animals.pdf

U.S. Fish & Wildlife Service. (2004, February). *ESA Basics: 30 years of protecting endangered species.* Washington, DC: U.S. Fish & Wildlife Service. Retrieved March 23, 2005, from http://endangered.fws.gov

Van Schaik, C. (2004). *Among orangutans: Red apes and the rise of human culture.* Cambridge, MA: The Belknap Press of Harvard University Press.

Waddell, C. (1996). Saving the Great Lakes: Public participation in environmental policy. In C. G. Herndl & S. C. Brown (Eds.), *Green culture: Environmental rhetoric in contemporary America* (pp. 141–165). Madison: The University of Wisconsin Press.

Werbach, A. (2005, January 13). Is environmentalism dead?: A speech on where the movement can and should go from here. *Grist Magazine.* Retrieved February 16, 2005, from http://www.grist.org/news/maindish/2005/01/13/werbach-reprint/

Wilson, E. O. (2000). Balikpapan Orangutan Society–USA fundraiser invitation [brochure]. Aptos, CA.

CHAPTER SEVEN

Substitution or Pollution? Competing Views of Environmental Benefit in a Gas-Fired Power Plant Dispute

Øyvind Ihlen
University of Oslo and Hedmark University College

During the late 1980s and the early 1990s, increasing numbers of companies tried to pass themselves off as being green, and as offering environmentally friendly products or services. Many of these attempts have been severely criticized as being forms of "greenwash," and instances in which businesses are portrayed as environmentally responsive but continue to ruin the environment have been highlighted (Bruno & Karliner, 2003; Greer & Bruno, 1996; Laufer, 2003; Lubbers, 2002; Rowell, 1997; Tokar, 1997; Utting, 2002b). The appropriation of environmental rhetoric has led to a call for a new environmentalism that exposes "corporate myths and methods of manipulation" (Beder, 1998, p. 12). It would, however, be naive to assume that all of the conflicts could be reduced to a simple polarity between green and nongreen, or that environmental organizations are a priori ethically superior to industry (Cantrill, 1993; Myerson & Rydin, 1996). Recent developments in international environmental politics have created an even more complex picture, and a fitting example is that the Kyoto protocol on carbon dioxide emissions supports the principle of cost-effectiveness and so-called flexible mechanisms, such as the trading of emission quotas. If a country does not reach its specified emission level target, it can buy quotas from others that have made larger cuts than required (see http://unfccc.int). As will be shown, this new international approach opens up new possibilities for industrial actors and poses new challenges for environmentalists.

This qualitative case study focuses on the debate that was initiated by the proposal of the Norwegian company, Naturkraft, to build gas-fired power plants (GPPs) on the basis that this would improve the environment. At face value,

Naturkraft built on an environmentally progressive idea. The company would export power from GPPs as a substitute for power from more polluting Danish coal-fired power plants.

Norwegian environmentalists, however, were enraged by the plans of Naturkraft, and labeled the proposals as a new form of greenwash. Their basic objection was that energy use should be curbed, and that the power from the GPPs would only be an addition to the electricity that is generated by the coal-fired power plants. Instead of reducing carbon emissions, building GPPs would lead to more pollution.

The environmentalists were relying on a domestic approach and a national action discourse that was rooted in a deontological ethic, whereas the proponents of GPPs were using an international approach and a "thinking globally" discourse rooted in a consequentialist ethic (Hovden & Lindseth, 2004).[1] The former believed that it is always wrong to emit carbon dioxide, whereas the latter thought that it would be acceptable to emit more carbon dioxide domestically if this led to lower global emissions. The two discourses were put to the test when the Norwegian parliament discussed GPPs, and the debate turned into the biggest environmental conflict in Norway during the 1990s.

The research question for the case study is: What rhetorical strategies did the actors in this conflict use, and how did the rhetoric contribute to the actors' success in the political arena with regards to the GPP plans? The study has wider relevance in that it demonstrates how environmental rhetoric must be adapted to a more complex political context in which domestic and internationally oriented approaches compete. In many instances, environmentalists cannot simply rely on labeling something as greenwash.

The next section deals with theoretical perspectives on environmental rhetoric and the rhetorical tools used in this study. It is followed by a section that details the methods that were employed. The subsequent analysis is divided into two parts: the first deals with the strategies that Naturkraft used to obtain its building permits, and the second analyzes how the environmentalists were able to force Naturkraft to postpone its plans. The last section summarizes and discusses the issue further.

THEORETICAL PERSPECTIVES

This chapter focuses on environmental rhetoric, which M. Jimmie Killingsworth (1996) defined as "a topic or a field of rhetorical practice concentrating on the hu-

---

[1] *Deontological ethics* could be defined as "any approach to normative ethics that denies that the rightness of an action depends on how it promotes intrinsically good consequences" (Gensler, 1998, p. 201). Religious rules and the declaration of human rights are both examples of deontological ethics. As for *consequentialist ethics*, the basic idea is that "we ought to do whatever maximizes good consequences" (Gensler, 1998, p. 201).

man relationship to the natural environment and dealing with the forms or systems of discourse arising from ethical and political disputes over environmental protection and developmental planning" (p. 225). Environmental rhetoric is not employed only by environmental organizations, it is also used by businesses that often neglect to improve their environmental behavior (Bruno & Karliner, 2003; Greer & Bruno, 1996; Hager & Burton, 1999; Lubbers, 2002; Rowell, 1997; Utting, 2002b).

Studies that use rhetorical theory have shown how even in the late 1960s, industry used the methods of shifting blame to consumers or highlighting its own environmental measures to be perceived as environmentally friendly (Brown & Crable, 1973). More sophisticated efforts to co-opt environmental rhetoric include the plastic industry's attempt to portray plastic as a natural resource that could be "born again." This amounts to what Burke would call "perspective by incongruity," or revising a perception by putting together two unrelated elements (cited in Paystrup, 1995). Often, however, the environmental rhetoric of business is short on arguments, and resembles a form of new managerial rhetoric that uses images, identification, and entertainment as its main tools (Feller, 2004; Sproule, 1988).

A few studies have been conducted that focus on competing claims in environmental conflicts. One study pointed out how the strategies of loggers and environmentalists mirrored and matched each other, with both parties relying on techniques of vilification and simplification (Lange, 1993). Other studies have shown how this mirroring has been extended to other areas. The Wise Use movement has co-opted the rhetoric of environmentalists, has proclaimed its members to be the "real" environmentalists, and has mimicked the structure and identity of the environmental movement (Peeples, 2005). However, little case-specific research has been conducted into the dynamic interaction and contest between actors arguing that their solutions benefit the environment the most. This case study aims to help fill that research gap, which is important because such conflicts are now occurring more frequently, and environmental debate increasingly concerns discourse that focuses on new information, new concepts, and new practices (Myerson & Rydin, 1996).

Discourses on new information typically concern the way in which that information is generated and presented, and can be exemplified by the debate on the conclusions of the Intergovernmental Panel on Climate Change (IPCC) on whether human-induced climate change is a fact or not. Oil companies such as ExxonMobil have campaigned against the conclusions of the panel (Rowell, 1997). The book, *The Skeptical Environmentalist* (Lomborg, 2001), is another example of the discourse on new information, in which the author tries to debunk what are called *environmental myths*.

Discourses on new concepts are exemplified by the debate on sustainable development. This concept has become very influential, because it is said to overcome the polarity between economic growth and environmental protection

(Myerson & Rydin, 1996). Some have argued that the concept has the potential to move environmentalists out of the fringe position that has been created by limiting "ecospeak" (Killingsworth & Palmer, 1992). Others have pointed to the potential for cooptation and the fact that a general adoption of the concept could drown out the diversity of perspectives or turn environmental issues into technological matters that avoid the need to make choices (Peterson, 1997; Tokar, 1997). Business and industry have used the phrase liberally, although it often means business as usual, rather than asking whether the business itself contributes to sustainability (Feller, 2004; Utting, 2002a).

The discourse on new practices is illustrated by the national action discourse and the "thinking globally" discourse (Hovden & Lindseth, 2004). In the late 1980s in Norway, it was the former that prevailed as politicians adopted the goal of the domestic stabilization of carbon dioxide emissions. In the 1990s, however, this approach was abandoned in favor of an international perspective. So-called flexible solutions were advocated, and it was argued that it would be more cost effective for Norway to finance measures in countries in which it was possible to obtain more environmental benefits for the same amount of money (Andresen & Butenschøn, 2001; Nilsen, 2001).

The case that is studied in this chapter primarily concerns discourse on new practices. Although the case predates the introduction of emissions-quota trading, the debate on such mechanisms is an important backdrop. Rather than relying on a new managerial rhetoric, however, the actors involved for the most part used traditional rhetorical tools. This invites analysis that builds on Aristotelian rhetoric (Aristotle, trans. 1991; Corbett & Connors, 1999), as well as the extensions of this tradition (Burke, 1969a, 1969b; Perelman & Olbrechts-Tyteca, 1971).

From Aristotelian rhetoric, I use the concept of *kairos*, which can be defined as the identification of the right moment for an utterance (Sipiora & Baumlin, 2002). I also focus on the three artistic proofs, *ethos, pathos,* and *logos,* that is, ethical appeal, emotional appeal, and appeal to reason. These proofs are linked to the rhetor, the audience, and the message, respectively. In a given discourse, "these are at all times coordinate [*sic*] and interact mutually, distinguishable but not separable from one another, although one may occasionally take precedence over the others" (Conley, 1994, p. 15). A third concept that I use from Aristotelian rhetoric is that of *topics,* defined as a stock of general lines of argument, such as comparison or relationship (Corbett & Connors, 1999).

As for the extensions of Aristotelian rhetoric, I first turn to Kenneth Burke and his notion of identification, creating a "we," as an important way of achieving persuasion. Different rhetorical strategies might be used to this effect, like for instance, when that the rhetor emphasizes the common ground between him or her and the audience (Burke, 1969a). The second Burkeian concept that will be used is the pentad. The pentad helps to analyze and understand the rhetorical strategies of a rhetor and how rhetorical patterns index, as well as construct, and embody the motives of the rhetor (Brock, 1999; Stillar, 1998). The crucial question is:

"What is involved, when we say what people are doing and why they are doing it?" (Burke, 1969a, p. xv). When conducting a pentadic analysis, the analyst focuses on the choices that the rhetor has made in describing a situation (the act, the scene, the agent, the agency, and the purpose), and tries to understand what this says about the perspective of the rhetor, what kind of possibilities he or she sees, and what he or she will eventually do. The basic idea is that symbol use tells us something about the motives of the rhetor and how he or she tries to structure or restructure the audience's perception of a situation.

Perelman pointed out that the starting point for argumentation is the agreement between the rhetor and the audience. A rhetor will attempt to transfer this agreement to his or her thesis, by, for instance, trying to secure a communion with the audience, quite like Burke's concept of *identification*. Whenever there is disagreement between the rhetor and the audience, this is either in relation to the status of the premises, the choice of premises, or their presentation. In debating the status of the premises, it is useful to focus first on what the audience perceives as reality, that is, facts, truths or presumptions. Another type of premises relates to what the audience considers to be preferable (Perelman, 1982; Perelman & Olbrechts-Tyteca, 1971). Focusing on the types of premises helps to assess the scope of the appeal of the rhetoric and give a more comprehensive analysis of the proofs that were identified with the help of Aristotelian rhetoric.[2]

The relevant rhetorical concepts are elaborated on in the analysis, not with the purpose of flag-posting, or merely showing that certain rhetorical tools have been used, but to point out how these tools function in the context of the case. Rhetoric strives to tell us how everything in an utterance should appear in the context of the text, and analyses of rhetoric should assess whether texts (in the broad sense) achieve what they are meant to achieve. In the case presented here, the goal of one side was to obtain building permits for GPPs and to build them, and the goal of the other side was to prevent Naturkraft from being given permits and afterward to stop the building of the plants altogether.

METHOD

This chapter focuses on the strategies of Naturkraft and of what so far have been simply called "the environmentalists." In the first phase of the conflict, the most important environmental actors were the organization Nature and Youth and the ad hoc umbrella organization that it helped to create—the Climate Alliance. The latter was put on ice during the second phase of the conflict when Nature and Youth initiated a new ad hoc organization, the Action Against GPPs, that recruited individuals as members. Multiple methods were used to analyze the strategies of these actors:

---

[2]I have discussed the integration of Aristotelian rhetoric and theoretical extensions more thoroughly in other works (Ihlen, 2002, 2004a, 2004b).

1. Close rhetorical analysis was undertaken of a press release, brochure, and report that were issued by Naturkraft, a white paper from the Climate Alliance, and a leaflet from Action Against GPPs. Quotes were translated from Norwegian to English.

2. Qualitative interviews with the President of Naturkraft, the leader of the Climate Alliance (who later fronted the Action Against GPPs), and the treasurer of Action Against GPPs (who also served as a board member). The interviews were taped, and transcripts of the paragraphs that were to be used were submitted to each interviewee for approval. No major changes were requested.

3. Archival research was carried out and unearthed strategy memos and an internal evaluation report of Action Against GPPs.

4. Analysis of pivotal quotes from the organizational actors in two large Norwegian dailies, which were tracked using online archives, were analyzed.

The use of multiple methods in this way (triangulation) helps to counter the potential problem of interviewees trying to justify their choices or promote their own role or the role of their organization in influencing the outcome of a conflict. The claims of interviewees ought to be treated with caution, and should be checked against evidence that is found in strategy documents, media coverage, or statements from other actors.

The basic problem that is associated with such a study is, nonetheless, the extrapolation from the existence of a certain type of rhetoric to decisions made by politicians. The best way to respond to this challenge may be to check for alternative explanations, to allow for additional explanatory factors, and to be careful not to overstate the findings.

ACT I: GETTING PERMITS

Naturkraft was established as a private enterprise in 1994. The company soon secured the support of the Minister of Trade and Energy, who had previously opposed similar plans for GPPs. Why this U-turn? It seems that Naturkraft's business plan and rhetorical strategy fitted the needs of the government, as it came to be perceived that a domestic approach to curbing climate change would be "too harmful" to the Norwegian economy.

An Environmental Ethos

Naturkraft tried to position itself as an environmentally friendly actor. Indeed, the very name Naturkraft, which can be translated as "nature power," connotes something clean and natural, rather than polluting or artificial. The choice of this name was a clear attempt to promote an environmental *ethos*, which is one of three means of persuasion that is identified in Aristotelian rhetoric. When an actor uses an ethical appeal, it is a textual strategy that aims at being perceived as trustworthy, intelligent, and knowledgeable (Aristotle, trans. 1991; Corbett & Connors, 1999).

Naturkraft (1995a) identified what it called "our joint environmental challenges," and called for cooperation between countries:

> Building a GPP in Norway will demand that several partners cooperate in order to arrive at the best solution. First and foremost there is talk of cooperation between the Nordic countries, to ensure that the power demand will be covered in a way that takes into consideration our joint environmental challenges. Here we are faced with a problem that no country can solve on its own. (p. 9)

Naturkraft used typical environmental rhetoric that aimed at identification through the emphasis of common ground and the use of the personal pronoun "we" to include both the rhetor and the audience. These are typical identification strategies, and prime ways of achieving persuasion (Burke, 1969b). This type of rhetoric is also an illustration of the "thinking globally" discourse and of how it purports to have an improved environment as its goal.

However, as will be shown, Naturkraft also used two other means of persuasion: emotional appeals, which is the rhetor's attempt to invoke emotions by the use of vivid descriptions or honorific or pejorative words; and logical appeals, which involve the use of inductive or deductive reasoning (Aristotle, trans. 1991; Corbett & Connors, 1999). The three types of appeals might be used alone or in combination, but Naturkraft's business idea was largely based on logos to establish itself as an environmentally sound project. More specifically, Naturkraft used *common topics*, that is, a "stock of general lines of argument that can be used in the development of any subject" (Corbett & Connors, 1999, p. 87), rather than *special topics* that belong to disciplines such as environmental science. Naturkraft initially argued from the common topic of relationship.

Logos and Relationship: Growth in Energy Demand

In its first brochure, Naturkraft acknowledged that the best environmental alternative was to curb energy use. Nevertheless, it described how the need for electricity had grown and was expected to increase further, and that supply would have to be increased. A rhetor that uses the topic of relationship attempts to provide an answer to the question why by pointing to cause and effect, antecedent and consequence, and contraries or contradictions. The rhetor "pursues this line of argument: given this situation (the antecedent), what follows (the consequence) from this?" (Corbett & Connors, 1999, p. 104). If a rhetor can establish such a connection, then it strengthens the case logically, and gains a rhetorical advantage. In Naturkraft's announcement, the company used tables with public figures to build on the ethos of science to support this argument: "A yearly increase of 1–2% is projected to last into the next decade. This means there is a need to supply approx.

3–7 billion kilowatt hours (TWh) of new electric energy into the Nordic power system . . . Thus, during the next decade new production capacity has to be introduced" (Naturkraft, 1995b, p. 9).

In Burkeian terms (Burke, 1969a), this scene explained the act, that of trying to build GPPs. The increase in demand would happen anyway, and it would be best if a "clean" alternative existed. This motive of the rhetor and the emphasis on scenic features direct attention away from the chosen activity of the protagonist, in essence, a different course of action than that which is indicated is made to seem impossible. The potential counterstrategy of the environmentalists would be to attack the underlying premises: that the energy demand should be met with higher production, and that the choice is between GPPs and coal-fired power plants. The latter is termed a *false dilemma*, as it could be argued that energy conservation, measures for energy efficiency, and commitment to renewable energy sources should be prioritized instead.

Logos and Comparison: "The Environmental Friendly Alternative"

At the outset, Naturkraft planned to offer energy from both GPPs and hydroelectric plants; it labeled this *environmentally correct* energy (Naturkraft, 1995a). That description rested on the common topic of comparison, whereby the rhetor compares two or more things according to their similarity or difference, superiority or inferiority (Corbett & Connors, 1999). One thing is understood in the light of another. Naturkraft (1994) compared its "product" with coal, oil, and nuclear power:

> Naturkraft . . . wants to exploit the superior environmental benefits of natural gas and hydroelectric power. By replacing power production based on coal and oil, the alternative of gas and hydroelectric power might result in considerable reductions in the emissions of CO_2 and NOx in the Nordic region. The production of gas power and hydroelectric power might also replace nuclear power if a decision is made for a gradual phasing out of nuclear power in the Nordic countries. (p. 1)

Here, Naturkraft outmaneuvered nuclear power by playing up the controversy surrounding its safety, and by stressing that the company would offer safe energy. As for coal and oil, such sources emit more carbon dioxide (CO_2), nitrogen oxides (NOx), and sulfur dioxide (SO_2) than natural gas. This premise is related to the structure of reality: it does not need much justification, as it is a universal and something with which a rational audience would agree. As it is paramount that a rhetor adapt to the audience, the use of such premises gives the argumentation an edge (Perelman & Olbrechts-Tyteca, 1971).

The comparative element in the rhetoric of Naturkraft is questionable, as it rested on a comparison with a specific selection of energy sources. Instead of positioning GPPs against renewable sources such as solar power, they were compared to other fossil fuels. However, hydroelectric power was neutralized, as it was included in the product of Naturkraft.

Logos and Relationship: The Substitution Argument

Naturkraft not only posited that natural gas was less polluting than other fossil fuels, but went one step further and tried to establish a cause-and-effect relationship for what happens when power that is based on natural gas is introduced to the market. To repeat: "By *replacing* power production based on coal and oil, the alternative of gas and hydroelectric power might result in *considerable reductions in the emissions* of CO_2 and NOx" (Naturkraft, 1994, p. 1, emphasis added).

It could be argued that the use of the word "replacing" is a form of strategic ambiguity, as it does not necessarily imply complete replacement. In interviews and media coverage, however, this substitution argument did not contain the ambiguity. The argument was based on a topic of relationship, which again rested on an *enthymeme*. Deductive formal reasoning presents premises that lead to a conclusion. An enthymeme, however, asks the audience to supply a missing premise (Corbett & Connors, 1999). The substitution argument asked the audience to supply the unstated major premise, "(all) environmental friendly energy sources with competitive power will replace polluting energy sources." The stated minor premise was that "Naturkraft offers competitive and environmentally friendly energy," and the conclusion that followed was that the energy from Naturkraft would replace more polluting sources.

This enthymeme is problematic for several reasons, and needs additional qualification. As for the environmental dimension, one would have to presuppose that every other country was looking for ways to reduce its emissions. This may be true, but it is not certain that it is true in all situations. In logical terms, it is not possible to distribute the middle term, *competitive and environmentally friendly energy sources*. The motives for choosing energy carriers are not necessarily rational or precise, and can be rooted in perceived security needs or psychological and economic motives (Eldegard, 1995). For example, a country with coal supplies might prioritize the maintenance of employment levels and the subsidizing or exemption of plants in the coal regions from carbon taxes. Thus, there is no economic iron law that can prove that the substitution argument is correct.

Another issue is the question of what the power from the GPPs would replace. Would GPPs help to curb the emissions that stem from current production and consumption, or would they only reduce the emission growth? The environmentalists argued that the former had to be the aim (Climate Alliance, 1996), but Naturkraft disagreed. The president of Naturkraft said that the company initially talked about being able to reduce the growth of emissions, but that at the same time, he believed that it could be documented that GPPs would also contribute to cuts in existing emissions, "but it was not up to us to close down existing coal-fired power plants" (Naturkraft president, Auke Lont, personal communication, June 25, 2003).

The most important premise for Naturkraft and the substitution argument in the mid-1990s was that the then impending international climate negotiations in Kyoto might entail the introduction of stricter regulations and economic measures, such as carbon taxes. If such taxes were made applicable to all carbon emissions, then this would favor GPPs over coal-fired power plants. "The moment you pose restrictions on the emissions, then the costs of removing these emissions from coal will be so high that you can compete when offering natural gas. And this is what we saw in the beginning of the 1990s, that stronger restrictions would be introduced" (Naturkraft president, personal communication, June 25, 2003). This, then, could be interpreted as Naturkraft's attempt to identify *kairos* (Sipiora & Baumlin, 2002), that is, the opportune moment to introduce an utterance, or in this case, a project. The problem for Naturkraft was, however, that no binding international agreements existed at the time that the company proposed to build the GPPs, and thus there was no system that would give the producer credit for environmental benefits. By building GPPs, Norway would increase its own emissions, gain financially on exports, and only hope that those exports would contribute to fewer emissions in the neighboring countries. A system for pollution trading was still some years away.

Environmental Protests

The environmentalists set up the Climate Alliance to protest against the plans of Naturkraft. The alliance included 25 organizations and political parties, including almost all of the youth parties. Nature and Youth ran the small secretariat. The activists that belonged to the Alliance conducted extensive lobbying and also published a counterpaper. The title of the preface of the counterpaper read: "No to GPPs—save energy." Beyond declaring their opposition and hinting at alternatives, it was also necessary for the opponents to point out exactly *why* they were opposed to GPPs: "Construction of GPPs in Norway may increase the threat of dangerous climate changes. What we have to consider is whether we should contribute to increased energy use in the Nordic region based on fossil fuels" (Climate Alliance, 1996, p. 3). The cause-and-effect argument was used with an enthymeme that stated, "building GPPs increases the likelihood of dangerous climate changes." The unstated premises here are that "all emissions of climate gases increase the likelihood of climate changes" and "GPPs emit climate changing gases." The leader of the Climate Alliance postulated that the Alliance specifically aimed at an emotional appeal with their use of the adjective, "dangerous," along with "threat" (Lars Haltbrekken, personal communication, April 22, 2003).

The basic rhetorical structure of the environmentalists mirrored that of Naturkraft in that they also used a scene/act ratio in which a scene description, in the rhetor's mind, dictates the action that follows (Burke, 1969a). However, the major element in this scene description was not the growth in energy demand, but the threat of climate change. More specifically, energy would have to be saved and people should not contribute to the growth in the use of fossil fuels. The environmentalists were sticking to the domestic approach from a deontological ethical basis—that it would always be wrong to increase emissions. As previously mentioned, those that advocated the international approach operated from a base of *consequentialism*, and saw increased emissions as acceptable if it led to good consequences, such as lower global emissions (Hovden & Lindseth, 2004).

However, the arguments of the Climate Alliance failed to impress a majority of the politicians. The Norwegian parliament, the Storting, supported the GPP plans by a 74 to 44 vote in June 1996 (Storting Deliberations, 1996), and a few months later, Naturkraft was granted building permits. The company's description of the scene (growth in energy demand), and the arguments about the new practice (the international approach) fit the needs of the government.

ACT II: FORCED POSTPONEMENT OF THE PROJECT

After Naturkraft received its building permits, an inner circle of environmental activists met to discuss a new strategy. They ended up establishing Action Against GPPs in February 1997, and used the structure and membership base of Nature and Youth as the backbone of the new organization. This marked a shift away from the lobbying efforts that both sides had pursued in the conflict thus far. The environmentalists now wanted to mobilize public opinion against GPPs and threaten to use civil disobedience. The organization would later boast of having a list of 1,000 people who were willing to participate in such actions. The topics of relationship and comparison were used to steal back the initiative from Naturkraft.

Relationship: The Threat of Civil Disobedience

Again, the environmentalists relied on a scene/act ratio (Burke, 1969a). The activists tried to argue that the situation demanded that they use civil disobedience to stop the construction of the plants. In making this threat, the environmentalists faced two challenges: that of legitimizing such actions, and that of making the threat seem real. The activists themselves acknowledged these challenges (Action Against GPPs, 1997a). A memorandum entitled "Civil Disobedience, Democracy, and GPPs" was written, and featured arguments based on three main themes that formed a situation that made civil disobedience unavoidable. The first was that building GPPs would lead to increased pollution that would affect the lives of the following generations. This relied on a combination of logical and emotional appeals. The premise was that climate change was largely a problem that would be felt by future generations, which struck a personal chord of responsibility, as it involved us all. The second theme was that increasing domestic emissions would go against international agreements to curb emissions. This argument could be deemed a logical appeal that builds on the premise that all agreements should be kept. It could, however, be argued that no binding agreements existed at the time, and thus that the reasoning was faulty. The final theme was that the GPP process had been undemocratic, as the government controlled the companies involved, and was both judge and jury in the treatment of the building permits. This was not only a logical appeal, but also attacked the ethos of the government and the legitimacy of the decision to build GPPs.

The activists also crafted a strategy that actively played up the connection to a previous environmental conflict—the battle over Alta. The Alta conflict took place in 1981, and concerned a hydropower project that became a leading symbol for environmentalists and the indigenous Sámi people, who felt that the area that had been proposed for damming belonged to them. The symbolic significance was increased further when Gro Harlem Brundtland, the Norwegian Prime Minister who saw the project through, later said that she regretted the decision as the development was "unnecessary" (Myklevoll, 1990).

Invoking the Alta conflict functioned as an important symbolic resource in at least two respects. It was a powerful emotional symbol for something that was widely regretted afterward, and it invoked memories of massive popular protest actions that totally dominated the public, political, and media agenda for a long period. The leader of Action Against GPPs used the media to stress the point: the Prime Minister "should learn a lesson from the Alta development. If he forces through construction, he will regret it in ten years time" (Andersen, 1997, p. 10).

To yield photo opportunities and to keep up the pressure, the activists announced plans for a protest camp in Øygarden, which was where Naturkraft had said that it would start construction of the first GPP in the summer of 1997. In preparing for this camp, the activists pitched a Sámi tent in Øygarden. This symbolic action was not lost on journalists, who gave the event extensive coverage (e.g., Furuly, 1997a).

Comparison: Pollution Like 600,000 Cars

According to several of the strategy documents that were written by environmental activists, Naturkraft had been able to dictate the way in which the issue was debated during the first phase of the conflict. The substitution argument was the focal point of the public debate, and the activists were therefore forced to relate to this complex argument. The activists shared the conviction that they had to simplify the issue, to make it tangible, and to redirect the debate to get people to discuss GPPs during lunch breaks. One later memorandum stated that "our task is to focus on the main conflict, and not dwell on detailed discussions. Other actors can do this . . . We have to be careful not to drown in details and compromises" (Action Against GPPs, n.d.). This follows much of the advice that is given to environmental advocates in general: "avoid intermediate or mixed positions" (Sandman, 1994). The treasurer/board member of Action Against GPPs used the phrase "dumbing down" to describe this change in the rhetoric (personal communication, April 14, 2003).

The solution that was adopted was either to circumvent the substitution argument, to stop relating to it, or to reshuffle the order of the arguments and to emphasize different aspects first (Leader of Action Against GPPs, personal communication, April 22, 2003). Both strategies functioned to move the focus elsewhere, and were well suited to media discourse, which has trouble focusing on too many arguments and complexities. The activists' argument attempted to establish that the GPP issue concerned pollution, and had some particular consequences: "This meant that the discussion of energy use was overshadowed, but in retrospect this was probably a wise move. It is easier for people to understand the problem with a climate in chaos, than the problem with high energy consumption" (Action Against GPPs, 1998, p. 19). According to the evaluation report, many of the activists argued that more emotional appeals should be used, as this could help to increase mobilization. The phrase, "dangerous climate change," was singled out as particularly fitting, because the activists expected the climate issue to dominate the agenda in the upcoming climate negotiations in Kyoto (Action Against GPPs, 1998). A leaflet was printed that contained the following lead: "Pollution changes the climate on earth. The government wants to build GPPs that increase this pollution. Climate changes cause greater extremes of weather, such as hurricanes, increased sea levels, and the spread of deserts" (Action Against GPPs, 1997b). The leaflet was organized around three pictures and three main arguments that were reflected in the subheadlines. The first was a picture that showed a child in a playpen, which was accompanied by the text, "Our descendants inherit the problems," and the second was a picture of an inflated globe with the subheading "[GPPs] Sabotage environmental cooperation," and the third was a comparison topic that used a picture of a car spewing out exhaust fumes, and the accompanying text "pollutes like 600,000 cars." To keep the comparison simple, the focus was on carbon dioxide, rather than on other emissions from the two sources.

The comparison argument was now repeated with such frequency that it irritated Naturkraft, and almost bored the environmentalists themselves: "You must repeat and repeat and repeat and repeat and repeat. And when you are growing dead tired yourself, *then* maybe someone will remember what you have said" (Leader of Action Against GPPs, personal communication, April 22, 2003).

Naturkraft thought that the comparison made by the environmentalists was irrelevant, and their main grievance was that cars could not be compared with electricity production. The most typical retort against comparison arguments is to point out the differences between the two entities that are being compared. Naturkraft, nonetheless, did not want to engage in a dialogue over the issue, and instead decided that they should continue their present course and repeat the substitution argument (Naturkraft president, personal communication, June 25, 2003).

The activists, however, felt that their case had gained momentum with the simplified, repetitive comparison between cars and GPPs. Soon, the media and more opposition politicians were using the pollution argument, which belonged to the national action discourse. The activists lamented that they had not chosen this strategy earlier on (Action Against GPPs, 1998). A poll in one of the main dailies showed that 44% of respondents were against GPPs, and that only 28% supported them. The opposition had also increased "qualitatively" as the Church of Norway had involved itself on the side of the protesters. It was now clearly impossible to isolate the GPP opponents as idealistic fringe elements, and the pressure on the government mounted (Furuly, 1997b; Nilsen, 2001; Vassbotn, 1997).

Retreat Alternative

The activists maintain that at this time they drew on important previous conflict experience, namely that it was necessary to provide the opponent with a means of retreat. It is claimed that the action group made a strategic choice to keep quiet while another actor, the environmental foundation, Bellona, provided the government with a third way. Bellona relaunched the idea that the carbon emissions from a GPP could be reinjected into empty oil reservoirs (Leader of Action Against GPPs, personal communication, April 22, 2003). The government seized on this technological alternative. In the beginning of May 1997, it urged Naturkraft to postpone construction:

> It is just recently that all the reactions against GPPs have been disclosed and that the issue has become more controversial. The government takes notice of the increased opposition of late. Because of this it is necessary to ask Naturkraft to postpone the building so that as much information as possible can be acquired. (Nygaard, 1997, p. 2)

After some hesitation, Naturkraft agreed to postpone construction. The actors on both sides of the conflict, commentators in the media, and scholars all agree that the postponement came about as a result of protests from the environmentalists (Naturkraft president, personal communication, June 25, 2003; Leader of Action Against GPPs, personal communication, April 22, 2003; Bonde, 1997; Hovden & Lindseth, 2004; Nilsen, 2001; Vassbotn, 1997). The same sources repeatedly pointed out that the ruling Labor Party most probably had the upcoming Storting election in mind, and did not want to conduct an election campaign at the same time as police and protesters were clashing over GPPs. In other words, the environmentalists had succeeded in involving enough people to make their threat seem credible. The proponents of the national action discourse had seemingly outmaneuvered the proponents of the "thinking globally" discourse through the clever use of rhetoric. At the time of writing, Naturkraft has not yet built its plants.[3]

CONCLUSION

This case study has moved beyond discussions of shallow attempts at greenwash to focus on the rhetoric of the proponents and opponents of GPPs. It has been shown how the former won a majority of politicians over, but also how the opponents were able to steal back the initiative and, at least temporarily, halt the building of the GPPs. The strategies that were used by both sides exemplify the discourses of national action and "thinking globally" (Hovden & Lindseth, 2004), and therein lies the wider relevance of the study. Perspectives that emphasize international approaches ("thinking globally" discourse) have prevailed in international climate politics. In this complex new setting, industrial actors can make relatively adequate claims about their offers of environmental alternatives. The major disagreement seems to be tied to the extent to which flexible mechanisms will be used, and how they will be constructed and implemented. However, proponents of domestic cuts (national action discourse) argue that the international approach advocates measures that come too little, too late.

Naturkraft was able to secure its building permits by cultivating an environmental ethos, for instance, through the use of identification, but also by arguing logically. A majority of the politicians agreed to the scene description of an increasing demand for energy that had to be met. Compared to other fossil fuel types, power from GPPs was without doubt a cleaner alternative. Naturkraft also seemed to convince the majority of the politicians that power from GPPs would replace power from more polluting sources. Arguing on the topics of relationship and comparison may function to strengthen ethos and give the upper hand because of their resemblance to basic cognitive schemes; if something can be established as an antecedent, then a consequence follows naturally. Furthermore, a specific phenomenon is often better understood when compared with other phenomena.

Naturkraft had identified an opportune moment at which to launch its business idea, as a shift in official Norwegian climate policy had occurred at the start of

---

[3]The case was complicated further when a minority government that consisted of GPP opponents took office in the fall of 1997. This government demanded that Naturkraft should reduce its carbon dioxide emissions by 90%. This requirement was put forward after a competing GPP project with "pollution free" technology was launched in 1998. In 2000, the minority government left office after a majority in the parliament instructed it to alter the emission demand. With a new government in place, Naturkraft was finally given a green light. By this time, however, the market situation had changed. The electricity price had dropped, and the owners considered the price for natural gas as too high (Ihlen, 2004b).

the 1990s. The government now favored an international approach, and it was quite possible that the international climate negotiations would introduce some kind of measures to curb emissions. Naturkraft thus fitted in with political need, and it seemed that the environmentalists had lost their "monopoly" on environmental rhetoric. The substitution argument, in particular, proved difficult to beat, and the Climate Alliance failed in its lobbying efforts.

The Action Against GPPs group was, however, able to build a huge and vocal opposition to the plans of Naturkraft. Less than a year after parliament had approved the plans, the government urged the company to postpone construction, and opinion polls reflected the negative public sentiment toward GPPs. The activists used the topics of relationship and comparison, the former to argue that the use of civil disobedience was unavoidable, and the latter to make the issue in question simple and tangible. The whole repertoire of ethos, logos, and pathos was put to use, the latter of which also tied to a previous large-scale environmental conflict. The activists made their threat physical by establishing a protest camp. A couple of strategic maneuvers seemed also to have had an influence. First, the activists circumvented the complex substitution argument, and simplified and repeated their basic contention, thus exploiting the limits of the media's *modus operandi*—the need for simplification and lack of space. "Dumbing down" their rhetoric seemed to work, and was a conscious move on the part of the environmentalists. The second maneuver was the offering to the government by the environmentalists of a retreat so that no face would be lost. Equally important, however, was the fact that the activists launched their threat in the period before the government started its election campaign and before the negotiations in Kyoto. This timing proved to be effective.

The efforts and success of the actors thus point to the insight of rhetoric that *kairos* is crucial. This study shows how both of the parties had to adapt to the new political setting, and how the mass media and domestic public opinion came into play and could not be ignored. The strategies that were employed mirrored each other in that both sides argued from the bases of relationship and comparison, and relied on particular scene descriptions that led to necessary actions and new practices. This made the actors seem rational, which thus strengthened their ethos. This study also shows how the use of pathos was crucial to the ability of the environmentalists to mobilize against the GPPs.

The study also points to potential courses of action in such situations. Actors need to identify the premises in the rhetoric of their opponents and to challenge them. When, for instance, a topic of relationship is used, it must be asked whether the description is correct, and the basic premises of the action that is called for should be scrutinized. As shown, for example, it is not necessarily a natural consequence that growth in energy demand should be met with growth in energy production. Similarly, when a topic of comparison is used, the inductive reasoning behind it should be examined to expose potentially faulty generalizations or poor analogies. Topics of relationship that use deductive reasoning, however, should be analyzed for fallacies such as undistributed middle terms or false dichotomies (Corbett & Connors, 1999).[4]

In addition to illustrating how rhetorical theory can help actors in a conflict, this study has also explained why certain traditional tactics are called for when struggles take place in the media arena: simplify, be concrete, repeat, and present your perspective or point before engaging with the arguments of the opponent. Such rhetoric may be needed to get the arguments across in the media arena, and to help arguments to stand out in the political arena with clear policy suggestions. The challenges are thus plentiful for industrial actors, politicians, activists, and those who wish to understand environmental rhetoric in general. This will be increasingly true as international environmental politics grow more complex.

REFERENCES

Action against gas-fired power plants. (n.d.). Strategy document [internal document]. Oslo, Norway: Author.

Action against gas-fired power plants. (1997a). Kampanjeplan [campaign plan] [internal document]. Oslo, Norway: Author.

Action against gas-fired power plants. (1997b). Miljø eller gasskraftverk? [Environment or gas-fired power plants?] [leaflet]. Oslo, Norway: Author.

Action against gas-fired power plants. (1998). *Evaluering av Fellesaksjonen mot gasskraftverk* [Evaluation of action against gas-fired power plants]. Unpublished manuscript. Oslo, Norway: Author.

Andersen, R. (1997, April 17). Het høst for Jagland [Difficult fall for Jagland]. *Dagbladet*. Oslo, Norway: Dagbladet.

Andresen, S., & Butenschøn, S. H. (2001). Norwegian climate policy: From pusher to laggard? *International Environmental Agreements, 1*(3), 337–356.

Aristotle. (trans. 1991). *On rhetoric: A theory of civic discourse* (G. A. Kennedy, Trans.). New York: Oxford University Press.

Beder, S. (1998). *Global spin: The corporate assault on environmentalism*. London: Chelsea Green Publishing Company.

Bonde, A. (1997, May 7). *Jagland vil utsette gasskraft-utbygging* [Jagland wants to postpone GPP building]. *Aftenposten*. Oslo, Norway: Aftenposten.

Brock, B. L. (Ed.). (1999). *Kenneth Burke and the 21st century*. New York: State University of New York.

Brown, W. R., & Crable, R. E. (1973). Industry, mass magazines, and the ecology issue. *Quarterly Journal of Speech, 59*(3), 259–271.

Bruno, K., & Karliner, J. (2003). *earthsummit.biz: The corporate takeover of sustainable development*. Oakland, CA: Food First.

Burke, K. (1969a). *A grammar of motives* (Rev. ed.). Berkeley, CA: University of California Press.

---

[4]It is important to emphasize that pointing to fallacies or using logos arguments do not result in machinelike adherence. Logos arguments and deduction of fallacies might function to strengthen a rhetor's point about how the opponent is mistaken, but have no imperative power. It is not a matter of creating black-or-white argument categories, because the context of the argument has to be taken into consideration as well. A rhetor will most often have several arguments, and use of one weak argument does not necessarily mean that a debate is lost (Jørgensen, Kock, & Rørbeck, 1994; Kock, 2004). Nonetheless, deduction of fallacies can be an important rhetorical means to weaken the opponents' rhetoric and help the rhetor in constructing his or her rhetoric in ways that avoid fallacy charges.

Burke, K. (1969b). *A rhetoric of motives* (Rev. ed.). Berkeley, CA: University of California Press.

Cantrill, J. G. (1993). Communication and our environment: Categorizing research in environmental advocacy. *Journal of Applied Communication Research, 21,* 66–95.

Climate Alliance. (1996). *Motmelding til St medl* [sic] *nr 38 (1995–1996). Om gasskraftverk i Norge* [Counter paper to Report to the Storting No. 38 (1995–1996). On gas-fired power plants in Norway]. Oslo, Norway: Author.

Conley, T. M. (1994). *Rhetoric in the European tradition* (Rev. ed.). Chicago: The University of Chicago Press.

Corbett, E. P. J., & Connors, R. J. (1999). *Classical rhetoric for the modern student* (4th ed.). New York: Oxford University Press.

Eldegard, T. (1995). *Gasskraft og klimapolitikk* [Gas power and climate policy]. *Sosialøkonomen, 49*(7/8), 2–10.

Feller, W. V. (2004). Blue skies, green industry: Corporate environmental reports as utopian narratives. *Environmental Communication Yearbook, 1,* 57–76.

Furuly, J. G. (1997a, May 10). *Gasskraftverk i Norge? Avblåser ikke miljø-aksjonene* [GPP in Norway? Environmental protest actions are not called off]. *Aftenposten.* Oslo, Norway: Aftenposten.

Furuly, J. G. (1997b, May 9). *Striden om gasskraftverk: Ikke flertall for gasskraftverk* [The GPP conflict: No majority for GPPs]. *Aftenposten.* Oslo, Norway: Aftenposten.

Gensler, H. J. (1998). *Ethics: A contemporary introduction.* London: Routledge.

Greer, J., & Bruno, K. (1996). *Greenwash: The reality behind corporate environmentalism.* Penang, Malaysia: Third World Network.

Hager, N., & Burton, B. (1999). *Secret and lies: The anatomy of an anti-environmental PR campaign.* Monroe, ME: Common Courage.

Hovden, E., & Lindseth, G. (2004). Discourses in Norwegian climate policy: National action or thinking globally? *Political Studies, 52,* 63–81.

Ihlen, Ø. (2002). Rhetoric and resources: Notes for a new approach to public relations and issues management. *Journal of Public Affairs, 2*(4), 259–269.

Ihlen, Ø. (2004a). Norwegian hydroelectric power: Testing a heuristic for analyzing symbolic strategies and resources. *Public Relations Review, 30*(2), 217–223.

Ihlen, Ø. (2004b). *Rhetoric and resources in public relations strategies: A rhetorical and sociological analysis of two conflicts over energy and the environment.* Oslo, Norway: Unipub Forlag.

Jørgensen, C., Kock, C., & Rørbeck, L. (1994). *Retorik der flytter stemmer: Hvordan man overbeviser i offentlig debat* [Rhetoric that moves votes: How to convince in public debate]. Copenhagen, Denmark: Gyldendal.

Killingsworth, M. J. (1996). Environmental rhetoric. In T. Enos (Ed.), *Encyclopedia of rhetoric and composition* (pp. 225–227). New York: Garland.

Killingsworth, M. J., & Palmer, J. S. (1992). *Ecospeak: Rhetoric and environmental politics in America.* Carbondale, IL: Southern Illinois University Press.

Kock, C. (2004). *Retorikkens relevans* [The relevance of rhetoric]. In Ø. Andersen & K. L. Berge (Eds.), *Retorikkens relevans* [The relevance of rhetoric] (pp. 17–29). Oslo, Norway: Sakprosa.

Lange, J. L. (1993). The logic of competing information campaigns: Conflict over old growth and the spotted owl. *Communication Monographs, 60,* 239–257.

Laufer, W. S. (2003). Social accountability and corporate greenwashing. *Journal of Business Ethics, 43,* 253–261.

Lomborg, B. (2001). *The skeptical environmentalist: Measuring the real state of the world.* Cambridge, England: Cambridge University Press.

Lubbers, E. (Ed.). (2002). *Battling big business: Countering greenwash, infiltration and other forms of corporate bullying.* Totnes, England: Green Books.

Myerson, G., & Rydin, Y. (1996). *The language of environment: A new rhetoric.* London: UCL Press.

Myklevoll, T. (1990, August 26). *Gro angrer på Alta-utbyggingen* [Gro regrets the Alta development]. *Dagbladet.* Oslo, Norway: Dagbladet.

Naturkraft. (1994). *Statoil, Statkraft, og Norsk Hydro etablerer Naturkraft a.s; Norsk el-eksport til Norden basert på gass- og vannkraft* [Statoil, Statkraft and Norwegian Hydro establish Naturkraft AS: Norwegian electricity export to the Nordic Region based on gas and hydroelectric power] [press release]. Lysaker, Norway: Author.

Naturkraft. (1995a). *Hva gjør vi med gassen som utvinnes i Nordsjøen etter stengetid?* [What do we do with the gas that is extracted in the North Sea after closing time?] [brochure]. Lysaker, Norway: Author.

Naturkraft. (1995b). *Melding om planlegging av tiltak: Gasskraftverk på alternative byggesteder: Kårstø, Kollsnes, Tjeldbergodden* [Advance notice of planning of venture: Gas-fired power plants at alternative sites: Kårstø, Kollsnes, Tjeldbergodden] [report]. Lysaker, Norway: Author.

Nilsen, Y. (2001). *En felles plattform? Norsk oljeindustri og klimadebatten i Norge fram til 1998* [A joint platform? Norwegian petroleum industry and the climate debate in Norway until 1998]. Norway: Centre for Technology, Innovation and Culture, University of Oslo.

Nygaard, O. (1997, May 9). *Gir etter for press* [Yields to pressure]. *Aftenposten*. Oslo, Norway: Aftenposten.

Paystrup, P. (1995). Plastics as planet-saving "natural resource": Advertising to recycle an industry's reality. In W. N. Elwood (Ed.), *Public relations inquiry as rhetorical criticism: Case studies of corporate discourse and social influence* (pp. 85–116). Westport, CT: Praeger.

Peeples, J. A. (2005). Aggressive mimicry: The rhetoric of wise use and the environmental movement. *Environmental Communication Yearbook, 2*, 1–17.

Perelman, C. (1982). *The realm of rhetoric* (W. Kluback, Trans.). Notre Dame, IN: University of Notre Dame Press.

Perelman, C., & Olbrechts-Tyteca, L. (1971). *The new rhetoric: A treatise on argumentation* (Rev. ed.; J. Wilkinson & P. Weaver, Trans.). London: University of Notre Dame.

Peterson, T. R. (1997). *Sharing the earth: The rhetoric of sustainable development*. Columbia, SC: University of South Carolina Press.

Rowell, A. (1997). *Green backlash: Global subversion of the environmental movement*. London: Routledge.

Sandman, P. M. (1994). Mass media and environmental risk: Seven principles. *RISK, 5*, 251–260.

Sipiora, P., & Baumlin, J. S. (Eds.). (2002). *Rhetoric and kairos: Essays in history, theory, and praxis*. Albany, NY: State University of New York Press.

Sproule, J. M. (1988). The new managerial rhetoric and the old criticism. *Quarterly Journal of Speech, 74*, 468–486.

Stillar, G. F. (1998). *Analyzing everyday texts: Discourse, rhetoric and social perspectives*. Thousand Oaks, CA: Sage.

Storting Deliberations. (1996). [No. 107, June 14, session 1995–1996]. Oslo, Norway: Stortinget.

Tokar, B. (1997). *Earth for sale: Reclaiming ecology in the age of corporate greenwash*. Cambridge, MA: South End Press.

Utting, P. (2002a). Corporate environmentalism in the South: Assessing the limits and prospects. In P. Utting (Ed.), *The greening of business in developing countries: Rhetoric, reality and prospects* (pp. 268–292). London: Zed Books.

Utting, P. (Ed.). (2002b). *The greening of business in developing countries: Rhetoric, reality and prospects*. London: Zed Books.

Vassbotn, P. (1997, May 11). *Jaglands retrett* [Jagland's retreat]. *Dagbladet*. Oslo, Norway: Dagbladet.

CHAPTER EIGHT

Bridging the North–South Divide: The Global Responsibility Frame at Earth Summit +5

Marie A. Mater
Houston Baptist University

The environment has gained prominence as an important political, moral, and social issue. Originally debated at local and national levels, environmental concerns have now become a routine theme of international and global discussions, similar to national security and economic issues. One of the most important global meetings was the United Nations Conference on Environment and Development (UNCED) in Rio de Janeiro in June 1992. This meeting, also known as the Earth Summit, was unprecedented in the sheer number and diversity of the participants. Heads of state of more than 100 countries, delegations from 178 countries, and representatives of more than 1,000 nongovernmental organizations (NGOs) gathered to consider global environmental problems and development issues.

Originally hailed as a historic opportunity for global cooperation on these issues, the Earth Summit quickly became the setting for a contentious debate between the nation-states and NGOs of the North and the South. Throughout the conference's preparatory meetings, and most of the conference itself, delegates from the North and the South were embroiled in a heated dispute. On one hand, the nation-states and NGOs of the North (particularly the United States and Greenpeace International) wanted an environmental regulatory presence in the South, while maintaining their own status quo. On the other hand, the Southern nation-states and NGOs (Malaysia and Third World Network) argued for economic development unhampered by Northern environmental restrictions and enhanced with Northern financial and technological assistance. Schwarz (1992) described the resulting atmosphere: "At times, talking to delegates from developed and developing nations before and during Unced (UNCED) left the eerie feeling that the two sides had been invited to different conferences" (p. 30).

At the end of the Earth Summit, the nation-states from both the North and the South did adopt *Agenda 21* (United Nations, 2004). *Agenda 21* is a comprehensive plan of action for human impacts on the environment. It provides guidance for the local, national and global levels. The main sections of the document include "Social and Economic Dimensions, Conservation and Management of Resources for Development, Strengthening the Role of Major Groups, and Means of Implementation" (United Nations, 2004). Each of these sections has approximately 10 chapters for more in-depth treatment of specific issues. According to the *Agenda 21* preamble, the responsibility for this plan lies first and foremost with nation-states. However, it also argues that the United Nations and other international, regional, and subregional organizations have a role to play. Finally, it calls for the participation of NGOs and the public. In fact, *Agenda 21* argues that the involvement of all social groups is critical for nation-state successes. It also calls for broad public participation in environmental decision making.

Five years after the Earth Summit, a special session of the United Nations General Assembly was held to review and appraise the progress made toward achieving the implementation of *Agenda 21*. This meeting became known as "Earth Summit +5" and was held June 23 through June 27, 1997, in New York at the U.N. headquarters. According to the Malaysian President of the General Assembly, Mr. Razali Ismail, the goal of the special session was to ". . . take a hard, honest and critical look at what has been done and what has not been done since Rio. . . . We need to recall and reemphasize the compact that brought about the Earth Summit" (United Nations Department of Public Information, 1996, para. 3). While highlighting sustainable development success stories, the meeting was also to identify why specific actions had not been taken and to suggest corrective actions.

Earth Summit +5 did not have the same attendance record as the first Earth Summit. This was because many Southern heads of state refused to attend the meeting and sent lower level government representatives instead. Notably absent was the outspoken Malaysian leader, Mahathir bin Mohamad, whose speech at the first Earth Summit condemned Northern efforts to impose environmental regulations on Southern nation-states as another form of colonialism. However, many Northern leaders did attend, and United States President, William Jefferson Clinton, personally presented a statement. Importantly, Earth Summit +5 provided the unprecedented opportunity for environmental NGOs from both the North and the South to also present 5-minute addresses to the General Assembly. This was unlike the first Earth Summit, where NGOs like Greenpeace International and Third World Network were not allowed to address the governmental delegations and had to submit written "Reports on Global Issues" in order to influence the construction of the final version of *Agenda 21* by the nation-state representatives.

Analysis of the discourse produced at Earth Summit +5 by important Northern and Southern actors suggests that *global responsibility* emerged as the masterframe of the *Agenda 21* discussions. The hegemony of the *global responsibility* masterframe seems to support Apel's (1987, 1991, 1993), Corrick's (1990), Moscovici's (1990), and Strydom's (1999a, 1999b, 2000, 2002) argument that we have entered an "era of responsibility," because of the unique nature of environmental problems. In this era of responsibility, the Habermasian two-track public sphere, which I now elaborate on, serves an important function because cooperative solutions to environmental problems must be achieved through communication. This communication occurs at all levels, from the local general public spheres to the global mediated and regulated public spheres.

This chapter reviews the environmental communication literature that analyzes global environmental conferences like the two Earth Summits. It then examines the environmental communication literature that utilizes Habermasian approaches and sets out an updated Habermasian theory to explain how conferences like the Earth Summits produce discourse. To interpret the environmental discourse produced by important Northern and Southern actors at Earth Summit +5, the cognitive discourse analysis method is used. The official statements presented by representatives of the United States, Malaysia, the Third World Network, and Greenpeace International are analyzed and interpreted to determine how these actors seemed to construct their understanding of *Agenda 21* in the wake of the North–South debate at the first Earth Summit. The impact of the discourse produced by these actors is then considered in the conclusion.

COMMUNICATION LITERATURE ON GLOBAL ENVIRONMENTAL CONFERENCES

Global environmental conferences like the Earth Summit and Earth Summit +5 have not been studied extensively by environmental communication scholars in the United States. The Earth Summit is briefly mentioned in Waddell's (1996) study of public participation in environmental policy regarding the Great Lakes; Brown and Herndl's (1996) analysis of the extreme environmental rhetoric of the John Birch Society; and Cox's (2004b) analysis of President Clinton/Vice President Gore and the rhetoric of U.S. environmental politics. NGOs' increased participation in global environmental conferences like the Earth Summit is also examined briefly in Tevelow's (2004) work on NGO capacities in different institutional contexts. President Bush's poor performance at the Earth Summit is mentioned several times in Carcasson's (2004) analysis of the American presidency and the framing of international environmentalism.

One of the more thorough examinations of the Earth Summit is Peterson's (1997) rhetorical analysis of President Bush's address. Peterson begins by describing the agreements (including *Agenda 21*) that were signed at the Earth

Summit. She then contrasts the United States' active participation in the 1987 *Montreal Protocol on Substances That Deplete the Ozone Layer*, with its lackluster participation at the Earth Summit. She then argues that President Bush portrayed a Lone Ranger role at the conference; and as a result, the response to his speech by other world leaders (especially Mahathir) was negative. Peterson also analyzes how *sustainable development* is defined using the themes of equity and cooperation. She concludes by arguing that "Despite attempts to work across political and disciplinary boundaries, the working definition of *sustainable development* that emerges in the *Rio Declaration on Environment and Development* fails to fundamentally alter the traditional perspective on development" (Peterson, 1997, p. 85).

Peterson and Pauley (2000) also examined President Bush's participation at the Earth Summit. Like the previous work, they begin by discussing the conventions and nonbinding statements that were signed by many nation-states, the lackluster participation of the United States, and the Lone Ranger role that President Bush played. Peterson and Pauley (2000) argued that although President Bush wanted the United States to lead global efforts, his was "a failed bid for leadership" (p. 81). Additionally, they examine President Bush's definition of nature "as commodity" and contrast this to one Southern definition of nature based on a "... holistic understanding of human life and its relationship with other life forms" (Peterson & Pauley, 2000, p. 87). They conclude by pointing out that the hostility between the North and the South is based on these different definitions of nature and that perhaps *sustainable development* indicates an awareness of possible alternatives to the market model.

ENVIRONMENTAL COMMUNICATION LITERATURE UTILIZING HABERMASIAN APPROACHES

In addition to the work on global environmental conferences, environmental communication scholars have also used Habermasian approaches to help them understand societal debates on environmental issues. Generally, these scholars fall into two groups: those that use Habermas' theory of the public sphere to explore the spaces in which environmental discourse is produced and those who use his theory of communicative, instrumental, and strategic action to examine the environmental discourse that is produced. There are, however, several (Daley & O'Neill, 1991; Karis, 2000) who use his work to do both.

Environmental communication literature that incorporates the theory of the public sphere to explore the spaces in which environmental discourse is produced relies on Habermas' (1962/1989) original theory presented in *The Structural Transformation of the Public Sphere*. A distinctive feature of this theory is the use of communication by actors in public spaces free of power. There are several studies in the environmental communication literature that use this notion of the public sphere. Cantrill's (1996) study of the Beartooth Alliance in Montana indicates that the rhetoric of grassroots groups creates a sense of community and limits their attempts to shape public policy. Todd's (2003) analysis of the Brazilian Amazon debate between the government and indigenous resistance groups argues that there is an emergence of local actors on the global level. Cox's (2004a) examination of the North American Free Trade Agreement and the Free Trade Area of the Americas finds that these types of agreements create barriers to the public sphere's demand for transparency, public participation, and democratic accountability. Finally, DeLuca's (1999) and DeLuca and Peeples' (2002) critiques of the public sphere take issue with the oral nature of the space in a time of mass communication and argue for supplementing it with the notion of "public screens" that can account for image events.

Environmental communication literature that uses the theory of communicative, instrumental, and strategic action to examine the environmental discourse that is produced relies on Habermas' (1981/1984, 1981/1987) examination of these ideas in *The Theory of Communicative Action*. Communicative action produces argumentative speech through public participation. Instrumental action, on the other hand, restricts speech to that which is produced by technical experts. Strategic action works to manipulate the discussion so as to achieve a desired end. Several studies in the environmental communication literature use these ideas to examine environmental discourse. Killingsworth and Palmer's (1992) chapter on environmental impact statements indicates that although the documents are open in theory to public discussion, in practice they are closed because only scientific opinions count. Ross' (1996) study of two Mohawk Indian controversies attempts to correct communicative action (and decrease strategic action) with the ethical notion of responsible care. Patterson and Lee's (2000) analysis of the Kingsley Dam relicensure in Nebraska claims that the instrumental process emphasized balance which hid the subjectivity of decision making and reduced the reasonable actor to the role of umpire. This led to a colonization of the public sphere. Plevin's (2000) examination of national environmental publications' use of guilt argues that this does not encourage communicative action, and may actually repress public participation. Kinsella's (2004) call for public expertise underscores the instrumental action that typically occurs in energy and environmental policy debates and suggests the public attain technical competence. Finally, Schwarze's (2004) survey of the United States Forest Service's public participation efforts in Minnesota finds that the instrumental action present eroded stakeholder confidence and trust in the service.

As mentioned previously, there are several environmental communication studies that use both Habermas' theory of the public sphere and his theory of communicative, instrumental, and strategic action to examine environmental discourse production. In Daley and O'Neill's (1991) comparison of press coverage of the *Exxon Valdez* oil spill, they go beyond Habermas' conception of a public space and argue that the mass media are the public sphere. In the case of the *Exxon*

Valdez, Daley and O'Neill (1991) contended that "Obviously, the public sphere was not monoideological; its discursive character was fluid, open, and complex. Numerous voices were able to articulate competing definitions of the situations and offer narratives that anticipated new consequences" (p. 53). They do point out, however, the press coverage favored governmental and industrial officials. Moreover, they argue that the individual focus on Captain Joseph Hazelwood ". . . did the ideological work of closing down an interrogation of the system power of technological control" (Daley & O'Neill, 1991, p. 49). In the end they concluded, "Native claims to nature, inseparable from culture, politics, and economics, appear to be outside the mainstream reportorial knowledge system. In this case, press coverage either ignored them or absorbed them into the feeble public of victimized fisherman" (Daley & O'Neill, 1991, p. 55). For Daley and O'Neill, the mass-mediated public sphere does two things. First, it excludes non-Western modes of thinking and arguing. Second, if non-Western modes of thinking and arguing do enter the public sphere, they get filtered through a Western lens.

Another environmental communication study that uses both Habermas' theory of the public sphere and his theory of communicative, instrumental, and strategic action is Karis' (2000) examination of the failure of two large projects in Adirondack Park in the early 1990s. Although he does not cite Habermas directly, he does use his theories to understand the importance of their defeat. Karis (2000) argued, ". . . environmental issues, perhaps more than any other technical/scientific matters, are precisely those issues that can provide for greater latitude for the inclusion of dialogue, and debate of various human values" (p. 229). For Karis, public debate over environmental issues means more inclusion, not exclusion. Additionally, Karis (2000) said, ". . . public dialogue over environmental matters can and should offer a 'corrective' to the problem of technological expediency . . ." (p. 229). He believes that communicative action can overcome instrumental action. Moreover, he believes the potential for public deliberations in society are increasing because of environmental problems.

DISCOURSE PRODUCTION AND ANALYSIS IN PUBLIC SPHERES

If, as Karis believes, we are entering a time period in which there is an increase in public deliberation on environmental issues, then Habermasian approaches to studying discourse production in the public sphere are useful. As illustrated, environmental communication scholars have used his theories to understand both the spaces in which discourse is produced and the actual discourse that is produced. Unfortunately, however, Habermas' revised thinking on the complex nature of the public sphere has been neglected. In this section of the chapter, then, Habermas' (1992/1996) two-track theory of the public sphere that is set out in *Between Facts and Norms* is introduced.

A Revision of Discourse Production in the Public Sphere

Habermas (1992/1996) argued that there are two types of public spheres: (a) general public spheres that develop spontaneously and provide a medium for unrestricted communication and (b) procedurally regulated ones (like parliaments), which try to produce cooperative solutions to practical questions. General public spheres not only include organized groups like civic organizations or social movements, but also include less organized groups such as informal gatherings at coffee shops or rock concerts. Moreover, abstract mass-mediated public spheres of listeners, readers, and viewers exist. General public spheres try to raise a "crisis consciousness" about issues like the environment. If they are able to get the attention of larger numbers of the public, then *problematization* takes place, in which the weight of public opinion requires that the regulated public sphere become involved (Habermas, 1992/1996). Once in the regulated public sphere, the issue becomes the subject of formal methods of deliberation required by that particular institution.

In the problematization phase, the general public sphere has to fulfill important expectations. It is like a warning system with sensors that not only detects and identifies problems, but also thematizes them and dramatizes them so that they become issues that must be taken up and dealt with by the procedurally regulated public spheres (Habermas, 1992/1996). In this role, the general public sphere can be viewed as ". . . a network for communicating information and points of view . . . [in which] the streams of communication are, in the process, filtered and synthesized in such a way that they coalesce into bundles of topically specified *public* opinions" (Habermas, 1992/1996, p. 360). This kind of public opinion has political influence that can be transformed into political power when it affects the beliefs and decisions of those in the procedurally regulated public sphere. In order to understand more thoroughly this important process of transforming differing positions on an issue into public opinion, we must revisit Habermas' earlier work on communicative action and discourse.

Habermas (1981/1984) defined communicative action as ". . . the interaction of at least two subjects [who] . . . seek to reach an understanding about the action situation and their plans of action in order to coordinate their actions by way of agreement" (p. 86). The objective of reaching understanding about a particular situation implies that the actors must be capable of behaving in a communicatively rational fashion. This means that the actors must be able to achieve their agreement on the situation (and the plan for action) through language. The mutual understanding of a situation that results from communicative rationality, then, can be examined in two related ways: (a) how it is defended in terms of validity claims and (b) how the actors relate to the world in terms of value spheres (Habermas, 1981/1984).

First, an actor must defend his or her position on an issue in a valid manner. This is done through "validity claims," which are based on Habermas' (1976/

1979) earlier identification of the universal conditions necessary for mutual understanding: reciprocal understanding, shared knowledge, mutual trust, and accord with one another. Therefore, the actor claims that his or her statement is *true* so that the audience will recognize and accept it as legitimate knowledge about the suggested action. Moreover, his or her statement is presented as *right* with respect to the normative context of the action. Finally, the actor presents himself or herself as *sincere* with regard to his or her intentions while discussing the action.

In the process of coming to a mutual understanding of a situation, the audience must accept all of the speaking actor's validity claims or make known his or her dissent. Habermas (1981/1987) explained that consensus cannot come about, for example, if the audience ". . . accepts the *truth of an assertion* but at the same time doubts the sincerity of the speaker or the normative appropriateness of his [or her] utterance . . ." (p. 121). When disagreement over a validity claim occurs, the participants move their communication to the argumentative level of discourse. At the argumentative level of discourse, actors must give good reasons for their validity claims. Moreover, these reasons can be criticized by the other actors. Discourse has three special characteristics: (a) it excludes all force, except the force of the better argument, (b) it is subject to special rules, and (c) its aim is to produce cogent arguments that are convincing (Habermas, 1981/1984).

Second, an actor produces his or her reasons or arguments in discourse based on his or her various relationships to the world. These relationships with the world are constituted in three ways, each according to the different value spheres:

1. The objective world (as the totality of all entities about which true statements are possible);
2. The social world (as the totality of all legitimately regulated, interpersonal relations);
3. The subjective world (as the totality of the experiences of the speaker to which he [or she] has privileged access). (Habermas, 1981/1984, p. 100)

In other words, the reasons given by an actor must be oriented to the world of nature, the world of collective norms, and the actor's inner world. These three worlds are interdependent and operate concurrently in any argument.

In communicative action, the process of reaching understanding generally occurs against a background of shared cultural understanding. This changes, however, when a situational definition cannot be agreed on because an actor is thematically stressing one of the three value spheres in a different way from the other actors. Consequently, the definition of situations is continually being revised by actors who interpret the world differently. "Stability and absence of ambiguity are rather the exception in the communicative practice of everyday life. A more realistic picture is . . . a diffuse, fragile, continuously revised and only momentarily successful communication" (Habermas, 1981/1984, p. 100). This is especially true in times of crisis, when actors are continually reworking their definitions of the political, moral, or social question that is the source of contention.

Cognitive Discourse Analysis of the Public Sphere

Discourse, then, is the means through which the public comes to understand a particular issue as it is constructed by various actors. Cognitive discourse theorist Klaus Eder, who worked closely with Habermas in the 1970s and 1980s, has developed a three-step method for analyzing both the texts and the strategies of actors as they present their definition of the situation in question in the general public sphere. It is important to analyze both the texts and the way they are presented in the public sphere. One way to begin to analyze these texts is to examine the frames at work within them. For Eder (1996b), frames are:

> . . . stable patterns of experiencing and perceiving events in the world which structure social reality . . . we use and apply frames in order to sort the world, thus reducing the continuous stream of events to a limited number of significant events. Frames give to these selected events an objective meaning, thus disregarding subjective differences and idiosyncrasies ascribed to individual persons. (p. 166)

Frames should be understood metaphorically not as a picture frame that brackets an actor's definition of the situation, but as structures that constitute an actor's definition of the situation.

In discourse production in the general public sphere, frames operate at three different levels. On the first level, "microframes" are found in the deep structure of a particular actor's text (Strydom, 2000, p. 84). The cognitive framing devices used for constructing these microframes are based on Habermas' value spheres of the objective world, the social world, and the subjective world mentioned previously. Eder (1996a) referred to these conceptions as the "real," the "right," and the "meaningful" (p. 206). The cognitive framing device of the "real" communicates economic, scientific, and technological facts. The cognitive framing device of the "right" communicates laws and moral codes. The cognitive framing device of the "meaningful" communicates ethical or value choices made for an actor's identity formation. These three cognitive framing devices are always present in an actor's text, but one of them is usually dominant.[1] On the second level, "mesoframes" are constructed by actors with their entire text (Strydom, 2000, p. 84). It is the frame that gets identified with a particular actor and their text in the general public

---

[1] In order to interpret the Earth Summit +5's special focus on time, Aristotle's (trans. 1991) identification of specific "times" as the topic of discussion in judicial, deliberative and epideictic speeches has been used to interpret time within Eder's conceptions of the "real," the "right," and the "meaningful" (p. 206). For example, references to the past involve the objective world (real), references to the future involve the social world (right), and references to the present involve the subjective world (meaningful).

sphere. On the third level, a "masterframe" emerges from the competing meso-frames that are released by each actor (Strydom, 2000, p. 84). This is the hege-monic frame that comes to construct the public's reality of a situation as a result of the problematization process.

In order to distinguish the different frames put forth by an actor in their text, an analysis of the deep structure is required. This is because the different micro-frames are not usually presented literally in a text. Instead they must be inferred from the idea elements found within it. Paulo Donati (1992), who collaborated with Eder, explains this process of interpretation:

> Since a frame is a known structure, the elements that are constitutive of it are implic-itly considered as "naturally" tied together. The consequence is that mentioning some elements—sometimes even one—is usually enough to "suggest" or to recall the whole set, as, for example, when a body lying on the ground near a knife immedi-ately suggests a whole homicidal plot, or even better when some elements become the accepted "symbols" of entire conceptual constructions, such as a red flag. (p. 141)

Fisher (1997) concurred with this explanation and clarified that these elements operate metonymically in which one part can stand for the whole. To reveal these idea elements within selected texts, the cognitive framing devices of the "real," the "right," and the "meaningful" are coded. For each of the actors' texts, the words, phrases, clauses, sentences, and/or paragraphs that correspond to a particular framing device are coded and labeled with what Gamson, Croteau, Hoynes, and Sasson (1992) called an "idea element," consisting of two or three words that sum-marizes an actor's argument. All of the idea elements present within the text are compiled and sorted to determine their stress, or emphasis in the text. The com-piled and sorted idea elements are then interpreted to determine the microframe being suggested by them.

Once the microframes are identified, the symbolic packaging of them must be examined. Eder (1996b) explained this process:

> Through symbolic packaging framing devices are "attached" to the social world, to social situations and to social actors . . . The frames in action-contexts cannot be un-derstood outside the specific symbolic package that gives them consistency, coher-ence and validity. Thus, the second step of discourse analysis involves a move from the level of analysis of cognitive structures to the level of *narrative* structures. (p. 168)

The next step, then, is to examine all of the cognitive framing devices and the microframes and interpret how they are communicated together within the same text. Generally, this involves a reference to a narrative or ideology that finds reso-nance within a given society (Fisher, 1997). Examples of symbolic packages in-clude capitalism and communism. Symbolic packages may be mentioned directly, or like frames, they may be inferred from the tapestry of cognitive framing devices and microframes present in the text.

Finally, cognitive framing devices and microframes are situated in the public spheres via the strategies of the actors producing them. Eder (1996b) claimed that the empirical social contexts of framing devices must be taken into account by examining the symbolic packaging organizing them. He also argued that framing strategies transform the microframes and the symbolic packages into mesoframes (Eder, 1996a). He does not, however, provide a method for studying the framing strategies of the actors.[2] Because the empirical context of a discourse is always already imbued with power, "rules" affect what actors can and cannot say (Foucault, 1972). Consequently, power influences the framing strategies and the symbolic packages available to actors. "Rules of exclusion" include the outright prohibition of an actor or the rejection of "unreasonable" or "false" arguments (Foucault, 1972, pp. 216–220). "Rules of internal control" comprise the role the discourse plays, the role of the author as a unifying principle and the organization of disciplines (Foucault, 1972, pp. 220–224). "Rules of speaking subjects" consist of the qualifications required of the speaker, the existence of an exclusive "fellowship of discourse," and the existence of an actor's allegiance to doctrine and the social appropriation of discourse (Foucault, 1972, pp. 224–227). In the following discussion of the interpretative results for each actor's text, the cognitive framing devices (given from left to right in the order of thematic stress), microframes, symbolic package, mesoframe, and framing strategy are presented as shown in Fig. 8.1.

To interpret the framing devices, microframes, symbolic packages, mesoframes, and framing strategies in the selected texts, the author used the following methodology. First, to interpret the framing devices used within a particular text, a surface-structure analysis of the semantics of the text was applied. The framing devices of the "real," the "right," and the "meaningful" were assigned color codes. For each of the texts, the words, phrases, clauses, sentences, and/or paragraphs that corresponded to a particular framing device were highlighted with the appropriate color and then labeled with an idea element consisting of two or three words, which summarized an actor's objective, social, and subjective positions. Next, these idea elements were compiled for each text according to whether they identified the framing devices of the "real," the "right," and the "meaningful." This compilation of idea elements was then used to interpret the order and stress of the framing devices within a text. This compilation was also used to infer the mesoframe of the entire text. The symbolic package was inferred from the narrative suggested by the actor and the text. The rules affecting the framing strategy of the actor were interpreted from the actor's political or social position and history in the global public sphere.

---

[2]Eder (1996b) himself admitted that the relationship between frames and framing strategies is undertheorized.

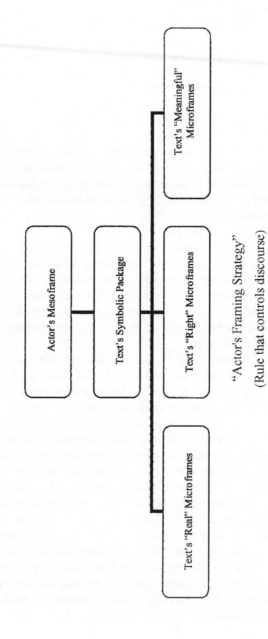

FIG. 8.1. Presentation of interpretative results.

INTERPRETATIVE RESULTS OF EARTH SUMMIT +5

The Official Statement of United States
President William Jefferson Clinton

The official statement of United States President William Jefferson Clinton at the Earth Summit +5 stressed the cognitive framing devices of what is "real" and what is "meaningful" in order to construct a frame of *global responsibility* within the symbolic package of a *global superpower* under the influence of the "rule of speaking subjects" in which the speaker is required to have certain qualifications (see Fig. 8.2).

President Clinton relied on the cognitive framing device of what is "real" throughout his speech. He began by referring to the United States' environmental record in *the past* in order to show that his administration has been actively working to achieve *Agenda 21* goals. For example, he stated, "Here in America, we have cleaned up a record number of toxic waste dumps . . . We have passed new laws to better protect our water and created new national parks and monuments. We have worked to harmonize our efforts for environmental protection, economic growth and social improvement . . ." (Clinton, 1997, para. 4). President Clinton also relied on numerous economic examples to show that implementation of *Agenda 21* relies on *economic instrumentalism*. He proclaimed that the United States will increase environmental financial assistance to the South:

> To help developing nations reduce greenhouse gas emission, I am pleased to announce that the U.S. will provide them with $1 billion in assistance over the next five years. These funds will go to programs that support energy efficiency, develop alternative energy sources and improve resource management to promote growth [that] does not adversely affect the climate. (Clinton, 1997, para. 16)

Additionally, President Clinton (1997) believed that private investment was important and stated, "The United States will continue to encourage private investment that meets environmental standards. . . . Common guidelines for responsible investment can greatly sustain growth in developing countries" (para. 17). President Clinton (1997) also had *faith in science and technology* and relied on environmental statistics and figures to make his case for *global responsibility*:

> The science is clear and compelling: We humans are changing the global climate. Concentrations of greenhouse gases in the atmosphere are at their highest level in more than 200,000 years and climbing sharply. If the trend is not changed, scientists expect the seas to rise two feet or more in the next century. (para. 7)

He believed that new technologies would help to solve the environmental problem of global climate change and said, "We must create new technologies . . . Many of

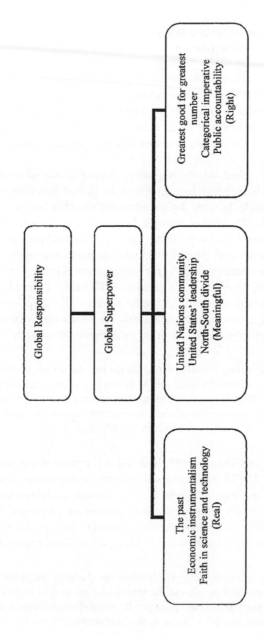

"Rule of speaking subjects-Speaker qualifications"

FIG. 8.2. President Clinton's interpretative results.

the technologies that help us meet the new air quality standards can also help address climate change" (Clinton, 1997, para. 12).

President Clinton also stressed the cognitive framing device of what is "meaningful" in order to establish his identity as the leader of the United States. He began by describing a *United Nations community* in which "... the nations of the world joined together around a simple but revolutionary proposition ..." (Clinton, 1997, para. 1). President Clinton believed that within this community, the *United States' leadership* should be an example to other nation-states. President Clinton (1997) first pointed out his environmental record in his own country: "Yesterday, I announced the most far-reaching efforts to improve air quality in our nation in 20 years, cutting smog levels dramatically, and for the first time ever, setting standards to lower levels of the fine particles in the atmosphere that form soot" (para. 5). He then went on to say that he would take action outside of the United States. "Today, I want to discuss three other initiatives we are taking to deal with climate change and to advance sustainable development here and beyond our borders" (Clinton, 1997, para. 15). Furthermore, he acknowledged that there is a *North–South divide*, but argued that every nation should still do its part: "No nation can escape this danger. No nation can evade its responsibility to confront it. We must all do our part—industrial nations that emit the largest quantities of greenhouse gases and developing nations whose emissions are growing rapidly" (Clinton, 1997, para. 8).

Finally, the cognitive framing device of what is "right" is also present in President Clinton's text, although it is not stressed as much. He used the microframe of *greatest good for greatest number* when he argued:

> In our area, the environment has moved to the top of the international agenda because how well a nation honors it will have an impact, for good or ill, felt across the globe. Preserving the resources we share is crucial not only for the quality of our environment and health, but to maintain stability and peace, within nations among them. (Clinton, 1997, para. 2)

President Clinton (1997) also relied on the *categorical imperative* by ending his speech with a quote from the *Bible*:

> ... "One generation passes away and another comes, but the earth abides forever." We must strengthen our stewardship of the environment so that when this generation passes, it will be a rich and abundant earth abode—and the coming generation will inherit a world as full and as good as the one we have known. (para. 19)

In order to accomplish this moral goal, President Clinton argued that there needs to be *public accountability*. For example, he argued the United States will "... bring to the Kyoto conference [on climate change] in December a strong American commitment to realistic and binding limits that will significantly reduce our emissions of greenhouse gases" (Clinton, 1997, para. 14).

President Clinton's microframes are used to create a powerful symbolic package and frame. President Clinton re-creates the specific image of the United States as the only *global superpower*—militarily, economically, and culturally. This symbolic package is very effective because it draws on the four decades of Cold War history, the emergence of the United States as one of the world's strongest economies, and the hegemony of American culture worldwide. As the only global superpower, the United States shoulders a great deal of responsibility. President Clinton framed the United States' environmental position as one of *global responsibility*. Consequently, he argued that the United States, as the undisputed world leader, should take on a good deal of responsibility for the state of the global environment.

The power inherent in the global public sphere influences this framing strategy in the form of the "rule of speaking subjects," in which the speaker is required to have certain qualifications. Because President Clinton was the leader of the last remaining superpower, he was expected to have the qualifications that are needed to play a central role in the regulated public sphere of a United Nations gathering. He was especially aware of the fact that President Bush was perceived to have abdicated this leadership position at the first Earth Summit. Consequently, President Clinton incorporated the microframes of *United Nations community* and *United States' leadership* and the symbolic package of a *global superpower* into his overall frame of *global responsibility*.

The Official Statement of Malaysian Minister of Science, Technology, and the Environment Dato' Law Hieng Ding

The official statement of the Malaysian Minister of Science, Technology, and the Environment, Dato' Law Hieng Ding, at Earth Summit +5 highlighted the cognitive framing device of what is "real" more than what is "right" and what is "meaningful" in the construction of his frame of *global responsibility* within his symbolic package of a *global community* because his framing strategy was also governed by the "rule of speaking subjects" in which an actor needs to have certain qualifications (see Fig. 8.3).

Dato' Law placed the most stress on the cognitive framing device of what is "real." As the Minister of Science, Technology, and Environment, Dato' Law (1997) has to have a strong *faith in science and technology*:

> ... [the] transfer of environmentally sound technologies is crucial to the success of Agenda 21. While we note that our experts are discussing this matter, we would urge that the TRIPs [Trade-Related Aspects of Intellectual Property Rights] agreement in WTO should be reviewed to facilitate the transfer of environmentally sound technologies. (para. 25)

Dato' Law linked this *faith in science and technology* with a belief in *economic instrumentalism*. He claimed that development assistance is needed for imple-

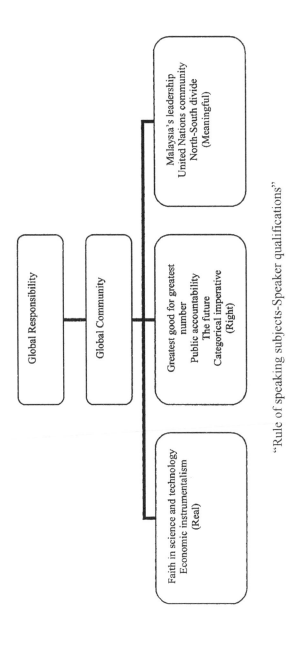

"Rule of speaking subjects-Speaker qualifications"

FIG. 8.3.   Dato' Law's interpretative results.

menting the goals of *Agenda 21.* "Finance is perhaps the most tangible indicator of the commitment of developed countries to the entire process of promoting sustainable development" (Law, 1997, para. 23).

The cognitive framing device with the second degree of stress is what is "right." Dato' Law called for the *greatest good for greatest number* in which "future cooperation must be based on a genuine partnership of shared values and common destiny" (Law, 1997, para. 19). Dato' Law (1997) recognized that this would not be easy, but he was ". . . confident that together we can achieve the political consensus necessary for this purpose" (para. 31). Like President Clinton, Dato' Law believed that *public accountability* was necessary. He argued that this should be achieved through the United Nations: "We would like to see the strengthening of the role and work of CSD and UNEP, with a clear definition of responsibilities. The CSD [Commission on Sustainable Development] should remain the premier body at the policy level. . . . UNEP [United Nations Environment Programme] should remain the international body to mobilize action and should therefore be strengthened" (Law, 1997, para. 29). According to Dato' Law, this *public accountability* should extend into *the future.* He explained: "Our message should be clear and precise for all to understand that our commitments are for real action and that people's well-being, and that of future generations, remain in the center of our deliberations. We support the sectors identified for future programmes of work" (Law, 1997, para. 21–22). Finally, Dato' Law (1997) relied on the *categorical imperative* and said that Malaysia's success as a rapidly developing country ". . . bestows a responsibility on us to help others" (para. 15).

The cognitive framing device with the least amount of stress in Dato' Law's statement is that which is "meaningful" for identity formation. Throughout his statement, Dato' Law pointed out *Malaysia's leadership* in the *United Nations community.* For example, he publicly acknowledged the role of Razali Ismail, the Malaysian President of the United Nations General Assembly: "Mr. President, it gives me great pleasure to see you in the Chair of this important meeting. Your leadership and untiring efforts in the cause of sustainable development is a great source of pride and encouragement to all Malaysians" (Law, 1997, para. 2). Furthermore, Dato' Law (1997) stated, "Malaysia's effort to protect the environment predates the Rio Summit. We have established clear policies and programmes to deal with environmental matters at the national, regional, and global levels" (para. 11). Dato' Law (1997) highlighted the fact that Malaysia had been able to do this in spite of the disadvantage of being a developing country:

> Like other developing countries, Malaysia has met its international obligations and commitments largely through its own resources and efforts. But this has not been easy for us. For a developing country like Malaysia, development is a priority. With limited domestic resources, meeting international commitments puts additional stress on our capacity to address development priorities. (para. 10)

Although Dato' Law (1997) identified Malaysia as a developing country in the previous quote, he made a plea for bridging the *North–South divide* that was prevalent at the Earth Summit: "Our experience over the past five years clearly shows that international assistance and cooperation is vital to achieving the goals of Agenda 21. We therefore need to begin anew by breaking away from the North–South divide, which has stalled so many important issues at the negotiation table" (para. 18–19).

Dato' Law's microframes come together to form a unique symbolic package and a frame that is the same as that presented by President Clinton. Dato' Law puts forth the image of a *global community* that must work together to achieve the goals set out in *Agenda 21*. This symbolic package is effective because it draws on current globalization trends (e.g., transportation, economics, and communication) that are bringing the various peoples of the world closer together. It is also a way to bridge the contentious *North–South divide* that was apparent at the first Earth Summit. Like President Clinton, Dato' Law ultimately framed his position as one of *global responsibility*. Unlike President Clinton, he believed that Malaysia had the potential to take on a leadership role within the *global community*.

Dato' Law's framing strategy was also influenced by the "rule of speaking subjects" in the form of the speaker qualifications necessary for participation in the United Nations' regulated public sphere. Because Dato' Law was not the prime minister of his country, he had to rely on Malaysia's presidency of the General Assembly and the country's traditional role as an outspoken Southern nation to provide him credibility in the political ritual of the United Nations. Consequently, he incorporated the microframes of *Malaysia's leadership*, *United Nations community*, and the *North–South divide* and the symbolic package of a *global community* when constructing his overall frame of *global responsibility*.

The Official Statement of Third World Network Director, Martin Khor

The official statement of Third World Network Director Martin Khor at Earth Summit +5 stressed equally the cognitive framing devices of what is "real" and what is "meaningful" in his construction of his frame of *global responsibility* within the symbolic package of *people power*. Power enters into the framing strategy as the influence of the "rule of speaking subjects" in which there is membership in a "fellowship of discourse" (see Fig. 8.4).

Khor relied on the cognitive framing device of what is "real" to point out the scientific facts about the state of the global environment and to argue that the problem is a result of the global economic system. He relied on a *faith in science and technology* to determine how fast the global environment is deteriorating: ". . . the world is rushing even nearer to the brink of ecological disaster. . . . As the forests and lands are mined, [and] the atmosphere is polluted" (Khor, 1997, para. 2).

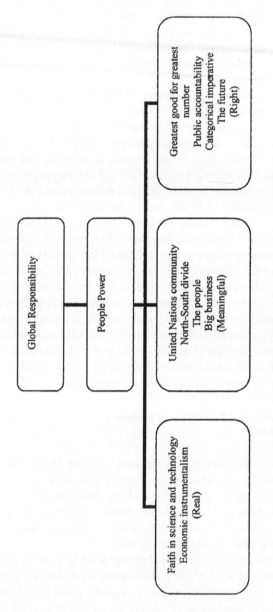

Global Responsibility

People Power

Faith in science and technology
Economic instrumentalism
(Real)

United Nations community
North-South divide
The people
Big business
(Meaningful)

Greatest good for greatest
number
Public accountability
Categorical imperative
The future
(Right)

"Rule of speaking subjects-Fellowship of discourse"

FIG. 8.4. Khor's interpretative results.

He then argued that global *economic instrumentalism* is the cause of the problem. Khor (1997) declared:

> The kind of globalization prevailing today . . . is rapidly spreading the same consumption and production patterns that we have already proclaimed unsustainable. It represents the growing power of big business that is increasing its monopoly of the economy and extending its reach to policymaking bodies. (para. 17–18)

Khor also stressed the cognitive framing device of what is "meaningful" in his speech. Like President Clinton and Dato' Law, he acknowledged a *United Nations community* made up of ". . . many of you present in this hall and in the outside rooms of this building, who are going against the status quo and pioneering the way ahead" (Khor, 1997, para. 15). Unlike Dato' Law, Khor (1997) promoted the *North–South divide* by relying on the dichotomies of developed/developing, North/South, and rich/poor throughout his statement. Significantly, Khor highlighted many examples of where *the people* have taken a stand against a *faith in science and technology* and *economic instrumentalism*. He began by saying ". . . we stand and salute the hundreds and thousands of local community leaders and the millions of ordinary people around the world, who have provided us the hope that something is being done to save the Earth" (Khor, 1997, para. 6). He then provided nine different examples (indigenous peoples, local communities, environmental activists, farmers, consumer movements, antigenetic engineering campaigners, women, NGOs, and environmental journalists) of groups of *the people* who have become the "real heroes" fighting for *Agenda 21* (Khor, 1997). At the same time, Khor (1997) also provided seven different examples (loggers, hazardous industries, commercial interests, agricultural establishments, tobacco industry, genetic engineering industry, and men) of *big business* that *the people* are resisting.

Khor (1997) used the cognitive framing device of what is "right" to bolster his use of the other two cognitive framing devices. He began by illustrating that the first Earth Summit attempted to achieve the *greatest good for greatest number*:

> Five years ago at Rio, global civil society looked at the Earth Summit as a source of hope for a new global partnership that would bring us back from the brink of ecological catastrophe and at the same time help developing countries and poor communities to develop in sustainable ways. (para. 1)

To achieve the *greatest good for greatest number*, Khor (1997) maintained that there needs to be more *public accountability* of *big business*. One example that he cited is the tobacco industry:

> We salute the consumers and consumer movements . . . who have taken the tobacco industry to court and forced it, in the United States at least, to admit its liability, to pay billions of dollars in compensation, and to agree to request that government regulate their behaviour. (para. 11)

Khor (1997) also relied on the *categorical imperative* that is implicit in the "Spirit of Rio":

> ... it is vital to reassert the principles at the heart of the Spirit of Rio: that the poor have the right to development, the rich have the duty to change their lifestyles and to help the poor, and that the common but differentiated responsibilities to save the Earth should be put into practice. (para. 24)

Finally, Khor (1997) concluded his speech with a reference to *the future*, specifically "... the future of the Earth" (para. 36).

Khor's microframes create a symbolic package and frame that one would expect from the director of an environmental social movement based in the South. Khor symbolically packages his framing devices within the narrative of *people power*. This package relies on the prevalent perception of Southern environmental activists resisting and struggling against corporations. It also taps into the long history of local resistance movements in the South during colonialism. Like the other two actors, Khor framed his position as one of *global responsibility*. In his case, however, he claimed that in the current global civil society the only group taking *global responsibility* is *the people*.

Power, in the form of the "rule of speaking subjects," influenced Khor's framing strategy because of his membership in a "fellowship of discourse." Khor is the director of a large Southern environmental NGO based in Malaysia and active in the general public spheres. Consequently, he had to champion the very notion of resistance and struggle inherent in his framing devices of *North–South divide, the people, big business, public accountability, categorical imperative*, and *the future* and symbolic package of *people power* when presenting his frame of *global responsibility*.

The Official Statement of Greenpeace International Executive Director, Dr. Thilo Bode

The official statement of Greenpeace International Executive Director, Dr. Thilo Bode, at Earth Summit +5 stresses equally the cognitive framing devices of what is "meaningful" and what is "real" in order to construct a frame of *global responsibility* contained within the symbolic package of *broken promises* under the influence of the "rule of speaking subjects" in which there is a "fellowship of discourse" (see Fig. 8.5).

Unlike the other actors, Bode significantly highlighted the cognitive framing device of what is "meaningful" for identity formation throughout his speech. Bode (1997) began by identifying Greenpeace as the representative of *the people* in the *United Nations community* when he said, "Greenpeace appreciates the opportunity and honor to address you, the world's sovereign governments, on behalf of millions of people worldwide" (para. 2). Like Khor, he promoted the existence of a *North–South divide* with his use of the dichotomy of developed/developing and

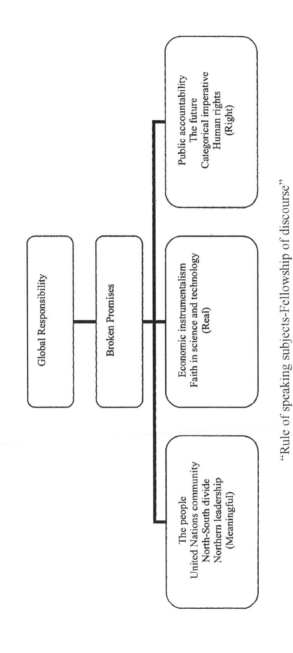

"Rule of speaking subjects–Fellowship of discourse"

FIG. 8.5.   Bode's interpretative results.

the specific references to countries or regions in the North such as the United States, Britain, Canada, and Germany and those in the South such as Amazonia, Congo, Papua New Guinea, and Brazil (Bode, 1997). Bode (1997) concluded by arguing for *Northern leadership* within the *United Nations community*:

> Industrialized countries cannot simply wash their hands of responsibility by making investments in the developing world. Nor can the developing world use the global environmental crisis solely as leverage for obtaining finance. . . . real leadership by industrialized countries is needed. (para. 18)

Bode's speech also relied heavily on the cognitive framing device of what is "real." Again, like Khor, he attacked the prevalence of *economic instrumentalism* by noting that governments have given in to "commercial interests" and that "Progress in protecting our environment will take more money. But money is not enough" (Bode, 1997, para. 4, 18). Bode's *faith in science and technology* is evident because of his use of scientific facts to describe environmental reality. For example, he presented the following scenario on climate change:

> Yet outside this building there is another reality. Carbon dioxide emissions have increased to unprecedented levels. Sea levels will rise so much that entire nations represented in this room may vanish. The frequency of extreme weather events such as storms which cause billions of dollars of damage have increased. (Bode, 1997, para. 9)

Bode (1997) then went on to argue, "More than three-quarters of the known reserves of oil, coal, and gas must remain in the ground if we are to avoid catastrophic climate disruption" (para. 12).

The cognitive framing device with the least amount of stress in Bode's speech is that which is "right." Bode (1997) noted that *United Nations community* must provide *public accountability* so that the goals of *Agenda 21* can be met:

> It has become fashionable to say that governments can do very little, and that all power now lies with unaccountable multinational companies and institutions in a newly globalised market. But let that not disguise the power and accountability which you, together, hold to impose environmental and social limits, controls and standards. (para. 20)

He also pointed out that *the future* was at stake because ". . . you have failed as yet to act. You have given in to commercial interests; you have put national interests above the welfare of future generations" (Bode, 1997, para. 4). Bode refers to the *categorical imperative* as well. He noted, "Whatever the promises you made at Rio, the condition of the world has worsened, in many cases at a faster rate than five years ago" (Bode, 1997, para. 6). Finally, Bode (1997) linked the environment with *human rights*:

Mr. President, the abolition of slavery, decolonisation, and the adoption of the International Declaration of Human Rights were defining points in human history— moments in which people and then nations took deliberate steps toward true humanity. Today we are at such a threshold again. (para. 17)

The unique combination of microframes in Bode's text produced a moralistic symbolic package of *broken promises* to convey the frame of *global responsibility*. Bode's symbolic package is potent because the moral authority of governments is being challenged. Bode believed that the governments of the world have "sold out" to *big business* and have broken their promises to *the people*. As a result, the *United Nations community* has the *global responsibility* to renew its promise of *public accountability* with regard to *Agenda 21*.

Like Khor, Bode was influenced by the "rule of speaking subjects" through membership in a "fellowship of discourse." Bode was the executive director of one of the largest Northern environmental NGOs that is very active in the general public spheres. Generally, Northern NGOs have focused more of their effort toward governmental lobbying than widespread local resistance. Because of this, Bode needed to use the cognitive framing devices of *Northern leadership*, *public accountability*, *categorical imperative*, and *human rights* with his *broken promises* package in order to frame his position as one of *global responsibility*.

CONCLUSION

The interpretative results of the political and social actors at Earth Summit +5 indicate that there are similarities in microframes and symbolic packaging present in the texts. For example, President Clinton, Dato' Law, and Khor all placed the most stress on the cognitive framing device of what is "real," which privileges the objective value sphere. When addressing this value sphere, all actors used the microframes of *economic instrumentalism* and *faith in science and technology*. In terms of the cognitive framing device of what is "meaningful," all actors used the microframes of *United Nations community* and *North–South divide* to discuss the subjective value sphere. Additionally, President Clinton, Dato' Law, and Bode used some version of the *leadership* microframe (e.g., *United States' leadership*, *Malaysia's leadership*, and *Northern leadership*). The two NGO actors, Khor and Bode, used the microframe of *the people*. With regard to the cognitive framing device of what is "right," all actors used the microframes of *public accountability* and *categorical imperative*. President Clinton, Dato' Law, and Khor also used the *greatest good for greatest number* microframe. Moreover, Dato' Law, Khor, and Bode employed the microframe of *the future*. Finally, Dato' Law and Khor employed symbolic packages that rely on the narrative of a global public sphere (*global community* and *people power*).

The interpretative results also suggest important differences among the micro-frames and symbolic packaging present in the texts. First, unlike the other actors, Bode placed the most stress on the subjective value sphere of what is "meaning-ful." Additionally, he is the only one who used the microframe of *human rights* in the social value sphere. In the subjective value sphere, Khor is the only actor who employed the *big business* microframe. In the objective value sphere, President Clinton is the only actor who utilized the microframe of *the past*. Interestingly, Dato' Law is the only actor who did not have a unique microframe of his own. Also important to note is the lack of the microframe of *the present*. In terms of symbolic packages, Bode's narrative of *broken promises* seems to point the finger at President Clinton's narrative of a *global superpower*.

Although the four actors used their microframes differently and symbolically packaged their frames in unique ways, they all put forth a common mesoframe of *global responsibility*. This was a shift from the problematic masterframe of *sustainable development* that had emerged from the first Earth Summit that was discussed by Peterson and Pauley (2000). Significantly, actors from both the North and the South agreed that *both* the North *and* the South must take responsibility for the state of the global environment. The constructed understanding of this shared sense of *global responsibility* allowed these actors to begin to bridge the North–South divide that they created at the first Earth Summit, and which had prevented real environmental change from occurring.

Perhaps the masterframe of *global responsibility* was able to accomplish this be-cause it constructs a new environmental understanding based on all three of Habermas' value spheres. The objective value sphere is suggested by the term *global* in that one can think of the "real" earth, the environment that can be stud-ied through science and changed through technology and economic development. The social value sphere is brought to mind by the term *responsibility* because there are "right" actions that need to be taken to interact with both the earth and other people. The subjective world is also evoked by the term *global* because one can re-flect on a "meaningful" new identity in a world that is increasingly intercon-nected. *Sustainable development*, on the other hand, privileges the objective value sphere because *sustainability* conveys awareness of the impact of economic activ-ity on the environment and *development* implies increased economic activity. The seemingly conflicting goals of environmental protection and increased economic development can both be justified with the concept of *sustainable development*.

Although a new masterframe emerged at Earth Summit +5, progress on achieving the goals of *Agenda 21* has not occurred as quickly as had been hoped. The report, *Global Challenges, Global Opportunity*, which was published prior to Earth Summit +10 held in Johannesburg, South Africa in 2002, indicates that al-though there has been some success, much work remains (United Nations De-partment of Economic and Social Affairs, 2002). The Earth Summit +10 focused on five specific areas of *Agenda 21*: agriculture, biodiversity, energy, health, and water. Examples of progress and decline for each of these areas include:

Agriculture: Food production and consumption are increasing, but agricultural expansion threatens other ecosystems.

Biodiversity: Protected forest areas in all regions are increasing, but the total of the world's forested areas continues to decline.

Energy: Renewable energy is growing on a small scale, but consumption of all types of energy is growing.

Health: The goal of a 50% reduction in child mortality rates due to diarrhoeal diseases has been achieved, but indoor air pollution is a major killer.

Water: Aquaculture is expanding to meet the growing demand for fish, but over one billion people still lack access to safe water. (United Nations Department of Economic and Social Affairs, 2002)

While progress on *Agenda 21* has not been fast paced, it is clear that the masterframe of *global responsibility* was still operating at the Earth Summit +10. In his address to the conference, Secretary-General Kofi Annan (2002) said:

> And if there is one word that should be on everyone's lips at this summit, one concept that embodies everything we hope to achieve here in Johannesburg, it is responsibility. Responsibility for each other—but especially the poor, the vulnerable, and the oppressed—as fellow members of a single human family. Responsibility for our planet, whose bounty is the very basis for human well-being and progress. And most of all, responsibility for the future—for our children, and their children. (para. 4–7)

Additionally, the final *Report of the World Summit on Sustainable Development* resolves to "... assume a collective responsibility to advance and strengthen the interdependent and mutually reinforcing pillars of sustainable development—economic development, social development and environmental protection—at the local, national, regional and global levels" (United Nations Department of Economic and Social Affairs, 2002, p. 1).

Perhaps one reason for the lack of progress at achieving the goals of *Agenda 21* with the *global responsibility* masterframe is the election of George W. Bush as President of the United States in 2000. According to the League of Conservation Voters (2003), President Bush has had one of the worst presidential environmental records ever. For example, he proposed an energy plan based on corporate interests and has pushed for drilling in the Arctic refuge. Additionally, he abandoned the Kyoto treaty on global warming and supports only minimal reductions of carbon emissions. He has proposed rolling back the Clean Air Act and limiting the authority of the Clean Water Act. Finally, his "Healthy Forests" initiative allows for extensive logging of trees. These proposals and the discussions surrounding them suggest that President Bush's discourse has stressed the cognitive framing device of what is "real" by incorporating the microframes of *economic instrumentalism* and *faith in science and technology*. More importantly, however,

President Bush's discourse seems to have also stressed the cognitive framing device of what is "meaningful" by introducing and privileging the *big business* microframe in the United States' presidential discourse. Future communication research utilizing the cognitive discourse analysis method should be able to track the continuing evolution (or devolution) of American environmental policy and its rationalization in global environmental conferences.

REFERENCES

Annan, K. (2002). *The Secretary-General Address to the World Summit on Sustainable Development Johannesburg, 2 September 2002.* Retrieved March 7, 2005, from the World Summit on Sustainable Development Web site: http://www.un.org/events/wssd/statements/sgE.htm

Apel, K.-O. (1987). The problem of a macroethic of responsibility to the future in the crisis of technological civilization: An attempt to come to terms with Hans Jonas' 'principle of responsibility' (W. Brown, Trans.). *Man and World, 20,* 3–40.

Apel, K.-O. (1991). A planetary macroethics for humankind: The need, the apparent difficulty, and the eventual possibility. In E. Deutsch (Ed.), *Culture and modernity: East–West philosophic perspectives* (pp. 261–278). Honolulu: University of Hawaii Press.

Apel, K.-O. (1993). How to ground a universalistic ethics co-responsibility for the effects of collective actions and activities? *Philosophica, 52*(2), 9–29.

Aristotle. (trans. 1991). *On rhetoric: A theory of civic discourse* (G. A. Kennedy, Trans.). New York: Oxford University Press.

Bode, T. (1997, June). *Address by Dr. Thilo Bode, Executive Director of Greenpeace International, to the Special Session of the General Assembly of the United Nations.* Retrieved January 9, 1999, from the United Nations Commission on Sustainable Development Web site: gopher://gopher.un.org:70/00/ga/docs/S-19/statements/gov/BODE.TEXT

Brown, R. L., & Herndl, C. G. (1996). Beyond the realm of reason: Understanding the extreme environmental rhetoric of the John Birch Society. In C. G. Herndl & S. C. Brown (Eds.), *Green culture: Environmental rhetoric in contemporary America* (pp. 213–235). Madison: The University of Wisconsin Press.

Cantrill, J. G. (1996). Gold, Yellowstone, and the search for a rhetorical identity. In C. G. Herndl & S. C. Brown (Eds.), *Green culture: Environmental rhetoric in contemporary America* (pp. 166–194). Madison: The University of Wisconsin Press.

Carcasson, M. (2004). Global gridlock: The American presidency and the framing of international environmentalism, 1988–2000. In T. R. Peterson (Ed.), *Green talk in the White House: The rhetorical presidency encounters ecology* (pp. 258–287). College Station: Texas A&M University Press.

Clinton, W. J. (1997, June). *President William Jefferson Clinton's address to the UN Special Session on Environment and Development.* Retrieved January 9, 1999, from the United Nations Commission on Sustainable Development Web site: gopher://gopher.un.org:70/00/ga/docs/S-19/statements/gov/CLINTON.TEXT

Corrick, B. J. (1990). An era of responsibility. *The Futurist, 24,* 60.

Cox, J. R. (2004a). "Free trade" and the eclipse of civil society: Barriers to transparency and public participation in NAFTA and the free trade area of the Americas. In S. P. Depoe, J. W. Delicath, & M. A. Elsenbeer (Eds.), *Communication and public participation in environmental decision making* (pp. 201–219). Albany: State University of New York Press.

Cox, J. R. (2004b). The (re)making of the "environmental president": Clinton/Gore and the rhetoric of U.S. environmental politics, 1992–1996. In T. R. Peterson (Ed.), *Green talk in the White House: The rhetorical presidency encounters ecology* (pp. 157–180). College Station: Texas A&M University Press.

Daley, P., & O'Neill, D. (1991). "Sad is too mild a word": Press coverage of the *Exxon Valdez* oil spill. *Journal of Communication, 41*(4), 42–57.

DeLuca, K. M. (1999). *Image politics: The new rhetoric of environmental activism.* New York: Guilford.

DeLuca, K. M., & Peeples, J. (2002). From public sphere to public screen: Democracy, activism, and the "violence" of Seattle. *Critical Studies in Media Communication, 19*(2), 125–151.

Donati, P. R. (1992). Political discourse analysis. In M. Diani & R. Eyerman (Eds.), *Studying collective action* (pp. 136–167). London: Sage.

Eder, K. (1996a). The institutionalization of environmentalism: Ecological discourse and the second transformation of the public sphere. In S. Lash, B. Szerszynski, & B. Wynne (Eds.), *Risk, environment and modernity: Towards a new ecology* (pp. 203–223). London: Sage.

Eder, K. (1996b). *The social construction of nature: A sociology of ecological enlightenment.* London: Sage.

Fisher, K. (1997). Locating frames in the discursive universe. *Sociological Research Online, 2*(3). Retrieved October 1, 1997, from http://www.socresonline.org.uk/2/3/4.html

Foucault, M. (1972). The discourse on language. In *The archaeology of knowledge and the discourse on language* (pp. 215–237) (A. M. Sheridan Smith, Trans.). New York: Pantheon Books.

Gamson, W. A., Croteau, D., Hoynes, W., & Sasson, T. (1992). Media images and the social construction of reality. *Annual Review of Sociology, 18,* 373–393.

Habermas, J. (1979). *Communication and the evolution of society* (T. McCarthy, Trans.). Boston: Beacon Press. (Original work published 1976)

Habermas, J. (1984). *The theory of communicative action: Vol. 1. Reason and rationalization of society* (T. McCarthy, Trans.). Boston: Beacon Press. (Original work published 1981)

Habermas, J. (1987). *The theory of communicative action: Vol. 2. Lifeworld and system: A critique of functionalist reason* (T. McCarthy, Trans.). Boston: Beacon Press. (Original work published 1981)

Habermas, J. (1989). *The structural transformation of the public sphere: An inquiry into a category of bourgeois society* (T. Burger & F. Lawrence, Trans.). Cambridge, MA: The MIT Press. (Original work published 1962)

Habermas, J. (1996). *Between facts and norms: Contributions to a discourse theory of law and democracy* (W. Rehg, Trans.). Oxford, England: Polity Press. (Original work published 1992)

Karis, B. (2000). Rhetoric, Habermas, and the Adirondack Park: An exemplum for rhetoricians. In N. W. Coppola & B. Karis (Eds.), *Technical communication, deliberative rhetoric, and environmental discourse* (pp. 225–234). Stamford, CT: Ablex.

Khor, M. (1997, June). *Speech by Martin Khor, Director, The Third World Network, at the General Assembly, UNGA Special Session on review of the implementation of Agenda 21.* Retrieved January 9, 1999, from the United Nations Commission on Sustainable Development Web site: gopher://gopher.un.org:70/00/ga/docs/S-19/statements/gov/MKHOR.TEXT

Killingsworth, M. J., & Palmer, J. S. (1992). *Ecospeak: Rhetoric and environmental politics in America.* Carbondale and Edwardsville, IL: Southern Illinois University Press.

Kinsella, W. J. (2004). Public expertise: A foundation for citizen participation in energy and environmental decisions. In S. P. Depoe, J. W. Delicath, & M. A. Elsenbeer (Eds.), *Communication and public participation in environmental decision making* (pp. 83–95). Albany: State University of New York Press.

Law, H. D. (1997, June). *Statement by H. E. Dato' Law Hieng Ding Minister of Science, Technology and the Environment of Malaysia on overall review and appraisal of the implementation of Agenda 21 at the 19th Special Session of the United Nations General Assembly.* Retrieved January 9, 1999, from the United Nations Commission on Sustainable Development Web site: gopher://gopher.un.org:70/00/ga/docs/S-19/statements/gov/DING.TEXT

League of Conservation Voters. (2003, November). *2004 presidential candidate profiles.* Washington, DC: League of Conservation Voters.

Moscovici, S. (1990). Questions for the twenty-first century. *Theory, Culture and Society, 7,* 1–19.

Patterson, R., & Lee, R. (2000). The environmental rhetoric of "balance": A case study of regulatory discourse and the colonization of the public. In N. W. Coppola & B. Karis (Eds.), *Technical communication, deliberative rhetoric, and environmental discourse* (pp. 235–250). Stamford, CT: Ablex.

Peterson, T. R. (1997). *Sharing the earth: The rhetoric of sustainable development*. Columbia: University of South Carolina Press.

Peterson, T. R., & Pauley, K. L. (2000). George Bush goes to Rio: Implications for U.S. participation in global environmental governance. In A. González & D. V. Tanno (Eds.), *Rhetoric in intercultural contexts* (pp. 67–90). Thousand Oaks, CA: Sage.

Plevin, A. (2000). Green guilty: An effective rhetoric or rhetoric in transition? In N. W. Coppola & B. Karis (Eds.), *Technical communication, deliberative rhetoric, and environmental discourse* (pp. 251–265). Stamford, CT: Ablex.

Ross, S. M. (1996). Two rivers, two vessels: Environmental problem solving in an intercultural context. In S. A. Muir & T. L. Veenendall (Eds.), *Earthtalk: Communication empowerment for environmental action* (pp. 171–189). Westport, CT: Praeger.

Schwarz, A. (1992, June 25). Back down to earth: Global summit fails to live up to ambitions. *Far Eastern Economic Review, 155*, 61–62.

Schwarze, S. (2004). Public participation and (failed) legitimation: The case of Forest Service rhetorics in the Boundary Waters Canoe Area. In S. P. Depoe, J. W. Delicath, & M. A. Elsenbeer (Eds.), *Communication and public participation in environmental decision making* (pp. 137–156). Albany: State University of New York Press.

Strydom, P. (1999a). The challenge of responsibility for sociology. *Current Sociology, 47*(3), 68–82.

Strydom, P. (1999b). The civilization of the gene: Biotechnological risk framed in the responsibility discourse. In P. O'Mahony (Ed.), *Nature, risk, and responsibility: Discourses of biotechnology* (pp. 21–36). London: Macmillan.

Strydom, P. (2000). *Discourse and knowledge: The making of enlightenment sociology*. Liverpool, England: Liverpool University Press.

Strydom, P. (2002). *Risk, environment and society: Ongoing debates, current issues and future prospects*. Buckingham, England: Open University Press.

Tevelow, A. (2004). Global governance and social capital: Mapping NGO capacities in different institutional contexts. In S. P. Depoe, J. W. Delicath, & M. A. Elsenbeer (Eds.), *Communication and public participation in environmental decision making* (pp. 223–233). Albany: State University of New York Press.

Todd, A. M. (2003). Environmental sovereignty discourse of the Brazilian Amazon: National politics and the globalization of indigenous resistance. *Journal of Communication Inquiry, 27*(4), 354–370.

United Nations. (2004, December 17). *Agenda 21—Table of contents*. Retrieved March 7, 2005, from the U.N. Department of Economic and Social Affairs Division for Sustainable Development Web site: http://www.un.org/esa/sustdev/documents/agenda21/english/agenda21toc.htm

United Nations Department of Economic and Social Affairs. (2002, August). *Global challenge, global opportunity: Trends in sustainable development*. Retrieved March 7, 2005, from the U.N. Johannesburg Summit 2002 Web site: http://www.johannesburgsummit.org/html/documents/summit_docs/criticaltrends_1408.pdf

United Nations Department of Public Information. (1996). *Earth Summit +5 overview*. Retrieved January 8, 1999, from the United Nations Commission on Sustainable Development Web site: http://www.un.org/ecosocdev/geninfo/sustdev/es&5broc.htm

Waddell, C. (1996). Saving the Great Lakes: Public participation in environmental policy. In C. G. Herndl & S. C. Brown (Eds.), *Green culture: Environmental rhetoric in contemporary America* (pp. 141–165). Madison: The University of Wisconsin Press.

CHAPTER NINE

The Rhetoric of the Columbia: Space as a Wilderness, a Miracle, and a Resource

Jane Bloodworth Rowe
PhD Communication
Adjunct of Old Dominion University
Norfolk, VA

> *I could then see my reflection in the window, and in the retina of my eye the whole earth and sky could be seen reflected . . .*
>
> —CNN.com (2003a)

Astronaut Kalpana Chawla, one of seven astronauts aboard the ill-fated space shuttle Columbia, spoke these words to a reporter on January 28, 2003, just 3 days before she and her colleagues were killed when the shuttle broke apart over Texas during its re-entry to the earth. Chawla's description of her view of the night sky through a space shuttle window seemed to suggest a personal, emotional connection with the earth and sky, which she saw reflected in her eye. Her colleague, Laurel Clark, also implied a sense of wonder when she used the word, "magical" (CNN.com, 2003b), to describe her reaction to the sight of the earth, stars, and a newly hatched moth.

Comments by Chawla, an aeronautical engineer, and Clark, a physician, raised questions about space travel and its implications for the cultural constructions of the human/nature relationship. Their communication, which seemed to focus on the beauty and magic of the natural world, was particularly interesting when contrasted with the televised address delivered by President George W. Bush just hours after the Columbia accident. Bush depicted the nonhuman world as dark and dangerous, using phrases such as "the fierce outer atmosphere of the earth" and "the darkness beyond our world" to describe outer space (PBS, 2003).

This study examines the language used in Bush's speech (PBS, 2003), as well as comments made by the astronauts in media interviews (CNN.com, 2003a;

CNN.com, 2003b) and in e-mails to family and friends (Associated Press, 2003a) in order to gain insight into the communicators' perception of the earth/sky/human relationship and the impact that these perceptions might have on environmentalism. The study revealed that, although the astronauts described the natural world as a source of life, light, and beauty, Bush described untamed nature as a dark wilderness that should be conquered for use by humans. This metaphor of nature as a resource, which Merchant (1980) traced to the early modern period, prevailed in Bush's eulogy for these astronauts. Further, Bush, with his ability to grab media attention and to frame arguments within prevailing metaphors, was the most likely to have the largest impact on American discourse surrounding the earth/sky/human relationship.

The February 1, 2003 speech, then, when considered in this context, appeared to have political implications. Bush, in eulogizing the astronauts, was also framing his argument for continued space travel within the prevailing metaphors of nature as a resource that Corbin (1992) said dominated American discourse surrounding nature during the 19th and 20th centuries. This metaphor, when extended to the space program, could serve to widen the nature/culture dichotomy, minimize the importance of the natural world to human survival, and undermine the contemporary environmental movement.

These communication artifacts were chosen for this research because, after a review of the media coverage surrounding this flight and subsequent disaster, I determined that this communication dealt most directly with space, human life, and the relationship between the earth, space, and the universe. The rhetoric surrounding the Columbia was particularly relevant to a study of the earth/sky/human relationship because of the potential impact of this accident on continued manned space exploration. The 1986 crash of the shuttle Challenger seemed to rally public support in favor of space travel: the National Aeronautics and Space Administration (NASA) budget doubled over 5 years after that accident. In 2003, however, changed technology and a different world order made public and Congressional reaction to the most recent accident seem less certain (Lawler, 2003). This time, political advocates of space travel had to convince a public that was one generation removed from the Apollo moon landings. As Rees (2002) noted, "nobody under 35 can remember the era when men walked on the moon" (2002, para. 2). Rees, an Astronomer Royal and Royal Society Professor at Cambridge University, noted in 2002 that younger Americans thought of aggressive space exploration as "a remote historical episode" (para. 2) that was linked to a Cold War rivalry. Americans had not established a permanent space colony, and had not even placed a human on the moon since 1972. His comments were an indication that, even prior to the Columbia disaster, there was speculation that the public enthusiasm for space travel was waning.

A changed attitude toward the potential benefits of space exploration was just one challenge that advocates faced after this disaster. The timing of the accident, was "bitterly ironic" (Lawler, 2003, para. 1) because it occurred just 48 hours prior to the scheduled release of NASA's budget. That agency was about to launch an ambitious new program that included returning manned spacecraft to the moon and even sending humans to Mars. The accident refocused the immediate goals to those of investigating the cause of the accident (Lawler, 2003, para. 4–5).

Soon after the crash, talk show hosts began to speculate that the NASA budget cuts that occurred from 1998 to 2000 might have indirectly contributed to the disaster, which was thought to be linked to a piece of foam that fell from the shuttle after takeoff (Bowyer, 2003). Pictures taken by the launch pad cameras, which might have revealed what happened to the affected left wing, were out of focus (Space Today Online, 2003). NASA, then, faced the daunting task of not only investigating the accident, but updating the cameras and other safety and monitoring equipment to minimize the risk of future disasters.

Changing technology, including the use of robots in outer space, also raised questions about the need for human space travel, particularly after the death of the seven astronauts focused attention on the inherent dangers of sending humans into outer space (Bowyer, 2003; Lawler, 2003). In addition, the country was facing a recession, the threat of terrorism, the possibility of war with Iraq, and the upcoming presidential campaign, and the demise of the cold war seemed to make space exploration less essential to the nation's security (Lawler, 2003).

The disaster, then, created a need for advocates of a manned space program to defend their position, and Bush's speech, delivered hours after the crash, offered an opportunity to examine the arguments for continued manned space exploration. These artifacts also communicated the euphoria, as well as the horror, of space exploration. The astronauts, speaking before the crash, were focused on the thrill of seeing the earth and stars from outer space. Their presumed purpose in communicating was to describe their experience to the public and share their thoughts on seeing the stars, earth, and sky from outer space. Bush, speaking after the disaster, reminded the nation of the dangers of space travel and focused on the need to continue manned space exploration despite the danger. He also lacked the direct experience with space that the astronauts had, and he focused less on a sensory description of the earth and space and more on the human desire for exploration and knowledge.

In order to examine these artifacts, I have delineated and clustered the groups of major metaphors in the communication by Bush, Chawla, and Clark for the purpose of determining what these metaphors revealed about the communicator's perception of the human/nature relationship. A metaphorical method of inquiry was particularly revealing because scholars (Corbin, 1992; Fernandez, 1972, 1974; Merchant, 1980; Muir, 1994) described the prevailing metaphors in different cultures, or during different historical periods, in order to explain the historical changes in the construction of the human/nature relationship. Corbin (1992) noted that metaphors could be particularly useful to contemporary scholars of environmental rhetoric because the views of nature that prevailed within a culture drove economic and political decisions regarding the environment.

A metaphorical study of the rhetoric surrounding space travel, then, seemed a natural extension of these previous studies and appeared well-suited for an inquiry that sought to determine the potential impact of space exploration on the environmental movement. As Burke (1969) noted, the metaphor is perspective, and metaphoric inquiry provided insight into a culture's beliefs, values, and behaviors. This research followed the procedures of metaphoric inquiry examined by Ivie (1994) and Lakoff (1995, 2002) in a study of political rhetoric used to rally a country to war (Ivie) and to justify one course of political action over another (Lakoff). These scholars identified the metaphors within an artifact, then clustered the metaphors in order to identify the major themes embedded in the communication and to determine what these themes revealed about the communicator and the culture in which the communication occurred.

First, I examine these artifacts for a general sense of the meaning and context. The metaphors used to describe the earth/sky/human relationship, whether explicitly or implicitly stated, are identified. Then, I group these metaphors into clusters in order to determine what the major themes are and to analyze what these themes reveal about the communicators' perception of the earth/sky/human relationship. I then discuss the possible implications of these metaphors on the perceptions of the human/nature relationship in American culture.

In the following section, I review the literature surrounding metaphoric inquiry, with particular attention to the metaphoric studies of environmental communication. In the second major section, I describe the historical context in which the Columbia disaster occurred. In the third major section, I analyze the metaphors in the communication by Clark and Chawla, then by Bush, and I examine the possible reasons for the conflicting, yet interrelated metaphors. Finally, I discuss the impact that these metaphors hold for the cultural construction of the human/nature relationship.

THE METAPHOR AS A METHOD OF INQUIRY

The metaphor, Hart (1997) noted, "has been the subject of much scholarly inquiry" (p. 146). The definition of the metaphor evolved over the centuries, from Aristotle's description of a literary device to Burke's (1969) view that the metaphor was perspective. For the purpose of this research, Burke's definition was accepted, but it was necessary to further explore metaphoric inquiry as a method of scholarly research.

Aristotle described the metaphor as a literary device that could be used to charm the reader or to clarify ideas (Aristotle, trans. 1954). However, during the 20th century, rhetoricians began to re-think this definition. Richards (1936) maintained that the traditional treatment of a metaphor as a literary device was too limiting because the skill to use and understand metaphors was an integral part of the human ability to use language. He described metaphoric thought as "those processes in which we perceive or think of or feel about one thing in terms of another" (p. 116). The metaphor, Richards (1929) said, is a shift, or a carrying over of a word from its normal use to a new one.

In the last half of the 20th century, scholars extended Richards's ideas further and described the metaphor as the foundation for much of human thought. Ivie (1994) maintained that shared metaphors enabled members of a culture to form values and beliefs and provided a common frame of reference that made communication possible. Lakoff (1993, 1995) indicated that metaphors sometimes permeated thought without being explicitly stated and were often implied in the ontological and empirical assumptions of intangible objects. Metaphors, although culturally based, could also be imposed by people who held political power (Lakoff & Johnson, 1980).

Foss (1996) agreed that "metaphor constitutes argument" (p. 361), while Ortner (1973) described metaphors as symbols used by members of a culture to organize experiences and devise a plan of action. Ortner (1973) and Foss (1996) proposed the study of metaphor in rhetorical research in order to gain insight into the major ideas that guided thought in a particular culture. Researchers (Ivie, 1994; Lakoff, 1995, 2002) also studied the metaphors within political arguments to determine how these arguments were framed and why they were effective.

Scholars also applied metaphorical analysis to study the perceived human/nature relationship within a particular culture. Fernandez (1972, 1974) used this approach in anthropological fieldwork to describe the connotations attached to various species of animals (1972), then to describe the entire human/nature relationship (1974). In the following paragraphs, I review the research surrounding the major metaphors that have been used to describe the human/earth/sky relationship.

A Metaphoric Construction of the Earth/Sky/Human Relationship

Muir indicated "metaphors of the environment have spanned the ages of human discourse" (1994, para. 3). A review of the existing literature dealing with the metaphors surrounding the earth and sky described a complex construction of the human/nature relationship, but some common themes prevailed. Scholars tended to juxtapose the metaphors of nature as a mother versus nature as a machine, and nature as a living organism versus nature as a collection of isolated physical bodies. Merchant (1980) and Corbin (1992) identified the major metaphors that historically defined nature in Western culture. Merchant (1980) delineated two major metaphors, including the organic and the mechanistic, whereas Corbin (1992) pinpointed four basic metaphors for nature, including the organic, agricultural, mechanistic, and the economic.

Before the Renaissance, humans, aware that their survival depended on the cycle of the growing season or the abundance of wild game, were more likely to view themselves as a part of the natural world. An increased interest in the natural sciences during the early modern period led to the conception that nature could be manipulated by and for the benefit of humans, who were distinct from the nonhuman world. Nature, when reduced to a collection of physical bodies that could be viewed through a microscope or telescope or quantified by experimentation, became more abstract and further removed from the human experience (Evernden, 1989; Schildknecht, 1995).

Nash (1989) traced the conception of nature as an object of human study and management in the English-speaking world to the 17th-century philosopher, Francis Bacon (1955), who, in *Novum Organum*, described "Man, as the minister and interpreter of Nature" (p. 107). He maintained that humans, through knowledge, could subdue and change nature. Corbin and Nash also wrote of the early American metaphor of nature as a wilderness that, if conquered for human use, could become an economic resource. Bacon (1909b) outlined this view when he advocated colonizing the American wilderness.

Later, as the American wilderness vanished, the metaphor of a nature that needed careful human management, pioneered by Theodore Roosevelt, began to prevail. The concept of *nature* as separate from human culture dominated the 20th century. Nature came to be defined as "the separate and wild province, the world apart from man to which he has adapted" (McKibben, 1989, p. 48).

A continued focus on science and the dawn of space exploration further complicated the metaphors for the natural world. Muir (1994) identified two major perspectives that prevailed in environmental discourse during the 20th century. These included the view that nature was an interconnected web and the perspective of the earth as a spaceship rotating among the stars. Evernden (1989) also spoke of the wide range of metaphors for the natural world, including the metaphors of nature as a collection of physical objects, and nature as a miracle.

During the last half of the 20th century, some environmentalists sought to reintroduce the concept of nature as an organic whole that included humans. Lovelock drew from the ancient Greeks, who referred to the living earth mother as *Gaia* to propose that the earth itself was alive, and this view influenced the ecofeminists and even some mainstream environmental groups such as the Sierra Club (Merchant, 1999; Nash, 1989). Some ecofeminists linked Mother Earth with the innate empathy for the totality of life that Merchant (1992) claimed that women felt. DeLuca (1999b) focused on what he described as a postmodern view of the natural world; community-based environmental justice groups (unlike the mainstream groups that defined nature as uninhabited wilderness) focused on a holistic nature that enveloped their communities and served as the foundation for human life and culture.

The Earth/Sky Relationship

The living *Gaia*, of course, referred to the earth, and, in fact, the terms *earth* and *nature* seemed synonymous in metaphors such as Mother Earth and Mother Nature. Twentieth-century observers drew a distinction between the earth and the space beyond the earth's atmosphere (Bateson, 1999; McLuhan, 1977). However, the relationship between the earth, the sky, and humans was a subject of myths and human inquiry for centuries. Some ancient cultures linked the sky with divinity. To the ancient Sumerians, the night sky was source of an "aboriginal religious experience" and "the eternal prototypes and models" for life on earth (Cahill, 1998, p. 49). Other cultures looked to the family as a source of metaphors for an interconnected cosmos. Polynesian and Aztec cultures used the metaphors of Mother Earth and Father Sky (DeCandido, 1998; Marck, 1996). Some Chinese cultures also used the triad of Earth, Heaven, and Man to organize the spiritual hierarchy and the relationship between gods, ghosts, ancestors, and humans (DeBernardi, 1992).

Scholars (McLuhan, 1977; Merchant, 1980) also wrote of the perceived relationship between the earth, sky, and humans in past and contemporary Western culture. Merchant (1980) described an "organic cosmology" (p. xvi) that she said prevailed in Europe during the Middle Ages. This construction placed the living earth at the center of the universe and the moon, stars and sun literally above, but revolving around this planet. Heaven, the dwelling place of God, was conceived of as an actual, literal space that occupied the outermost sphere. This view was consistent with Lewis' (1967) description of the earliest definition of nature as the external world of objects, encompassing everything but God. Merchant further argued that, in the medieval view, the human body existed as a microcosm of the entire universe, and the individual's physical and mental well-being were influenced by the stars and planets.

However, the discovery of the telescope displaced the earth as the center of the universe and placed humans firmly in control of the earth. Copleston (1993) and Merchant (1980) referred to the mathematical and systemic view of nature that developed as the result of Renaissance philosophy and science. The mysterious relationship between the earth, sky, and heaven yielded to the belief that the earth was one planet among many in a solar system, and this planet was governed by humans rather than by astronomical occurrences.

During the 20th century, technological advances allowed humans to view the earth from outer space and, some observers asserted, further minimized the earth as home, mother, and center of the universe. Communication scholars, including McLuhan (1977) and DeLuca (1999a) indicated that the concept of nature, at least as it had historically been defined, may have ended with the growth of space travel. Photographs of the earth, taken from outer space, made this planet seem remote. The abstract concept of a nonhuman world that functioned as a system (McLuhan, 1977) replaced the concrete, life-giving world of plants, animals, soil, and water. The possibility of exploring and subduing space for human use sent the message that humans were not dependent on the earth for sustenance. Once viewed as the center of the cosmos, the earth was now reduced to "tiny globe viewed from outer space" (Muir, 1994, para. 9). Merchant's organic cosmos described the earth and sky as interconnected, while Bateson (1999) and McLuhan (1977) linked the earth with nature and placed space outside of nature.

A new metaphor, that of the earth as a spaceship, developed with space travel. Pictures of the planet from outer space depicted "a tiny earth floating against the backdrop of the universe" (Muir, 1994, para. 9), and this metaphor became "a popular and powerful tool for shaping human conceptions of the earth and of our role as crew of the ship" (Muir, 1994, para. 9). This planet had become a ship adrift in a void of darkness, kept on course only by its human crew. As Muir indicated, this view became problematic for two reasons. First, the idea of the earth as floating in, but separate from, the universe caused this earth to appear vulnerable to the natural forces of the universe that surrounded it. Secondly, under this metaphor, the earth existed for, and was dependent on, a human crew who steered the tiny, vulnerable ship, saving it from aimless and reckless wandering.

The metaphor of the earth as a spaceship implied that it was "clearly 'our' ship" (Muir, 1994, para. 12) and even indicated a sense that we not only controlled it, but we created it. In fact, space travel tended to send the message that, while the earth was vulnerable and guided by humans, humans were now less dependent on the earth because they could one day flee to a space colony if increased technology caused a "global catastrophe" that made this planet inhabitable (Rees, 2002, para. 11). The mechanistic metaphor that Merchant (1980) traced to the early modern period finally evolved into the conception of the earth as a disposable commodity.

Previous research, however (McLuhan, 1977; Muir, 1994), focused on the possible effects of the mediated images of the earth as viewed through space, whereas the ancient cultures focused on the view of the sky as seen from the earth. The comments made by Chawla and Clark afforded an opportunity to focus on the metaphors embedded in their first-hand accounts of the earth and sky viewed from a space shuttle window. Their rhetoric indicated that space travel, which allowed the public the opportunity to vicariously experience the view of the cosmos from outer space, could tend to support the organic, holistic universe that Merchant (1980) described. Advocates of space travel, however, must also demonstrate that the potential benefits to human culture outweigh the risks and expense (Rees, 2002). President Bush, perhaps in an effort to justify further space exploration, described a universe over which humans, bent on conquering the darkness, would one day assume complete control.

In the following section, I briefly review the events following the Columbia crash in order to explain that the disaster underscored the crucial need for space advocates to defend and justify continued manned space travel.

THE COLUMBIA DISASTER, THE INVESTIGATION, AND THE SPACE PROGRAM

The Columbia, the oldest shuttle in NASA's orbital fleet, was built in the late 1970s and was flown on its first mission in 1981. Prior to the 2003 flight, it had flown on 27 previous missions and undergone extensive refurbishing three times: in 1991 to

1992, from 1994 to 1995, and then again from 1999 to 2001. On the 2003 flight, the crew aboard the Columbia conducted 80 experiments, including medical experiments aimed at developing a treatment for cancer. The flight was, prior to its re-entry to the earth's surface, successful, and the crew members had collected information about the growth of prostrate cancer cells and experimented with creating a weak, low-soot flame that could reduce air pollution on earth.

In addition to Clark and Chawla, the crew members included payload commander, Michael P. Anderson, mission specialist, David Brown, commander, Rick Husband, pilot, William C. McCool, and payload specialist, Ilan Ramon, Israel's first astronaut. On February 1, 2003, the shuttle entered the earth's atmosphere at about 8:16 a.m. in preparation for its landing, but NASA lost contact with the shuttle at about 9 a.m. Video recordings showed multiple vapor trails as the shuttle appeared to break apart, and, when NASA failed to establish contact with the Columbia by 9:16 a.m., officials were dispatched to search for debris (Space Today Online, 2003).

Americans learned, through the media, the grisly account of the astronauts' remains being collected and identified through DNA analysis. In addition, most, but not all of the scientific data the crew members had collected was also lost. The results of some of the experiments had already been transmitted to earth, and some of the worms grown aboard the Columbia in order to test a synthetic nutrient solution were found alive in the debris (Space Today Online, 2003). An investigating board was appointed to determine the cause of the accident, and attention focused almost immediately on the foam that fell from the shuttle 82 seconds after the January 16 liftoff (Bowyer, 2003).

Still, both President Bush and NASA director, Sean O'Keefe, pledged that neither the disaster nor the extensive investigation would deter this country from pursuing its manned space program. Presidential spokesman, Ari Fleischer, speaking 2 days after the disaster, told reporters that, although it would be premature to speculate about when NASA would return to outer space, President Bush was committed to space exploration (Associated Press, 2003b).

Bush and O'Keefe assured the country that astronauts would return to space as soon as possible (Associated Press, 2003b), but NASA became the target of negative publicity in the months to come. In August 2003, an investigating agency confirmed that the falling foam had indirectly caused the shuttle disaster because it damaged the wing and allowed excessive heat to enter the shuttle. Further, the agency blamed the shuttle crash on a general lack of attention to safety within NASA (NPR, 2003). The Columbia Accident Investigating Board also noted "a lack of an agreed national vision for space flight" (NPR, 2003) and raised questions about NASA's priorities and its organizational culture (which was said to squelch criticism). The investigating board's report, with its reference to a lack of a national vision, also questioned the nation's commitment to an ambitious manned space program.

The Columbia disaster and the political climate in which it occurred raised questions about space travel. Was space an extension of nature, and the logical next step for human exploration? What impact did space travel have on the perceived relationship between nature and culture? In the following section, I examine the communication artifacts surrounding the Columbia flight in order to answer these questions.

ANALYSIS OF METAPHOR USE

When the metaphors in the communication by Chawla, Clark, and Bush were examined in clusters, one consistent theme emerged. Space was viewed as a part of nature. The space beyond the earth's atmosphere was described with the same metaphors that historically described the natural world. From a close reading of these artifacts, I delineated these major perspectives: nature as light versus nature as darkness, nature as mother and nurturer versus nature as fierce, untamed wilderness, and nature as a miracle versus nature as a machine and a resource. The perspective of nature as powerful and active was also evident in Chawla's and Clark's communication, whereas President Bush attempted to minimize nature's power and present the natural world as an obstacle that could be overcome and a resource that could be manipulated and used. In the following paragraphs, I further explore communication by Chawla and Clark, then examine President Bush's rhetoric in order to elaborate on the themes embedded in these artifacts.

Nature as Light, Life, Home, Miracle, and Self

Muir (1994) and Bateson (1999) indicated that pictures of the earth taken from outer space enhanced the human's perceived dominance over an objectified, abstract nature and widened the nature/culture dichotomy. However, the astronauts, who had direct, sensory experiences with space and the earth as viewed from space, often focused on the light visible from the stars, moon, and earth-based lightning as well as on the power that was implied in these images. Clark described the "incredible sights" including "lightning spreading over the Pacific, the Aurora Australis lighting up the entire visible horizon with the cityglow of Australia below, the crescent moon setting over the limb of the Earth . . ." (Associated Press, 2003a). She linked both the earth and the sky to light and action. When viewed from outer space, the electric lights radiating from the city were also a part of the light, life, and power that radiated from this planet.

Clark's observation, then, seemed to narrow the nature/culture dichotomy. "Even the stars have a special brightness," she said, speaking of the sky viewed from her perspective in outer space (Associated Press, 2003a). Her description of the universe was more akin to the organic cosmos described by Merchant (1980) than it was to the posters, described by Bateson (1999) that depicted the earth as an isolated body spinning in a void of darkness. Clark's "incredible sights" were part of a holistic universe that included lightning, the earth-based city lights, and the moon and stars. Chawla, however, expressed the holistic metaphor most explicitly when she described the reflection of the "whole earth and sky" in her eye (CNN.com, 2003a). Unlike the detached observer who saw a one-dimensional earth and space on a poster, Chawla was a part of this universe, and its light and power became a part of her.

Chawla and Clark, then, communicated their direct, sensory observations using language that implied a familiar relationship between humans and the cosmos. Their comments seemed reminiscent of the fertility metaphors surrounding the earth/sky relationship (DeCandido, 1998; Marck, 1996), which maintained that a familial relationship between the earth, sun, and sky was responsible for life on earth. The astronauts' language was replete with themes of a nature that served as both a life-giving force and as the physical location of human life and culture.

Clark implied the metaphor of nature as both life and place when, describing the moth hatching aboard the space shuttle, she said, "life continues in lots of places, and life is a magical thing" (CNN.com, 2003b, para. 7). The concept of reproducing plants under manipulated circumstances that was described by Bacon in "The new Atlantis" finally resulted in attempts to grow genetically engineered crops in space (Cohen, 1999). Observers, ranging from the 17th-century poet Andrew Marvel (1974) to modern activists, have criticized cultivated plants and animals that were reproduced in laboratories or anywhere outside of their native habitat. Clark, however, spoke of all life as a "magical thing" (CNN.com, 2003b, para. 9), when she described the sight of a moth hatching in space. "There was a moth in there," she said, "and it still had its wings crumpled up, and it was just starting to pump its wings up" (CNN.com, 2003b, para. 9). Unlike poets and contemporary critics, Clark seemed not to make a distinction between indigenous and cultivated species. Rather, she looked at all of life, even that life hatched under artificial conditions in outer space, as a miracle that defied comprehension. Moths in outer space were an extension of the miracle of life, not the replacement for life on earth that Rees (2002) indicated space travel could provide.

Clark saw the miracle of life extended beyond the earth's atmosphere, but she also expressed a special sense of belonging to her own planet, and, specifically, to her hometown. She referred to "our magnificent planet Earth," implying a sense of belonging and identity with place and the grounding of all life in the natural world (Associated Press, 2003a). DeLuca (1999b) and Bell (1994) spoke of the attachment to place that some rural residents felt for their homes and communities, which they saw as being grounded in nature. Clark, here, extended that sense of attachment to community to include the entire planet, but she also spoke affectionately and excitedly of seeing her Wisconsin hometown from high above the earth's surface. "Magically, the very first day we flew over Lake Michigan and I saw Wind Point clearly" (Associated Press, 2003a, para. 5), she said.

Clark was not speaking of the abstract, remote, mediated earth described by McLuhan (1977), Muir (1994) and Bateson (1999). Her "magnificent planet . . ." (Associated Press, 2003a) was concrete, physical, and experienced directly rather than through a photograph. It generated sights as spectacular as lightning and the Aurora Australis and as commonplace and comforting as midwestern home-towns. She described mountains, plains, and rivers, but she also described the man-made city lights and small towns; human life and human activity were a part of a holistic earth.

Clark also described nature as forceful; she frequently used active tense verbs that described the power of the natural world. She spoke of the moth's wings "pumping" (CNN.com, 2003b), "lightning spreading," "the Aurora Australis lighting," "the crescent moon setting over the limb of the earth," and "rivers breaking through tall mountain passes" (Associated Press, 2003a, para. 3). In this e-mail to her family and friends, she also spoke of the earth as the habitat and life-giving force for humans when she referred to "the continuous line of life extending from North America, through Central America and into South America" (Associated Press, 2003a, para. 3). The sense of nature as both place and power was extended when she spoke of "the vast plains of Africa and the dunes of Cape Horn. . . ." This was not a small, vulnerable, objectified planet but, was, instead, the earth that included powerful rivers, vast plains and huge mountains such as Mount Fuji, which, even when viewed from outer space ". . . does stand out as a very distinct landmark" (CNN.com, 2003b, para. 3).

Clark also explicitly spoke of humans as a part of a universe controlled by a forceful nature to which the humans must adapt. As a physician, she was intensely interested in the effects of a gravity-free environment on the human body. "The first couple of days you don't always feel too well," she said, because the astronaut must "adjust" to fluids shifting and learn "to fly through space without hitting things or anybody else" (CNN.com, 2003b, para. 1). She did adjust, however, and felt "wonderful" (CNN.com, 2003a, para. 1) after 2 weeks in space. Clark was em-powered not so much by the technology that allowed her to transcend space but by the human's ability to adapt to the changing conditions of the universe.

Chawla also referred to "our planet" (ABCNews.com, 2003, para. 7) when she said, "Just looking at Earth, looking at the stars during the night part of Earth; just looking at our planet roll by . . . is like living a dream." Here, she described a sense of belonging to the earth, which she placed in the cosmos and firmly linked with both the stars and humans. Her sense of belonging to both the earth and the entire universe belied the desire to explore, conquer, and own that was implied in some rhetoric about space travel. Williamson (2003) noted that the discussion sur-rounding tourism and real estate sales in spaced implied that humankind was "in-tent on making the space environment part of our domain" (p. 47). Chawla, how-ever, seemed contented to merely look at, rather than own, the earth and sky, and she seemed to further imply that she was connected to, rather than in dominion over, the universe.

Chawla also linked space travel with "a dream—a good dream" (ABCNews .com, 2003, para. 7), implying that she was traveling in space not only for the pur-suit of knowledge but for the fulfillment of a long-time goal. She had dreamed of flying since she was a child in India, and pursued this career goal although it conflicted with the role traditionally assigned to Indian women.

McLuhan (1977) and Bateson (1999) predicted that space travel would radically alter the view of the nonhuman world by replacing the concept of nature as a physical force with the abstract idea of ecology. Evernden (1989) also contrasted the "abstract," or modern (p. 162) concept of nature that existed as a collection of objects that functioned as an ecosystem to a concrete nature that included individual members of a species and specific places. Chawla and Clark spoke of this physical world of hometowns and moths, and the human was firmly grounded in this world.

Evernden (1989) further observed that, despite this modern notion of the natural world as a system, the metaphor of nature as a miracle had survived because direct observers knew a world that included the individual members of species, the "frogs and mourning doves" that he labeled as the "realities" (p. 162). Clark, trained in the natural sciences, was aware of the modern concept of nature as a collection of inert objects that functioned according to a set of laws. Yet her description of life as "magical" (CNN.com, 2003b) suggested the metaphor of nature as a miracle. Nature existed in moths and midwestern towns, and these were all one part of a holistic, miraculous universe.

As Evernden (1989) suggested, this metaphor of nature as a miracle grew out of direct experience. Despite their demanding work schedules, Chawla and Clark took time to observe, and marvel at, the stars and moths. Clark noted, "much of the time I'm working back in Spacehab and don't see any of it. Whenever I do get to look out, it is glorious" (Associated Press, 2003a, para. 5). Her reference to the constellation Orion as "my friend" (CNN.com, 2003b, para. 6) implied a personified nature that defied the modernist construction of nature as a laboratory and was reminiscent of the construction of the cosmos held by members of ancient cultures (Cahill, 1998; DeCandido, 1998).

Chawla, who would sometimes miss sleep so that she could watch the earth out of the space shuttle window, was described as "like a poet" whose descriptive language could move her audience to tears (CBSNEWS.com, 2003, para. 11). She spoke of the "awe" that the sight of earth "inspires" and the "good thoughts that come to mind when you see all that" (ABCNews.com, 2003, para. 7). Chawla and Clark's comments seemed closely aligned with those of the 19th-century British romantic poet, Wordsworth (1954a), who spoke of his wonder at seeing the rainbow. Wordsworth encouraged the "spontaneous wisdom" (1954b) that a day in the woods brought as a break from the "barren leaves" of "science and art" (1954b). The astronauts, through their contemplation of nature, gained a sense of the miracle and wonder of life and the beauty of the cosmos. Nature, in this context, was a teacher rather than an object of study, and her lessons were learned ontologically, from direct experience and observation, rather than through study and experiments.

This direct experience placed the astronauts into the universe and made them a part of the natural world. The idea of a unified nature, with humanity as a microcosm of the universe, was stated most explicitly by Chawla when she described the earth and sky reflected in her eye (CNN.com, 2003a). She spoke of nature as an extension of self, a metaphor that led to the feeling of kinship with the nonhuman world.

Nature as a Wilderness, a Resource and a Machine

In contrast to the views expressed by the astronauts, President Bush spoke of space as untamed wilderness, entirely separate from human culture. The president was perhaps thinking of pictures that depicted the earth as an isolated globe spinning in a dark void when he referred to "mankind's" willingness to penetrate the "darkness beyond our world" (PBS, 2003, para. 5) despite the inherent dangers. He linked this darkness with danger and presented untamed nature as reluctant to yield to humans. He described the "fierce outer atmosphere of the earth" (PBS, 2003, para. 3) as if implying that the nonhuman world was a fierce beast that must be conquered in order for the human's goal of discovery and conquest to continue.

Bush referred to the "inspiration of discovery and the longing to understand" (PBS, 2003, para. 5) as the reason for the human desire to penetrate the chaotic darkness. This implied that humans studied nature in order to pursue knowledge, but nature, to him, seemed to exist as an inert object of study rather than as the source of inspiration and beauty that Chawla and Clark described. The natural world also existed entirely separate from human culture; humans were led into the dark wilderness in order to discover, explore, and claim the territory as their own. He used the term *journey* (PBS, 2003, para. 5) literally and figuratively to describe the journey into space and the pursuit of knowledge. Nature became a path that led to a destination, which in this case seemed to be complete knowledge and power.

When Bush spoke of the "journey" (PBS, 2003, para. 5) into the "darkness" (PBS, 2003, para. 5), he was also implying the mechanistic metaphor that Merchant (1980) and Corbin (1992) described. Nature was a wilderness that had no innate value of its own, but, under human management, this wilderness could be modified in order to function as a machine. It was not the active nature that Clark depicted when she described the lightning, moon, and rivers. In fact, Bush used no active tense verbs at all to describe the nonhuman world, but instead referred to it as something that the Creator made and the human navigated. Quoting from the prophet Isaiah, Bush referred to a powerful God who "brings out the starry hosts one by one" and "calls them by name" (PBS, 2003, para. 5). The stars were firmly under the control of this personified God, who, like humans, assigned names to the nonhuman world. This human ability to use language and name physical objects separated the nonhuman world from both the Creator and the human. Lewis (1967) noted that words exist as a symbol for the actual object and the ability to use language and assign names empowers and separates the human from the object that is named.

Bush, then, described a nature that was at once distinct from humans and penetrable to them because of technology and human intelligence (PBS, 2003, para. 3). He also described this nature as under the control of a God who created humans in His own image and granted them authority over the earthly species by assigning Adam the task of naming them (Genesis 2:19, New King James Version).

Bush, in direct contrast to Clark and Chawla, refuted the metaphor of nature as home when he indicated that the astronauts "did not return safely to Earth, yet we can pray that all are safely home" (PBS, 2003, para. 7). This reference presumed that his audience shared his construction of the physical world as a temporary abode for righteous humans, who, after death, would enter their rightful home in a place beyond the universe. The natural world was a resource that could be used temporarily but that righteous humans would one day no longer need.

The metaphor of nature as a resource, then, had its roots in the Judeo–Christian worldview. It was a cornerstone of the early modern period when European explorers were intent on establishing colonies on other continents as well as during the 18th and early 19th century when American settlers were pushing westward across this continent. In fact, philosophers and poets often associated wilderness with evil, waste, and void (Hobbes, 1974; Spenser, 1974). The ocean, in particular, was described by Renaissance poets as dark, dangerous, and alien to humans (Spenser, 1974). In Bush's rhetoric, the space beyond the earth's atmosphere had replaced the ocean as the dark, dangerous, alien frontier through which humans, blessed with reason, courage, and innate goodness, must travel if they were to continue their exploration and conquest. Bush, when he spoke of the humans' ability to penetrate "the fierce outer atmosphere" (PBS, 2003, para. 3) because of their "courage, daring and idealism" (PBS, 2003, para. 3) expressed the early modern view that humans, blessed with the ability to reason and the courage to act, would prevail against an irrational nature.

During the Renaissance, scientific study increased the human's feelings of empowerment and further erased the spiritual connotations for nature that had survived into the Middle Ages (McKibben, 1989; Merchant, 1980). The idea of using nature as a venue for study, linked to the metaphor of nature as a resource, was strongly advocated by the early modern philosophers, including Bacon and Locke, in order to encourage activities such as growing exotic plants under controlled circumstances (Bacon, 1909a) and taming the American wilderness (Bacon, 1909b; Locke, 1963; Merchant, 1980). Certainly, the perceived need to dominate nature through knowledge was advocated by Bacon (1955), who claimed that natural philosophy could enable humans to "regain their rights over nature, assigned to them by the gift of God" (p. 135). Bacon (1955) linked the study of science to religion, and even politics, when he added that humans would govern "by right reason and true religion" (p. 135).

The ability to reason, Locke (1963) said, was the distinction between the human and the nonhuman and this ability, endowed to humans by God, allowed them to dominate other species. Humans had the God-given right to divide, and claim as their own, part of the land. "The Earth, and all that is therein, is given to Men for the Support and Comfort of their being," he wrote (p. 328). Locke's (1963) perspective became justification for colonization. God, he said, never intended for any land to remain uncultivated because unimproved land was useless (p. 333). Captain John Smith (1986), on a mission to colonize the New World, praised Virginia's resources, decried the lack of management that had left the wilderness "overgrowne with trees," and noted that "good husbandry" could correct the situation (p. 151).

Smith's views seemed typical of that of early Americans. Nash (1989) and DeLuca (1999a) observed that the American settlers viewed untamed wilderness as a beast to be conquered and natural resources as inexhaustible. The early 20th-century conservationists, however, also promoted the view of nature as a resource (Pinchot, 1947), and Corbin (1992) wrote that this perspective had prevailed so long in American environmental discourse that many Americans assumed that a "reasonable, scientific approach" (p. 205) to the human/nature relationship was the only sensible one. President Bush, then, implied the familiar metaphor, embedded in American culture (Corbin, 1992) of the natural world as irrational, untamed, and in need of human management in order to make it fit for human use. The astronauts spoke of the natural world as life, light, and a miracle.

In the following paragraphs, I explore the possible differences for the contrasting metaphors in these communication artifacts, with an emphasis on the possible use of the resource metaphor as a political tool.

Conflicting Metaphors: Nature as a Resource and Nature as a Miracle

President Bush's rhetoric was familiar to Americans, not only because it described space as a frontier to be conquered, but because it had its roots in Judeo–Christian religion. Cahill (1998) spoke of the distinction between the heavens and the earth that arose with the development of the Hebrew religion, which replaced the constellations, storm gods, and fertility goddesses of the ancient religions with a single personified God. Bush's universe, existed, both literally and metaphorically, under Heaven and distinct from humans. Under this worldview, the humans who had been created in God's own image had the God-given right and, in fact, obligation to subdue the nonhuman world (Genesis 1:27, New King James Version).

Clark and Chawla, by contrast, placed the spiritual back into the physical realm with their references to life as magical and human life as a part of this magic. Differences in religion, then, could account for some of the contrasts in the metaphors found in these artifacts. Bush, a Protestant, was speaking from within a Judeo–Christian worldview. Clark, a Unitarian, was likely to have been influenced by such thinkers as William Ellery Channing, an early 19th-century Unitarian minister, and Ralph Waldo Emerson, who emphasized the study of natural poetry over that of natural science and who thought that humans could approach God through nature, which held a "divine aura" (Cherry, 1980, p. 1). Chawla, a Hindu, would likely have thought of the natural world as a reflection of God and a source of inspiration for humans (Sharma, 2002).

Gender differences could also partly explain the differences in the rhetoric between the female astronauts and Bush. Ecofeminists used gender as a means of describing the conflicting views on the nature/culture relationship. The natural world, Merchant (1980, 1992) maintained, was historically defined as feminine in Western culture, and women enjoyed a special relationship with nature because of the mutual qualities of fertility and nurturance. Bell (1994) also spoke of the feminine interest in the nurturing side of nature, and added that men generally were more interested in nature as an aggressor and a competitor.

The obvious differences in experience also account for some of the conflicting constructions in these artifacts. The earthbound Bush was denied the opportunity to experience space directly, so he never saw Orion from outer space or the earth and sky reflected in his eye. He referred only to the "darkness" (PBS, 2003, para. 5) of an almost impenetrable atmosphere because, except for photographs depicting the space around the earth as dark and unformed, he had no way to know what the sky looked like from above the earth's surface. Experiencing the nonhuman world directly, Evernden (1989) noted, was crucial to an understanding of the concept of nature as a miracle.

President Bush's ontological relationship with the universe was different, then, as was his audience and his purpose in speaking. Chawla and Clark, speaking in midflight, were doubtless filled with the euphoria of the space expedition. They communicated with the public through the media, and they focused on describing the sights that they saw and sharing their thoughts about this experience with this public.

Bush, of course, was addressing the public directly, rather than in a mediated interview, and his presumptive purpose was to eulogize the astronauts and comfort the nation in the hours after the disaster. Eulogies, Hart (1997) said, followed a predictable pattern in that they "tell a selective history" (p. 124) designed to highlight the deceased's accomplishments. President Bush, in fact, framed the deceased astronauts as martyrs in the human quest for "discovery" (PBS, 2003, para. 5). "These astronauts knew the dangers, and they faced them willingly, knowing they had a high and noble purpose in life," he said (PBS, 2003, para. 3).

The use of emotionally charged language, which made this speech memorable, was also predictable in a eulogy. Bush was no doubt conscious of the historical significance of his speech, which was somewhat reminiscent of Reagan's eulogy for the victims of the 1986 Challenger accident. Reagan, borrowing from a World War II era poem written by a young American pilot, described the victims as having "slipped the surly bounds of earth to touch the face of God" (Garton, 1994, para. 1). This was not unlike Bush's reference to the "comfort and hope" (PBS, 2003, para. 6) that he said existed far beyond the visible skies. Again, there was the spatial connotation associated with space travel. God lived in a heaven that existed beyond the universe, and humans, through spiritual flight, could transcend the physical universe to arrive "safely home" (PBS, 2003, para. 7). Bush, then, like Reagan, used an analogy of flying beyond the universe to arrive at a heavenly home in order to comfort the grieving families and an anxious nation.

It was interesting to note that, although the astronauts made frequent references to the earth in their e-mails (CNN.com, 2003a; CNN.com, 2003b), President Bush made minimal reference to the earth, noting only that the astronauts did not return here but still arrived home. The concept of soaring beyond the mundane, gravity-bound earth to a spiritual home was embedded in Christian tradition. This journey, although a metaphysical one, was often depicted in spatial terms, with heaven always described as "up," or beyond the skies and clouds. Reagan, speaking from within the Christian tradition, described the Challenger astronauts as escaping the bounds of earth and flying, literally and metaphysically, to a sanctified space where they could reach out and touch God. His references to the afterlife were, presumably, meant at least in part to comfort the grieving families and nations, as were Bush's references to the astronauts arriving home. Still, it was important to remember that the 20th century brought humans the ability to overcome the earth's gravity and travel, literally, into the heavens. The ascent into the heavens that Reagan described was first "made physically by a fiery rocket," then "metaphysically to heaven" (Garton, 1994, para. 25). The Judeo–Christian tradition, then, minimized the importance of the earth as home and emphasized the limitations of an earthly life, while increased technology gave humans the power to overcome these limitations and moved humankind a step closer, spatially and metaphysically, to God. President Bush, then, spoke of the astronauts as having escaped the confinement of life in the physical world in order to soar to a home high above the visible sky.

The timing and purpose of President Bush's speech, then, was different from that of the astronauts' communication, and President Bush's eulogy, predictably, spoke of the courageous efforts to overcome the challenges of life in the physical plane. However, one cannot overlook the possible political implications when examining President Bush's choice of words. His speech, which emphasized his resolve to continue the space program, was particularly pertinent because of the ambitious plans his administration had for space travel. After the disaster and the subsequent report criticizing NASA and questioning the nation's commitment to a manned space program, President Bush and NASA were faced with the task of selling an ambitious space program to Congress and the American public. In January 2004, Bush unveiled details that included returning the space shuttles to flight, developing new spacecraft by 2008, completing the International Space Station designed for long-term research by 2010, and returning to the moon by 2020, with the intention of sending manned spacecraft to Mars and beyond (The White House, 2004).

Bush was also facing the 2004 presidential elections, and he was no doubt aware that the space program could become an issue in the race. Democratic presidential candidate, John Kerry, asserted that the space program should focus on improving life on earth rather than establishing long-term research stations in space (Space.com, 2004). Bush's plans, however, seemed aimed at making outer space an extension of America's domain (Associated Press, 2003b; Williamson, 2003). As Williamson (2003) indicated, space colonization even hinted at the possibility that space could be converted into real property that would enhance the wealth of some Americans. This was an extension of the economic metaphor of nature as a resource (and the earth as real estate) that drove American colonization and western expansion.

Lakoff (2002) argued that this resource metaphor was deeply ingrained in the American consciousness and was linked to both the Judeo–Christian religion and the conservative worldview that he said prevailed in contemporary American politics. "The resource metaphor (consider the term 'natural resources') assumes that whatever is in nature is, and should be, part of a human economic system" (Lakoff, 2002, p. 214). Under this view, God intended for humans to use nature for their benefit (Lakoff, 2002), and righteous humans were granted the moral strength "to stand up to" (Lakoff, 1995, para. 31) evil. Humans, then, under this metaphor, had a mandate to continue with the "high and noble purpose" (PBS, 2003, para. 3) of exploring and subduing the space beyond the earth's atmosphere, which God had clearly intended to enhance human knowledge and wealth. The need to glorify both God and country justified continued space exploration. As if in anticipation of public and Congressional questioning about the proposed budget and the shuttle accident, President Bush promised that "the Cause" (PBS, 2003, para. 5) would continue, an assertion that he repeated 2 days later at a speech at the National Institutes of Health in Bethesda, Maryland (Associated Press, 2003b). The eulogy, then, provided him with an opportunity to rally public support for this cause before the NASA budget requests and the 2004 presidential elections.

IMPLICATIONS OF CASE STUDY

Corbin (1992) referred to the "kaleidoscope of constructions for nature" (p. 21) through which humans have historically viewed the human/nature relationship. Certainly, the metaphors implied in the rhetoric by Bush, Chawla, and Clark ranged from the organic metaphor that Merchant (1980) described to a metaphor of nature as dark wilderness and nature as a resource that existed only for human use. Despite assertions by Bateson (1999) and McLuhan (1977) that space travel changed the way humans perceived their relationship with the natural world, the artifacts examined in this study reflected the metaphors that historically defined this relationship. Nature was extended to include the space beyond the earth's atmosphere.

However, these artifacts confirmed the existing literature on space travel on one key point, and this point was central to an understanding of the possible effects of space travel on environmentalism. Bateson (1999) complained that space exploration sent a message that it was all right to destroy the earth because humans could simply push onward to exploit outer space. This seemed particularly pertinent when considered in the light of President Bush's position on environmental issues (for which he received a failing grade from the League of Conservation Voters, which ranks elected officials and candidates on their environmental record), including the promotion of increased drilling for oil in wilderness areas, the attempt to deregulate environmental controls on industry, and the failure to support international treaties on climate change (Barringer, 2004; League of Conservation Voters, 2003; Lakoff, 2002). Although the resource metaphor had also served as the foundation of the conservation movement (Pinchot, 1947), President Bush seemed not to refer to the exhaustible resource that Pinchot maintained should be carefully preserved and managed. Rather, he was reflecting Bacon's view of an inexhaustible nature that, by all rights, belonged to humans to use, colonize, change, and exploit. Space was a natural extension of the earth and the next logical step for humans, and, like the 17th-century American continent that Smith described, the space beyond the earth's atmosphere seemed to hold the promise of unlimited resources.

By contrast, Clark and Chawla's interest in nature's spiritual, nurturing, and aesthetic qualities, although still perhaps anthropocentric, tended to place humans back in nature and undermine the perceived distinction between the humans and nature. Their rhetoric suggested that space travel could enhance concept of a living, organic nature that served as the foundation for what DeLuca (1999b) described as a postmodern environmental movement that included rural residents, community activists, and others who spoke outside the mainstream construction of nature as an object of scientific study and management.

However, the astronauts, dependent on mediated interviews, lacked Bush's power to take their rhetoric directly to the American public. They were also speaking as NASA employees, rather than as political or religious leaders, and their description of the natural world was outside of the mainstream construction of nature and therefore less likely to resonate with Americans immediately. Lakoff (1993, 1995, 2002) spoke of the rhetorical power that communicators gained when they framed their arguments within existing metaphors. The metaphors embedded in the astronauts' communication, then, seemed unlikely to have much of an impact on the public consciousness, particularly because they lacked the power to repeat them frequently and directly to the public.

In contrast, President Bush was reinforcing existing metaphors and framing an argument that was already consistent with the American values of religion and continued economic development. Lakoff (1995) spoke of the "carefully constructed conservative rhetoric" (para. 125) that had been repeated so often, and was so ingrained in American thought that it was accepted without question.

Bush, speaking from the presidential podium, had the political power and the ability to communicate the metaphor of space as a frontier to be conquered and a resource to be used, and he could communicate directly to the American public and Congress. These metaphors, then, were likely to drive the decision-making process. Space exploration, as posited by President Bush, is promoting an anthropocentric view of the natural world, at least in part because of the need to justify this exploration within the existing dominant metaphors. Further, this rhetoric is, as some observers (Bateson, 1999) predicted, undermining the metaphors of nature as home and mother that are promoted by some environmental groups (Merchant, 1980, 1992).

This study, then, extended the body of literature surrounding the metaphors embedded in environmental rhetoric and demonstrated that the same resource metaphor that assumed that the earth was part of the economic system was also driving the political rhetoric surrounding space exploration.

REFERENCES

ABCNews.com. (2003, February 1). Kalpana Chawla: A heroine in two nations. Retrieved from wysiwyg://14http://abcnes.go.com/secti...s/DailyNews/shuttle_chawlabio030201.html

Aristotle. (Trans. 1954). *Rhetoric* (W. Rhys Roberts, Trans.). New York: The Modern Library.

Associated Press. (2003a, February 3). E-mail from Astronaut Laurel Clark on day before disaster. Retrieved May 20, 2003, from wysiwyg://8/http://www.space.com/missionlaunches/clark_email_030203.html

Associated Press. (2003b, February 3). Bush says space program will go on. Retrieved September 9, 2003, from http://www.jsonline.com/news/gen/feb03/115672.asp

Bacon, F. (1909a). The new Atlantis. In C. W. Eliot (Ed.), *Bacon, Milton's prose, Thos. Browne: Vol. 3. The Harvard classics* (pp. 168–223). New York: P. F. Collier.

Bacon, F. (1909b). Of plantations. In C. W. Eliot (Ed.), *Bacon, Milton's prose, Thos. Browne: Vol. 3. The Harvard classics* (pp. 85–87). New York: P. F. Collier.

Bacon, F. (1955). *Novum organum*. In R. M. Hutchins (Ed.), *Francis Bacon: Great books of the Western world, Series 30* (pp. 107–195). Chicago: Benton.

Barringer, F. (2004, September 14). The 2004 campaign: The issues, Bush's record: New priorities in environment. *The New York Times*, p. A1.

Bateson, M. C. (1999). We are our own metaphor. *Whole Earth, 14*. Retrieved May 4, 2003, from Expanded Academic Index.

Bell, M. (1994). *Chiderly: Nature and morality in a country village*. Chicago: University of Chicago Press.

Bowyer, J. (2003, February 4). "NASA points left, the media right." *National Review Online*. Retrieved September 9, 2004, from http://www.nationalreview.com/nrof_comment/comment-bowyer020403.asp

Burke, K. (1969). *A grammar of motives*. Berkeley: University of California Press.

Cahill, T. (1998). *The gifts of the Jews: How a tribe of desert nomads changed the way everyone thinks and feels*. New York: Doubleday.

CBSNEWS.com. (2003, February 4). A closer look at Columbia's crew. Message posted to wysiwyg://44/http:/www.cbsnews.com/stories/2003/02/04/columbia/main53967.shtml

Cherry, C. (1980). *Nature and religious imagination*. Philadelphia: Fortress Press.

CNN.com. (2003a, February 2). India mourns space heroine. Message posted to http://www.cnn.com/2003/TECH/space/02/01/shuttle/col

CNN.com. (2003b, February 4). Astronaut Clark: Life is a magical thing. Message posted to wysiwyg://33/http:/www/cnn.com/2003.US/02/01/sprj.colu.profile.clark/index.html

Cohen, P. (1999, April 24). Flight of the soya bean. *New Scientist*, p. 24.

Copleston, F. (1993). *A history of philosophy* (Vol. 3). New York: Doubleday.

Corbin, C. (1992). *Discourse on nature: Rhetoric from the forests of Western Montana*. Unpublished manuscript, University of Iowa.

DeBernardi, J. (1992). Space and time in Chinese religious cultures. *History of Religions, 31*(3), 242–269. Retrieved May 6, 2003, from Expanded Academic Index.

DeCandido, G. A. (1998). Mother earth, father sky: Pueblo Indians of the American Southwest. *Booklist, 95*(1), 11. Retrieved May 6, 2003, from Infotrac.

DeLuca, K. M. (1999a). *Image politics; the new rhetoric of environmental activism*. New York: The Guilford Press.

DeLuca, K. M. (1999b). The possibilities of nature in a postmodern age: The rhetorical tactics of environmental justice groups. *Communication Theory, 9*(2), 189–215.

Evernden, N. (1989). Nature in an industrial society. In J. Angus & S. Jhally (Eds.), *Cultural politics in contemporary America* (pp. 151–164). New York: Routledge.

Fernandez, J. W. (1972). Persuasion and performances: Of the beast in every body . . . and the metaphors of everyman. *Daedalus, 101*, 39–61.

Fernandez, J. W. (1974). The mission of metaphor in expressive culture. *Current Anthropology, 15*, 119–134.

Foss, S. (1996). *Rhetorical criticism: Exploration & practice* (2nd ed.). Prospect Heights, IL: Waveland.

Garton, C. (1994). Slipping the surly bonds [Ronald Reagan's use of lines from John Gillespie Magee's poem after the 1986 Challenger accident]. *ANQ, 7*(3). Retrieved September 9, 2004, from Infotrac.

Hart, R. P. (1997). *Modern rhetorical criticism*. Boston: Allyn & Bacon.

Hobbes, T. (1974). Leviathan. In M. H. Abrams, E. T. Donaldson, H. Smith, R. M. Adams, S. H. Monk, L. Lipking, & G. H. Ford (Eds.), *The Norton anthology of English literature* (3rd ed., Vol. 1, pp. 1642–1650). New York: Norton.

Ivie, R. L. (1994). The metaphor of force in prowar discourse. In W. L. Nothstine, C. Blair, & G. A. Copeland (Eds.), *Critical questions: Invention, creativity, and the criticism of discourse and media* (pp. 259–285). New York: St. Martin's Press.

Lakoff, G. (1993, January 29). *The contemporary theory of metaphor*. Retrieved December 1, 2000, from lakoff@cogsci.Berkeley.EDU

Lakoff, G. (1995). Metaphor, morality and politics. *Social Research, 62*(2), 177–214.

Lakoff, G. (2002). *Moral politics: How liberals and conservatives think*. Chicago: University of Chicago Press.

Lakoff, G., & Johnson, M. (1980). *Metaphors we live by*. Chicago: University of Chicago Press.

Lawler, A. (2003, February 7). Shuttle disaster puts NASA plans in tailspin. *Science, 299*(5608), 796–798. Retrieved September 9, 2004, from Expanded Academic Index.

League of Conservation Voters. (2003). 2003 presidential report card. Message posted to http://www.lcv.org/Files/getfile.cfm?id=1656

Lewis, C. S. (1967). *Studies in words*. London: Cambridge University Press.

Locke, J. (1963). In P. Laslett (Ed.), *Two treatises of government* (Rev. ed.). New York: Cambridge University Press.

Marck, J. (1996). Was there an early Polynesian 'sky father'? *The Journal of Pacific History, 31*(1), 8–27. Retrieved May 6, 2003, from Expanded Academic Index.

Marvel, A. (1974). The mower, against gardens. In M. H. Abrams, E. T. Donaldson, H. Smith, R. H. Adams, H. Monk, L. Lipking, G. H. Ford, & D. Deices (Eds.), *The Norton anthology of English literature* (Vol. 1, p. 1292). New York: Norton.

McKibben, B. (1989). *The end of nature*. New York: Random House.

McLuhan, M. (1977). The rise and fall of nature. *Journal of Communication, 27*(4), 80–81.

Merchant, C. (1980). *The death of nature: Women, ecology and the scientific revolution.* San Francisco: Harper & Row.

Merchant, C. (1992). *Radical ecology: The search for a living world* [Revolutionary Thought/Radical Movements Series]. New York: Routledge.

Muir, S. A. (1994). The web and the spaceship: Metaphors of the environment. *ETC: A Review of General Semantics, 51*(i2), 145–153. Retrieved May 6, 2003, from Expanded Academic Index.

Nash, R. F. (1989). *The rights of nature: A history of environmental ethics.* Madison: University of Wisconsin Press.

NPR. (2003, August 26). Columbia: The search for answers: Report blames NASA culture, management for tragedy. Message posted to http://spaceflightnow.com/shuttle/sts1071status.html

Ortner, S. B. (1973). On key symbols. *American Anthropologist, 5,* 1338–1346.

PBS. (2003, February 1). The president on the Columbia [broadcast of speech by President Bush]. *Online NewsHour.* Retrieved May 10, 2003, from wysiwyg://16/http:www.pbs.org/newshour/bb/science/columbia/bush_2-1html

Pinchot, G. (1947). *Breaking new ground.* Seattle: University of Washington Press.

Rees, M. (2002, May 20). The last spaceship from Earth: In space, machines have proved more useful than men. But we should one day revive human spaceflight to ensure the survival of the species. *New Statesman, 131*(4588), 4–6. Retrieved January 31, 2005, from Expanded Academic Index.

Richards, I. A. (1929). *Practical criticism.* San Diego: Harcourt.

Richards, I. A. (1936). *The philosophy of rhetoric.* London: Oxford University Press.

Schildknecht, C. (1995). Experiments with metaphor: On the connection between scientific method and literary form in Francis Bacon. In Z. Radman (Ed.), *From a metaphorical point of view: A multidisciplinary approach to the cognitive content of metaphor* (pp. 27–50). Berlin and New York: W. de Gruyter.

Sharma, A. (2002). *Modern Hindu thought: The essential texts.* Oxford: University Press.

Smith, J. (1986). A map of Virginia. In P. L. Barbour (Ed.), *The complete works of Captain John Smith* (pp. 120–190). Chapel Hill: University of North Carolina Press.

Space.com. (2004, June 16). Kerry criticizes Bush for space vision. Retrieved September 9, 2004, from http://www.space.com/news/kerry_report_040616.html

Space Today Online. (2003, February 1). Seven astronauts died: Tragedy of Space Shuttle Columbia. Retrieved September 9, 2004, from http://www.spacetoday.org/SpcShtls/ColumbiaExplosion

Spenser, E. (1974). The faerie queene. In M. H. Abrams, E. T. Donaldson, H. Smith, R. A. Adams, S. H. Monk, L. Lipking, G. H. Ford, & D. Deices (Eds.), *The Norton anthology of English literature* (3rd ed., pp. 548–552). New York: Norton.

Spirit filled life Bible: New King James version. (1991). Nashville: Thomas Nelson.

The White House. (2004, January 14). President Bush announces new vision for space exploration program. Retrieved September 9, 2004, from http://www.whitehouse.gov/news/releases/2004/01/2004114-3.html

Williamson, M. (2003). Space ethics and protection of the space environment. *Space Policy, 19*(1), 47–53.

Wordsworth, W. (1954a). My heart leaps up when I behold. In G. B. Harrison (Ed.), *Major British writers: Vol. 2* (p. 91). New York: Harcourt.

Wordsworth, W. (1954b). The tables turned. In G. B. Harrison (Ed.), *Major British writers: Vol. 2* (p. 38). New York: Harcourt.

CHAPTER TEN

Internet Use and Environmental Attitudes: A Social Capital Approach[1]

Jennifer Good
Brock University

BACKGROUND: INTERNET USE AND THE ENVIRONMENT

Based on a myriad of indicators, the case can be made that human activity is having a detrimental impact on the diversity and sustainability of the natural environment (United Nations Programme Annual Report, 2003; Millennium Ecosystem Assessment, 2005). As a medium with the potential to disseminate information and facilitate interaction relatively inexpensively across time and space, the Internet has provided many environmentalists with the hope that it will be a valuable tool in the struggle for sustainable human activity.

In what follows, an overview of some of the literature that explores the relationship between the Internet and the environment is offered. What becomes clear is that although the Internet has become a tool in the environmental movement, the question of "To what end?" has only begun to be explored.

Early research provided a broad overview of Internet use in the name of the environment. For example, Rittner's (1992), *Ecolinking: Everyone's Guide to Online Environmental Information,* looked at the basics of environmental information available online and how to access that information. Similarly, Feidt and Roos (1995) offered an overview of online environmental resources such as environmentally themed Usenet groups and World Wide Web (WWW) sites; the authors concluded with the following prophetic statement:

---

[1]An earlier version of this chapter was presented as a top paper at the 2004 National Communication Association (NCA) convention in Chicago, IL.

The Internet contains a wealth of environmentally related information, much of it of considerable significance. The future information specialist will not be able to perform effectively without access to this resource. Although problems in gaining access to the Net and its resources abound and are likely to continue for some time, it is a wellspring that cannot be ignored. (p. 23)

More recent research explored the Internet as providing a great place for depositing and accessing specific environmental information (i.e., specific environmental topics and specific regions). For example, South (2001) offered an overview of the Internet as a source of information about toxicology. *The Bulletin of the American Society for Information Sciences* (see Environment information systems, 1995) provided a synopsis of the Internet-based aspects of the Environment Information System (EIS) in sub-Saharan African. Bullard (1998) looked at how Internet resources were being used in environmental education with an exploration of Internet use to facilitate teaching about Local Agenda 21 (the United Nations program to foster local initiatives for sustainable development) in an undergraduate classroom. Bullard found that although the Internet had a "large volume" of readily available material about Agenda 21, students had to be taught how to critically judge the appropriateness of online information.

The role of the Internet in environmental/resource management has also been investigated. Kay and Christie (2001) explored the role of the Internet in coastal management by cataloging all integrated coastal management Web sites. They found 77 such Web sites worldwide and concluded that there is "enormous potential" for the Internet in coastal management and that the medium is particularly well suited for projects between developed and underdeveloped coastal communities. The researchers, however, tempered their enthusiasm for this technology that they felt was still in its infancy: "[T]here remain significant research questions requiring further analysis before the full transformative potential and the possible impacts of such transformation on coastal management can be fully assessed" (Kay & Christie, 2001, p. 157).

Guerin's (2001) exploration of the role of the Internet in soil contamination and remediation offered a similar mix of enthusiasm and hesitation. By highlighting active listservs, Guerin stressed the important role the Internet played in facilitating participation and information sharing. Geurin (2001) also pointed out, however, that the listservs must be chosen carefully because they can overload people with unnecessary information and waste people's time. The mix of opportunity and caution was also present in Kangas and Store's (2003) look at Internet use and forest planning. Their focus was on use of the Internet for "teledemocracy" (i.e., participation in forest planning via the Internet) and although they concluded that interactive computer technology had potential in this capacity, they also cautioned that it should not replace any existing ways in which people become involved with such issues.

Two watershed management papers, Ratza (1996), who looked at the Great Lakes Information Network, and Voinov and Costanza (1999), who developed a case study of Maryland's Patuxent River, stressed the importance of the Internet as a tool for information sharing. "We argue that it is not the amount and quality of information that is crucial for the success of watershed management, but how well the information is disseminated, shared and used by the stakeholders. In this respect the Web offers a wealth of opportunities . . ." (Voinov & Costanza, 1999, p. 231).

Researchers have also looked at the Internet as a tool for environmental organizing and activism. Some early writings, such as Zelwietro (1995), expressed concern regarding whether the goals of the environmental movement and the "reality" of the Internet were compatible. So, although Zelwietro offered the example of the success of computer networking by the Native Forest Network to organize against Hydro Quebec's Great Whale Hydroelectric Project, he went on to comment that it is "too early to predict whether the commercial and the anti-commercial worlds [on the Internet] will be able to co-exist" (Zelwietro, 1995, p. 16). Like Zelwietro, Meisner (2000) offered a positive example of environmental Internet activism—that of the U'Wa people's coordination of an international defense of their ancestral lands that "wouldn't have happened without the Internet" (Meisner, 2000, p. 34). However, also comparable to Zelwietro, Meisner (2000) was cautious in his optimism about the technology and offered the following: "There is much to be hopeful about, but realism and awareness of the risks and limitations of these [Internet-based] approaches must constantly inform e-activism" (p. 38).

Three years after his initial work in the area of environmental organizations and the Internet, however, Zelwietro's (1998) tone was more optimistic. Based on a survey study of 135 environmental organizations in 10 countries, Zelwietro found that 68% of the groups had online access and that this access was related to an increase in interaction with other organizations, number of campaigns, number of information requests, and number of members (all of these elements compose what Zelwietro calls *politicization*). He concluded by saying that "[t]he Internet has provided an unprecedented opportunity for this population [of environmental organizations] to promulgate their messages to a wider and more diverse audience, as well as providing diverse groups a forum to debate and formulate action plans" (Zelwietro, 1998, p. 54).

Other researchers have shared Zelwietro's positive assessment of the role that the Internet can play in environmental organization and activism. Pickerill's (2001), British-based, environmental Internet research highlighted that the Internet facilitated speed, cost-effectiveness, interactivity, freedom of expression, and global reach for environmental organizations—as well as a shift in the stereotype of environmentalists as being "antitechnology." Similarly, Weeks (1999) explored the role of the Internet in the World Wildlife Fund's successful campaign to raise awareness about the tiger; Harkinson (2000) used the example of the Internet-based Conservation Action Networks (CAN) to illustrate how effective the Internet can be in environmental campaigns (i.e., he offered the example of the role the CAN played in the cancellation of a Russian whale hunt after a large scale, 4-day e-mail/fax campaign). Harkinson also highlighted that these "massive effects" can be achieved by the Internet, necessitating much less effort than traditional environmental activism required of the public.

This notion of the Internet facilitating environmental change with less effort, or while individuals carry on with "life as usual," has been explored by other researchers. For example, Oko (2000) wrote about how environmentally inclined music lovers could help an environmental cause merely by tuning in to Net Aid, an Internet-based international concert. Additionally, Howard (2001) highlighted that one can be "environmentally friendly" by merely using the right Internet Service Provider (i.e., EcoISP), which donates money to environmental organizations. Similarly, Gardyn (2001) and Bogo (1999) illustrated ways in which one can shop online and help environmental organizations and causes (e.g., at the Ecomall, 2005, Web site, where the mission is to "offer our visitors the inspiration and the resources to begin a more sustainable, natural, [and] environmentally-aware lifestyle," www.ecomall.com). Gardyn (2001) highlighted one respondent thusly, "With no time to shop and even less time to volunteer, she was thrilled to find a way she could contribute [to environmental organizations] by 'selfishly shopping for myself'." (p. 31).

Not all researchers, however, are so pleased with the notion that the Internet facilitates this sense of doing good for the environment while at the same time facilitating and encouraging life as usual. In their exploration of the Internet and environmental justice, the area of environmental work that focuses on environmental degradation in the most economically disadvantaged areas, Dordoy and Mellor (2001) pointed out that the Internet is most advantageous to the privileged groups in society (White, English-speaking, men) and it is in the hands of these people that Internet expertise resides. The Internet's increasing domination by powerful commercial interests means that "environmental organizations . . . will use new technology . . . but this is only within the context of political 'business as usual' where existing power structures retain their dominance" (Dordoy & Mellor, 2001).

So, although the Internet can clearly be a valuable tool in communicating about, organizing around, and managing environmental issues, the question of whether Internet use is, broadly speaking, part of the problem or part of the solution, is unclear, despite the issue being raised in a number of studies. A framework is needed within which to explore this question. In what follows, such a framework is proposed and explored.

INTERNET USE AND ENVIRONMENTAL SOCIAL CAPITAL

The Pew Internet and American Life Project began in 1999 (reports available as of May 2000) with funding from the Pew Charitable Foundation. Since its inception, the project has conducted over 35 studies (based on phone interviews and online surveys), exploring the Internet and social change. According to the Pew Internet and American Life Project, as of February 2004, approximately 65% of Americans had Internet access and 40% of Americans had been online for more than 3 years, and although there continues to be a digital divide, the Internet is increasingly a place "for all." For example, men (61%) and women (58%) are online in almost identical percentages, and although there are some differences in age (72% of 18- to 29-year-olds are online versus 24% of those 65 years old plus), ethnicity (62% White are online versus 45% Black and 52% Hispanic), community type (63% of suburban people are online, 60% of urban people, and 50% of rural people), income (43% of those earning less than $30,000 per year are online versus 84% of those earning $75,000 or more) and education (22% of those with less than a high school education versus 82% with more than a college education), every day an average of 64 million Americans go online (Pew Internet and American Life Project, 2005).

Also according to the Pew Internet and American Life Project (2005), the most common online activity is sending e-mail (53% of those with Internet access do this every day) followed by using a search engine to find information (30% of those with Internet access do this every day). In fact, "the Internet has become a mainstream information tool" (Pew Internet and American Life Project, 2005). The Pew Internet and American Life Project research highlights that when people go online, they expect the Internet to be a source of information about health care, government agencies, news, and shopping, and, according to the BBC, the Internet is also "the best place to go for news of the environment" (cited in Kirby, 2002). This finding comes from a poll conducted by the BBC, AOL/Time Warner, and Microsoft, with results based on over 25,000 respondents from 175 countries (Kirby, 2002).

So, it is clear that large numbers of people are making use of the Internet, and it is also clear that some of those people are making use of the Internet for purposes related to the environment (although it is safe to assume that many are not). What is less clear is whether use of the Internet has been "good" for how people understand their relationship with the natural environment. In order to think about this, the concept of *social capital* is valuable.

Robert Putnam is perhaps best known for popularizing of the concept of social capital. His 1995 article, "Bowling Alone: America's Declining Social Capital," and subsequent book, *Bowling Alone: The Collapse and Revival of American Community* (2001), detailed a long list of examples that illustrated how Americans are participating less in civic society (voting, church going, labor-union involvement, parent–teacher association participation, women's and fraternal organizations, volunteering) than they were 50, even 10, years ago (Putnam, 2001). Putnam talks about social capital as the "features of social organization such as networks, norms, and social trust that facilitate coordination and cooperation for mutual benefit" (Putnam, 1995, p. 67). A similar definition is offered by Shah, McLeod, and Yoon (2001): "[Social capital] is the resources of information, norms, and so-

cial relations embedded in communities that enable people to coordinate collective action and to achieve commons goals" (p. 467).

Putnam (1995, 2001) saw social capital as the essence and foundation of democracy. More specifically, Putnam (1995) highlighted that successful outcomes
related to "education, urban poverty, unemployment, the control of crime and
drug abuse, and even health" (p. 66) are dependent on social capital. However,
Putnam may have left out an essential item in his list: the idea that successful outcomes related to environmental degradation are dependent on social capital.
Without the "networks, norms, and social trust that facilitate coordination and
cooperation for mutual benefit" (p. 67) that Putnam (1995) proposed as essential
ingredients of social capital, organizing and working for environmental change
would be impossible. And the case can be made that only through environmental
change can we as a species continue to inhabit the planet; there is perhaps nothing
more mutually beneficial than that.

Putnam (1995) offered five explanations for why social capital has been declining.[2] The first two that he presents, he promptly rejects. First, he talks about
women entering the workforce and not having time for the civic activities with
which they were once very involved (Putnam discredits this hypothesis by pointing out that men also left civic activities during this time period). Second, Putnam
proposes that perhaps people are uprooting themselves too often to get involved
with the community (this he discredits by pointing out that since 1965, people
have been more, not less, likely to stay put). However, Putnam (1995) retained
three other explanations as plausible: fewer "married, middle-class parents" due
to "fewer marriages, more divorces, fewer children, lower real wages" and the ar-
rival of the big box stores which replace "community-based enterprises by out-
posts of distant multinational firms" (p. 75). Additionally, Putnam (1995)
claimed that television radically changed social capital through the "technical
transformation of leisure [time]" (p. 75). As Putnam (1995) pointed out, "Time-
budget studies in the 1960s showed that the growth in time spent watching television dwarfed all other changes in the way Americans passed their days and nights"
(p. 75).

It would seem that the case for television as a destroyer of social capital could
be made fairly easily. However, making such a case is not the intention of this
chapter (although television use, and its relationship to Internet use, is now ex-

---

[2]Based on his original list of five, Putnam (2001) further developed the list of "plausible suspects"
that could be blamed for the decline in social capital. The list is: busyness and pressure; economic hard
times; the movement of women into the paid labor force and the stresses of two-career families; residential mobility; suburbanization and sprawl; television, the electronic revolution, and other technological changes; changes in the structure and scale of the American economy, such as the rise of chain
stores, branch firms, and the service sector, or globalization; disruption of marriage and family ties;
growth of the welfare state; the civil-rights revolution; the sixties (most of which also happened in the
seventies), including: Vietnam, Watergate, and the disillusion with public life; the cultural revolt
against authority that included sex, drugs, and so forth (Putnam, 2001, p. 187).

plored).[3] This chapter intends to pursue an aspect of what Putnam highlights in the "What is to be done?" section of his 1995 paper where he ponders what is the "impact . . . of electronic networks on social capital?" (p. 76). Putnam is skeptical about the potential of electronic networks in this context, and others have shared his skepticism. Shah, Kwak, and Holbert (2001) highlighted two main reasons that researchers have linked social capital and media/electronic network use: first, time displacement (i.e., that use of the media/electronic networks might occupy time that would otherwise have been spent on civic activities); second, content (i.e., that the media/electronic networks might contain messages that would dissuade individuals from engagement). Putnam's (1995) feeling about the possible relationship between social capital and media use/electronic networks is based on a "hunch . . . that meeting in an electronic forum is not the equivalent of meeting in a bowling alley . . . but hard empirical research is needed" (p. 76). Researchers over the past few years have begun to produce such hard empirical research. The current research adds to this evolving collection.

Two widely cited studies (Kraut et al., 1998; Nie & Erbring, 2000) found relationships between Internet use and decreasing face-to-face interaction with people as well as a decreasing sense of personal well-being, both of which are related to social capital. Other research, however, has come to very different conclusions. Such studies have found "online tools are more likely to extend social contact than detract from it" (Howard, Rainie, & Jones, 2001, p. 397) and "Internet use increases participatory capital. The more people are on the Internet and the more they are involved in online organizational and political activity, the more they are involved in offline organization and political activity" (Wellman, Haase, Witte, & Hampton, 2001, p. 450).

In two studies of Internet use and political participation/efficacy/knowledge, one analyzing secondary data from the 2000 National Election Survey (Nisbet & Scheufele, 2002) and the other analyzing primary survey data collected in New York State (Scheufele & Nisbet, 2002), limited effects were found for the relationship between Internet use and political participation. The findings, however, must be understood in context. For example, the National Election Survey allowed for a limited measure of Internet use, that is, whether respondents had access to the Internet and if they did, whether they had "seen any information about the current [2000] election campaign on the Web or Internet" (Nisbet & Scheufele, 2002, p. 16). When Internet use was operationalized in more detail, that is, using Internet political information seeking, Internet nonpolitical information seeking, and Internet use for entertainment (Scheufele & Nisbet, 2002), nonpolitical information seeking was moderately positively predictive of political efficacy[4] ($\beta = .12$,

---

[3]Cultivation theory provides an excellent starting place for making this case (see Shanahan & Morgan, 1999).

[4]Efficacy was operationalized using a 5-point Likert scale and the statements, "People like me don't have any say about what the government does," and "Sometimes politics and government seem so complicated that a person like me can't really understand what's going on."

$p < .05$) and Internet use for entertainment was a stronger negative predictor of political efficacy ($\beta = -.13$, $p < .01$) as well as a moderate negative predictor of political knowledge ($\beta = -.13$, $p < .05$).

The aforementioned results suggest that it may not be *overall* use of the Internet that is related to political participation/efficacy/knowledge, but rather *specific* uses of the Internet. This was exactly the conclusion that Shah, Kwak, et al. (2001) came to: "[R]elationships between Internet use and the production of social capital must be viewed as more provisional—dependent on the motives individuals bring to their use of the World Wide Web ... how much time people spend on-line is less important than what they are doing" (p. 154). In their analysis of secondary data, the 1999 DDB Life Style Study, Shah, Kwak, et al. (2001) broke Internet use into product consumption, information exchange, financial management, and social recreation. A similar study by Shah, McLeod, et al. (2001) made use of a pooled dataset from the 1998 and 1999 DDB Life Style Studies and operationalized Internet use as information exchange, financial management, and participation in a chat room or online forum. In both of the aforementioned studies, only use of the Internet for information exchange was positively associated with the criterion variables trust in people and civic participation (Shah, Kwak, et al., 2001); civic engagement, interpersonal trust, and contentment (Shah, McLeod, et al., 2001). In contrast to the positive relationship between Internet use for information exchange and social capital, Shah, Kwak, et al. (2001) found that use of the Internet for social recreation was negatively related to the three criterion variables of civic engagement, interpersonal trust, and contentment.

Given that both studies relied on cross-sectional analysis, causal relationships cannot be assumed. As Shah, Kwak, et al. (2001) pointed out, "It is not unreasonable to believe that individuals who are engaged in civic life come to use the Internet for information exchange in order to fulfill their preexisting motivations" (p. 155). Regardless, whether the Internet facilitates existing civic interests or encourages new interests, research seems to suggest a relationship between Internet use for informational purposes and higher levels of social capital, manifest in increased political efficacy (arguably an important facet of social capital).

Although research has been done that speaks to Putnam's question of the relationship between electronic networks and social capital from the perspective of the relationship between Internet use and political attitudes, no research has looked at the relationship between electronic networks and social capital from the perspective of Internet use and attitudes about the natural environment (and, as has been proposed, attitudes about the natural environment, like attitudes about politics, can be understood as a facet of social capital).

Therefore, based on the above research, the following research questions are proposed:

RQ1: Is use of the Internet for informational purposes positively related to environmental awareness and concern?

RQ2:   Is use of the Internet for social/recreation purposes negatively related to environmental awareness and concern?

As has been pointed out, one of Putnam's (1995) key explanations for the erosion of social capital in the United States is "the technological transformation of leisure," or the way in which television invaded our lives and displaced our "social externalities" (p. 75). In addition to time displacement, television brought content into our lives that was not necessarily civically productive. The relationship between television and environmental attitudes has been a research topic for cultivation theory researchers over the years with findings indicating that heavy viewing of prime-time television can be predictive of decreases in environmental concern (Good, 2005; Good & Shanahan, 2005; Shanahan & McComas, 1999). This raises the possibility that Internet use might be related to environmental attitudes in the absence of television (i.e., the Internet's displacement of time that would otherwise be spent watching television). Based on this possibility, a third research question is proposed:

RQ3:   How is Internet use related to television viewing?

What these hypotheses omit is the reality that although there are very different intensities, and kinds, of Internet use, there are also very different people who make use of the Internet. Within the context of attitudes about the environment, there are, broadly speaking, those Internet users who already have an interest in, and concerns about, the natural environment (people we might refer to as *environmentalists*) and those who do not necessarily share such interests or concerns (people we might refer to as the *general public*). Given this, a fourth and final research question about these two populations is:

RQ4:   Do environmentalists and the general population differ in the relationship between Internet use and attitudes about the environment?

METHODS

Data Gathering

In order to explore the relationship between Internet use and environmental attitudes for the general population and environmentalists, two random samples were collected. The first was a national sample of 1,000 names and addresses that was purchased from Survey Sampling, a company that creates random lists based on the white page telephone directories and supplemented with other proprietary information sources. These respondents are referred to in the rest of the chapter as the *general sample*. The second sample was a list of 1,000 names and addresses, provided free of charge by the National Parks Conservation Association, randomly selected from a database of the organization's 500,000 members (members are people who have donated $35 or more in the last 18 months). These respondents are referred to in the rest of the chapter as the *environmentalist sample.*

Sixty-eight names and addresses from the general survey and 11 from the environmental were removed from the overall sampling frame because they were returned as undeliverable. There were 314 general surveys returned (19 of which could not be used) and 492 environmental surveys returned (7 of which could not be used). Therefore, the response rate for the general list was 34% (295 usable surveys) and the response rate for the environmental list was 49% (485 usable surveys). These response rates are in keeping with current expectations for survey response rates (Teitler, Reichman, & Sprachman, 2003). Further discussion of response rates can be found in the Limitations section at the end of the chapter.

The mean age for the general sample is 56 and the mean age for the environmentalist sample is 63. Fifty-one percent of the environmentalist sample is female while the general sample is weighted toward men, with 31% female (this is also discussed in the Limitations section at the end of the chapter). Ninety-seven percent of the general sample and 98% of the environmentalist sample have English as their first language. The general sample has "Some college" as the mean level of schooling while the environmentalist sample has "A 4-year college education" as the mean. The income for the two lists is similar—the mean for both fell in the $36,000 to $55,000 range. The majority of the respondents from the environmental list are either "Professionals" (44%) or "Retired" (45%), while the general list has a similar percent of "Professionals" (45%) with 22% "Retired" and 14% in "Manual or Clerical" positions. Respondents from both lists reside predominantly in the "Suburbs" (54% for the general sample and 57% for the environmentalist sample). The general sample resides in "Rural" areas (26%) and "Urban" areas (20%); this is reversed for the environmentalist sample: "Urban" (25%) and "Rural" (18%).

Measures and Reliabilities

Environmental Attitudes. The New Environmental Paradigm (NEP) scale (Dunlap, VanLiere, Mertig, & Jones, 2000; Dunlap & VanLiere, 1978) has been used to measure environmental attitudes in numerous studies (see Dunlap et al., 2000, for an overview). The NEP was originally published in 1978 and was retested and reconfigured by Dunlap et al. in 2000 in order to provide a more comprehensive scale with gender-neutral language and balanced statements (i.e., the original scale had very few antienvironmental statements). The new scale consists of 15 statements that are responded to on a 5-point Likert scale from "Strongly disagree" to "Strongly agree." The scale can be divided into five facets of an "ecological worldview": reality of limits to growth, antianthropocentrism, fragility of nature's balance, rejection of exemptionalism (i.e., rejection that humans are exempt from the rest of the environment) and possibility of an eco-crisis (Dunlap et al., 2000).

The NEP scale, therefore, taps into fundamental beliefs regarding the state of the natural environment and, at least implicitly, the need for environmental change. Strong beliefs that the natural environment is finite, has inherent value outside of the value to humans (and yet humans are an integral part), is fragile, and is on the brink of crisis, would arguably be at the foundations of what people would need to believe before they would be willing to participate in environmental change (Dunlap et al., 2000). Therefore, if the Internet is providing, "The resources of information, norms and social relations . . ." that "enable people" to come to an ideological place where they are interested in "coordinate[ing] collective action and achieve common goals" (Shah, McLeod, et al., 2001, p. 467) related to the environment, then the case can certainly be made that the Internet is positively related to social capital. This is not to say, however, that a positive relationship between Internet use and the NEP would support some sort of "social capital is necessarily good for the environment" argument.[5]

The reliability for these 15 statements in the current study is $\alpha = .84$ (Dunlap et al., 2000, found $\alpha = .83$). Based on this strong reliability and other tests of internal consistency, Dunlap et al. (2000) suggested treating the scale as a single construct. This is what has been done in the current study.

Internet Use.   Quantity of Internet use was measured with three open-ended questions that asked about the number of hours the Internet is used each day at work, each weekday at home, and over the weekend at home.

Type of Internet use was measured in two ways. First, general Internet use was measured by asking subjects to rank their top five types of online Internet activities (10 examples of Internet use were given: "E-mail," "Looking for/reading news," "Looking for/reading sports," "Surfing the WWW," "Chatting with friends," "Chatting with people met online," "Games," "Shopping," "Downloading pictures, music, etc.," "Other"). These rankings were weighted such that the activity ranked first was given a value of 5, the second ranked use was given a value of 4, and so on.

Second, Internet use specific to the environment was measured by combining responses to two Likert statements ("I use my computer online to get information about the environment" and "I exchange thoughts and ideas about the environ-

---

[5]Putnam places a "progressive" valence on the concept of social capital and the same has been done in this chapter. This is not to say, however, that given his definition of social capital, ". . . networks, norms, and social trust that facilitate coordination and cooperation for mutual benefit," other less progressive attitudes and behaviors could not be understood as being "high" in social capital. For example, consider members of the Wise Use movement. This is a movement that is often called "antienvironmental" and, according to the Policy Objectives of the Wise Use Movement document on the Wild Wilderness (2005) Web site, it has two basic tenets: "all constraints on the use of private property should be removed" and "access to public land should be unrestricted" (www.wildwilderness.org/wi/wiseuse.htm). Individuals who consider themselves adherents to the Wise Use movement no doubt have networks, norms, and a trust (at least among themselves) that allow for coordination and cooperation in the name of mutual benefit. This chapter does not examine this particular "take" on environmental social capital, but its exclusion should not imply that this is an unreasonable understanding of the concept. That said, it is logical to assume that the vast majority of those who score high on the NEP scale would fall into the "proenvironmental" rather than the "antienvironmental" category.

ment with others when I'm online") with responses to three open-ended questions ("These are the environmental and/or nature-themed WWW sites that I like to visit"; "These are the online and/or nature-themed things that I like to do online"; "These are the other environmental and/or nature-themed things that I like to do online," where no response received zero, one response received one and more than one response received two). The reliability for this scale was $\alpha = .57$. See Table 10.1 for summary statistics.

Television Viewing.   Quantity of television viewing was measured using a six-statement Likert scale (ranging from "Strongly agree" to "Strongly disagree") with statements such as "I watch less television than most people I know" and "One of the first things I do in the evening is turn on the television." Shrum, Burroughs, and Rindfleisch (2005) developed the scale and found that it had a high reliability in an initial study ($\alpha = .87$) and a slightly lower reliability on a subsequent study ($\alpha = .78$). The scale had a single dimension with an average loading of .72 in the first study and .69 in the subsequent study. The scale performed similarly in this study with high reliability ($\alpha = .86$) and a single dimension with an average loading of .77.

RESULTS

The first research question asks about the relationship between Internet use for information and environmental attitudes; the second research question asks about the relationship between the use of the Internet for social/recreation purposes and environmental attitudes. In order to get an overview of the relationship between environmental attitudes and Internet use, a regression model was run with the NEP scale as the dependent variable and the following independent variables: demographic variables (sex, age, income, education, area where one lives, language)[6] in the first block and each of 10 online activities: "E-mail," "Looking for/reading news," "Looking for/reading sports," "Surfing the WWW," "Chatting with friends," "Chatting with people met online," "Games," "Shopping," "Downloading pictures, music, etc.," "Other," in the second block.

The online activities of "Looking for/reading sports" ($\beta = -.20$, $p < .01$), "Chatting with friends" ($\beta = -.14$, $p < .05$) and "Looking for/reading news" ($\beta = -.14$, $p < .05$), were found to be negative predictors of environmental attitudes (i.e., greater use of the Internet for these activities was related to lower scores on the NEP). Means for Internet use and environmental attitudes were also calculated and there is a significant negative relationship between Internet use and scores on the NEP for the general sample (i.e., as time spent on the Internet increases, scores on the NEP decrease). See Table 10.2 and Table 10.3.

---

[6]These six demographic variables were used as controls in all regression and correlation analyses.

TABLE 10.1
Summary Statistics: General and Environmentalist Samples

Variable	Mean		Standard Deviation		Maximum		Minimum	
	General	Environmentalist	General	Environmentalist	General	Environmentalist	General	Environmentalist
Weekly Internet use (based on a combination of weekday and weekend hours)	3	2.4*	4	2.8	32	16	0	0
Internet use for environmental purposes (2-statement scale; 1–5 point Likert + 3 yes/no questions)	4	4.4**	2	2.6	11	16	2	0
NEP scale (15-statement scale; 1–5 point Likert scale)	50.5	59.3***	11.2	9.3	74	75	21	29
Weekly television viewing (based on a combination of weekday and weekend hours)	7.6	6.8*	5	5.1	30	40	0	0

Note. 2-tailed *t* test: * = difference $p < .05$; ** = difference $p < .01$; *** = difference $p < .001$.

TABLE 10.2
Regression Analysis: Internet Use and Environmental Attitudes

Independent Variable	Dependent Variable	β	t	p <
(General sample)				
Internet for sports	Environmental attitudes	−.2	−3.1	.01
Internet for chatting (with friends)	"	−.14	−2.1	.05
Internet for news	"	−.14	−2.2	.05
(Environmentalist sample)				
Internet for environmental info.	"	.14	2.6	.01

Notes.

- All regressions were run with the following in the first block: sex (1 = male, 2 = female), age, education (1 = Jr. high/middle school to 6 = postgraduate work), income (1 = less than $10,000 to 5 = more than $76,000), language (1 = English as first language, 2 = other first language), and area lived in (1 = urban, 2 = suburban, 3 = rural). Gender is also positively related with NEP in each of the above analyses (i.e., women are more likely to have environmentally positive attitudes).
- The following were run in the second block: Internet for sports, news, sports, chatting with friends, chatting with strangers, games, shopping and downloading (see p. XXX for more details on how this analysis was run).
- Only significant results are presented here.
- R^2 adjusted (each independent variable was run in a separate regression): Sports = .1; Chatting (friends) = .08; News = .08; Environmental information = .03.

TABLE 10.3
Means: Internet Use and Environmental Attitudes

General Sample		Environmentalist Sample	
Internet	Environmental Attitudes (NEP)*	Internet	Environmental Attitudes (NEP)
High (6.6 hrs/wk)	47.8	High (5.4 hrs/wk)	59.2
Medium (1.7 hrs/wk)	50.8	Medium (1.5 hrs/wk)	60
Low (barely use)	53.3	Low (barely use)	58.6

Note.   * ANOVA of differences between scores significant for the general sample ($p < .01$) but not for the environmentalist sample; NEP score ranges: 21–74 (general sample); 29–75 (environmentalist sample).

The third research question asked about the relationship between time spent watching television and time spent on the Internet. The results indicate that there is a positive correlation between television and Internet hours for the general public ($r = .27, p < .001$). In particular, "Downloading pictures, music, etc.," was positively related to television viewing for the general public ($r = .17, p < .01$). See Table 10.4.

TABLE 10.4
Partial Correlations Between Television and Internet Use

	General Sample	Environmentalist Sample
Overall Internet use/TV	$r = .27, p < .001$	ns
E-mail/TV	ns	$r = -.14, p < .01$
Internet news/TV	ns	$r = -.1, p < .05$
Surfing for information/TV	ns	$r = -.1, p < .05$
Downloading (pictures, music, etc.)/TV	$r = .17, p < .01$	ns

The fourth research question asks about potential differences between Internet use and environmental attitudes for the general population versus Internet use and environmental attitudes for environmentalists. Although the online activities of "Looking for/reading sports," "Chatting with friends," and "Looking for/reading news" were negative predictors of environmental attitudes for the general population, the only significant Internet use predictor of environmental attitudes for the environmentalist sample was the composite measure of "Use of the Internet for environmental purposes" ($\beta = .14, p < .01$). In other words, greater use of the Internet for environmental purposes was related to higher scores on the NEP for the environmentalist sample. And although there is a significant negative relationship between Internet use means and mean scores on the NEP for the general sample, there was no significant relationship between Internet use means and mean scores on the NEP for the environmentalist sample.

Finally, although overall Internet use was positively correlated with television viewing for the general sample, these activities were not related for the environmentalist sample. An exploration of the relationships between specific Internet activities and television viewing showed that "E-mail" ($r = -.14, p < .01$), "Looking for/reading news" ($r = -.1, p < .05$), and "Surfing the WWW" ($r = -.1, p < .01$) were all negatively related to television viewing for the environmentalist sample, whereas these relationships were not related for the general public (there was a positive correlation between "Downloading pictures, music, etc." and television viewing for the general sample). The tables listed have results for both samples.

DISCUSSION

Robert Putnam's (1995, 2001) writings on declining social capital in the United States have served as catalysts for a variety of research that has explored the relationship between mediated communication and social capital. Some of that research has focused on Internet use and implications for social capital, and it is into this category that the current research falls.

As Shah and her colleagues highlight (Shah, Kwak, et al., 2001), the broad reasons for assuming that there might be a relationship have had to do with time displacement (i.e., that perhaps the consumption of mediated communication was
happening at the expense of other "social activities") and content (i.e., that perhaps what was being consumed was not positively related interest in and concern
about the "social good"). Research that has focused on the relationship between
Internet use and politics (i.e., political participation, efficacy, and knowledge) as a
way of exploring the Internet and social capital, has found that the relationship
between the Internet and social capital is very much dependent on how the
Internet is being used. General Internet use for information (i.e., not specifically
political information use) is related to increased political efficacy, whereas
Internet use for recreation is related to decreased political efficacy (Scheufele &
Nisbet, 2002). Internet use for information exchange has also been positively associated with civic participation (Shah, McLeod, et al., 2001) as well as civic engagement, trust, and contentment—all considered to be aspects of social capital (Shah,
Kwak, et al., 2001). Internet use for social recreation, however, has been negatively
associated with civic engagement, trust, and contentment (Shah, Kwak, et al.,
2001).

In this chapter, it has been proposed that in the same way that the relationship
between Internet use and social capital can be manifest through politics, the relationship between Internet use and social capital can be manifest through attitudes
about the natural environment. And the results indicate that this may in fact be
the case. For the general sample, two of the recreational Internet uses, "Looking
for/reading sports" and "Chatting with friends," did show negative relationships
with environmental attitudes (i.e., the higher the Internet use for "Looking for/
reading sports" and "Chatting with friends," the lower the score on the NEP).
However, a type of Internet use that we might expect to fall into the *information*
category, "Looking for/reading news," was also negatively related to environmental attitudes. For the environmentalist sample, only one use of the Internet—"For
environmental purposes"—is significantly related to environmental attitudes
(i.e., the greater the use of the Internet for environmental purposes, the higher the
scores on the NEP).

Given that the data are cross-sectional, there are no conclusive comments that
can be made about causality (i.e., when people make use of the Internet they may
be reinforcing or changing their beliefs and ways of being in the world), although
it is easier to offer a possible explanation for the finding that Internet use "For environmental purposes" and environmental attitudes are related than to try and explain why use of the Internet for sports, news, and chatting with friends would be
negatively related to environmental attitudes.

It is reasonable to assume that those who are already involved with environmental issues (i.e., in the case of this research, the environmentalists who are
current members of an environmental organization), turn to the Internet for information. Whether those who are "more" environmentally concerned are that

much more likely to turn to the Internet for information and to get involved, or whether Internet use "causes" heightened environmental awareness and involvement (perhaps due to the surprising ease with which information can be obtained, and involvement can be undertaken), is not possible to ascertain from these data. What is clear is that there is some sort of relationship between Internet use for environmental purposes and proenvironment attitudes.[7] If we understand the environment as an issue related to social capital, then the Internet seems to be playing a positive role in the development, or at the very least, maintenance, of social capital.

For the general sample, we can surmise about the negative relationship between Internet use for sports and lower scores on the NEP. Perhaps sports-related Internet use displaces activities that would otherwise have taken place outdoors—and it is not unreasonable to propose that time spent outdoors can be related to positive attitudes about "the outdoors" or the environment (see R. Kaplan & S. Kaplan, 1989). As for the negative relationship between chatting with friends online and lower NEP scores, perhaps online chatting displaces face-to-face conversations that would have been richer, fuller, and more likely to help with the analysis of difficult topics such as environmental issues (i.e., Kraut et al.'s, 1998, notion that Internet use can replace strong personal face-to-face ties with weak impersonal Internet ties).

But what about the negative relationship between "Looking for/reading news" on the Internet and lower scores on the NEP for the general sample? The initial assumption was that news on the Internet was "information" (i.e., as opposed to social/recreational Internet use) and would be positively associated with environmental attitudes (i.e., based on Internet and social capital studies by Shah, Kwak, et al., 2001, and Shah, McLeod, et al., 2001). One explanation might be that the online news sources are often associated with television news sources (i.e., CNN, Fox) and serve essentially as mirrors for the television news stories. We do know from television news content analyses that environmental issues are more likely to be covered on the news than in fictional programming. About 10% of news coverage had an environmental focus in the mid-1990s compared to about 2% in entertainment programming (McComas, Shanahan, & Butler, 2001; Shanahan & McComas, 1997, 1999). We also know that the coverage of environmental news tends to be cyclical (Downs, 1972) and sensational (see Shanahan & McComas, 1999, for an overview of studies). In other words, although we might want to put "news" together with "information," it may very well be more similar to "entertainment."

---

[7]This finding distinguishes the current study from Scheufele and Nisbet's (2002) research in that they found that nonpolitical online information seeking was positively related to political efficacy, whereas online political activities were not related to political efficacy. The current research found that online environmental activities were positively related to NEP scores, whereas general online information seeking was either unrelated to NEP scores, or, in the case of the general sample, negatively related.

Another way in which Internet use could be related to environmental attitudes is through the Internet's displacement of television viewing. Previous research has shown a relationship between heavy viewing of television and a lack of environmental concern (Good & Shanahan, 2005; Shanahan, 1993; Shanahan & McComas, 1999; Shanahan, Morgan, & Stenbjerre, 1997). The results from this study indicate that, overall, there is a positive correlation between television and Internet hours for the general sample, but not for the environmentalist sample.

An exploration of the relationship between specific Internet activities further indicates differences between the samples: "E-mail," "Looking for/reading news," and "Surfing the WWW" were all negatively related to television viewing time for the environmentalist sample, whereas "Downloading pictures, music, etc.," was positively related to quantity of television viewing for the general sample. It would appear that the environmentalist sample is making use of the Internet for information purposes and that this use of the Internet is displacing television viewing. The general sample is making use of the Internet for entertainment purposes and this use is being added to television viewing (interestingly, a "uses and gratifications of television versus the Internet perspective" might cause us to predict otherwise). For example, researchers have found that the Internet is more likely to be used as an information medium and television as a medium for entertainment (Kaye, 1998).

A logical prediction, therefore, might be that when the Internet is in fact being used for information, as in the case of the environmentalist sample, television would continue to be turned to for entertainment. However, if the Internet is being used for entertainment, we might predict that such activity would displace television. That said, it would seem that environmentalists are turning to the Internet for environmental purposes and that these uses are displacing television. Both of these elements might help to explain why Internet use is related to higher NEP, and thus an increase in social capital, for those already involved in environmental issues: the Internet is providing a new source of information and connection with others, while decreasing the time available for television's cultivation of environmental apathy. Those in the general sample, however, are adding their Internet time to their television viewing, which means that television's cultivation of environmental apathy can continue to take place, and that Internet use may be displacing other activities (perhaps outdoor activities, interpersonal interaction, civic involvement, etc.) that might be related to environmentally positive attitudes.

LIMITATIONS AND FUTURE RESEARCH

One concern with the research presented here relates to the response rates. As was mentioned earlier, the response rates in this chapter—environmentalists responded at a rate of 49% whereas the general sample represents a 34% response rate—are in keeping with current expectations for survey response rates, response rates that have been declining steadily in the past few years (Teitler et al., 2003). That said, these response rates do leave the data susceptible to criticism of nonresponse bias. Research has shown that such things as the number of contacts (Schaefer & Dillman, 1998), level of personalization (Schaefer & Dillman, 1998) and incentives (Church, 1993) can have a significant positive impact on response rates. However, the resources were not available in the current study for these endeavors. Additionally, although it is clear that nonresponse bias is a problem, research such as Teitler et al. (2003) highlighted that even when the resources are available and response rates are very high, "significant nonresponse bias remains" (p. 137). The authors point out that research in this area needs to focus less on the money needed to sway individuals to become involved with survey research and more on learning "about the cognitive process underlying survey participation in today's world" (p. 137).

Another data-related concern is that given the relatively high mean age of the survey population, Internet use would be low. However, 64% of the environmentalist sample and 67% of the general sample made some use of the Internet— numbers that, according to The Pew Internet and American Life Project (2005), are very much in keeping with national averages for Internet use. Additionally, only 31% of the general sample was female. This is not in keeping with national averages for Internet use, which estimate that overall Internet use is evenly split between men and women (The Pew Internet and American Life Project, 2005). Future research could focus on possible differences between specific demographic categories in terms of how they use the Internet, watch television, and think about the natural environment.

A question might also be raised about what scores on the NEP scale tell us about the individual's actual environmental proclivities. Numerous studies that have made use of the NEP scale have found links between scores on the scale and environmental behaviors. For example, Roberts and Bacon (1997) found that the self-report of purchasing "environmentally friendly" products, reducing energy use, and recycling were all related to various aspects of the NEP. Other researchers (Ebreo, Hershey, & Vining, 1999; Vining & Ebreo, 1992) have also found the NEP to be related to recycling behavior. Schultz and Zelezny (1998) looked at the NEP (as well as other attitudinal measures) and five proenvironmental behaviors (recycling, using public transit, conserving energy, conserving water, purchasing environmentally friendly products) in Mexico, Nicaragua, Peru, Spain, and the United States. The researchers found that scores on the NEP were related to a composite environmental behavior score (minus the public transit) for Mexico, Spain, and most strongly, the United States.

In general, the research presented here is offered as foundational research into an unstudied area. As such, these initial glimpses are not intended to supply definitive conclusions or provide large sweeping generalizations, but rather to impart important information for the construction of future questions and the framing of future research. In addition to investigating differences between demographic

subgroups, future researchers could do time and activity use comparisons between self-identifying environmentalists who make extensive use of the Internet
and those who do not. Are those who do not use the Internet more likely to be involved with "on the ground" environmental activities? When environmentalists
do make use of the Internet, what exactly are they doing? How does this detailed
picture of individual usage patterns relate to actual involvement with environmental issues and proenvironment behaviors in day-to-day life?

CONCLUSION

Of all forms of mediated communication, the Internet could arguably hold the
most potential as an enhancer of social capital. The Internet gives us the capacity
to readily communicate with others and to access a huge wealth of information.
However, as indicated by the following statement, Putnam (2001) is not convinced: "The Internet is a powerful tool for the transmission of information
among physically distant people. The tougher question is whether that flow of information itself fosters social capital and genuine community" (p. 172). That said,
Putnam (2001) does offer that the "growth of telecommunication, particularly the
Internet" is one of the "clearest exceptions to the trend toward civic disengage-
ment" (p. 180).

One concern that Putnam (2001) has regarding the Internet, and modern communication technology generally, is that it has become easy to make a "show" of
being an involved citizen, without actually being involved. The example that
Putnam highlights is the environmental movement. "[Those who belong to environmental organizations] are valued supporters and genuine rooters for environmentalism as a good cause, but they are not themselves active in the cause. They
don't see themselves as movement foot soldiers in any sense . . ." (Putnam, 2001,
p. 158). Therefore, although the research presented here seems to offer the possibility, at least, that the Internet can facilitate environmental social capital for those
who have a predisposition for such things, perhaps use of the Internet for environmental purposes by "environmentalists" has become a crutch. Perhaps those same
people who sit surfing the Web used to be out interacting with real, diverse, people, helping to create real change. "[T]he Internet can be used to reinforce real,
face-to-face communities, not merely displace them with a counterfeit 'virtual
community' . . . [however, to] build bridging social capital requires that we transcend our social and political and professional identities to connect with people
unlike ourselves" (Putnam, 2001, p. 411).

In the end, Putnam hopes that his exploration of social capital has created debate about how to make institutions "social capital friendly." I too hope that this
initial foray into Internet use as a tool for environmental social capital will encourage others to continue the exploration.

REFERENCES

Bogo, J. (1999). Shop 'til you drop. *E Magazine, 10*(4), 48–51.

Bullard, J. (1998). Raising awareness of local agenda 21: The use of Internet resources. *Journal of Geography in Higher Education, 22*(2), 201–210.

Church, A. (1993). Estimating the effects of incentives on mall survey response rates: A meta-analysis. *Public Opinion Quarterly, 57*(1), 62–79.

Dordoy, A., & Mellor, M. (2001). Grassroots environmental movements: Mobilization in an information age. In F. Webster (Ed.), *Culture and politics in the information age: A new politics?* (pp. 167–182). New York: Routledge.

Downs, A. (1972). Up and down with ecology: The "issue attention cycle." *The Public Interest, 28,* 38–50.

Dunlap, R. E., & VanLiere, K. D. (1978). The "new environmental paradigm": A proposed measuring instrument and preliminary results. *Journal of Environmental Education, 9,* 10–19.

Dunlap, R. E., VanLiere, K. D., Mertig, A. G., & Jones, R. E. (2000). Measuring endorsement of the New Ecological Paradigm: A revised NEP scale. *Journal of Social Issues, 56*(3), 425–442.

Ebreo, A., Hershey, J., & Vining, J. (1999). Reducing solid waste: Linking recycling to environmentally responsible consumerism. *Environment and Behavior, 31*(1), 107–135.

Ecomall. (2005). Accessed June 6, 2005, at www.ecomall.com

Environment information systems in sub-Saharan Africa: An Internet resource. (1995). *Bulletin of the American Society for Information Science, 21*(4), 24–25.

Feidt, W., & Roos, C. (1995). Environmental information resources on the Internet. *Bulletin of the American Society for Information Science, 21*(4), 22–23.

Gardyn, R. (2001, January). Saving the earth, one click at a time. *American Demographics,* 30–33.

Good, J. (2005). *Television, materialism and environmental attitudes.* Unpublished manuscript.

Good, J., & Shanahan, J. (2005). *Television, the Internet and environmental attitudes: A GSS study.* Unpublished manuscript.

Guerin, T. (2001). Environmental monitoring in soil contamination and remediation programs: How practitioners are using the Internet to share knowledge. *Journal of Environmental Monitoring, 3*(3), 267–272.

Harkinson, J. (2000, July/August). Online and active: Green groups get results from the Internet. *E: The Environmental Magazine, 11*(4), 12–14.

Howard, B. (2001). EcoISP. *E Magazine, 12*(6), 11–12.

Howard, P., Rainie, L., & Jones, S. (2001). Days and nights on the Internet: The impact of a diffusing technology. *American Behavioral Scientist, 45*(3), 383–404.

Kangas, J., & Store, R. (2003). Internet and teledemocracy in participatory planning of natural resources management. *Landscape and Urban Planning, 62*(2), 89–101.

Kaplan, R., & Kaplan, S. (1989). *The experience of nature: A psychological perspective.* New York: Cambridge University Press.

Kay, R., & Christie, P. (2001). An analysis of the impact of the Internet on coastal management. *Coastal Management, 29*(3), 157–181.

Kaye, B. (1998). Uses and gratifications of the World Wide Web: From couch potato to Web potato. *The New Jersey Journal of Communication, 6*(1), 21–40.

Kirby, A. (2002). Internet 'best' for green news. *BBC News.* Retrieved October 16, 2002, from http://news.bbc.co.uk/2/hi/science/nature/2290380.stm

Kraut, R., Patterson, M., Lundmark, V. Kiesler, S., Mukopadhyay, T., & Scherlis, W. (1998). Internet paradox: A social technology that reduces social involvement and psychological well-being? *American Psychologist, 53*(9), 1017–1031.

McComas, K., Shanahan, J., & Butler, J. (2001). Environmental content in prime-time network and TV's non-news entertainment and fictional programs. *Society and Natural Resources, 14,* 533–542.

Meisner, M. (2000). E-Activism: Environmental activists are using the Internet to organize, spoof and subvert. *Alternatives, 26*(4), 34–38.

Millennium Ecosystem Assessment. (2005). Going beyond our means: Natural assets and human well-being (Statement of the MA Board). Retrieved October 4, 2005, from www.millenniumassessment.org/eu/index.aspx

Nie, N., & Erbring, L. (2000). Internet and society: A preliminary report. Retrieved April 8, 2003, from http://www.stanford.edu/group/siqss/press_release/internetstudy.html

Nisbet, M., & Scheufele, D. (2002). *Internet use and political participation: Political talk as a catalyst for on-line democracy*. Manuscript submitted for publication.

Oko, D. (2000). Rockers honor the earth. *E Magazine, 11*(1), 18–19.

Pew Internet and American Life Project: Recent trends. (2005). Retrieved January 8, 2005, from http://www.pewinternet.org/trends.asp.

Pickerill, J. (2001). Weaving a green web: Environmental protest and computer-mediated communication in Britain. In F. Webster (Ed.), *Culture and politics in the information age: A new politics?* (pp. 142–166). New York: Routledge.

Putnam, R. (1995). Bowling alone: America's declining social capital. *Journal of Democracy, 6*(1), 65–78.

Putnam, R. (2001). *Bowling alone: The collapse and revival of American community*. New York: Touchstone.

Ratza, C. (1996). The Great Lakes Information Network: The region's Internet information service. *Toxicology and Industrial Health, 12*(3–4), 557–561.

Rittner, D. (1992). *Ecolinking: Everyone's guide to online environmental information*. Berkeley, CA: Peachpit Press.

Roberts, J., & Bacon, D. (1997). Exploring the subtle relationships between environmental concern and ecologically conscious consumer behavior. *Journal of Business Research, 40*, 79–89.

Schaefer, D., & Dillman, D. (1998). Development of a standard e-mail methodology: Results of an experiment. *Public Opinion Quarterly, 62*(3), 378–397.

Scheufele, D., & Nisbet, M. (2002). Being a citizen online: New opportunities and dead ends. *The Harvard International Journal of Press/Politics, 7*(3), 55–75.

Schultz, P. W., & Zelezny, L. (1998). Values and proenvironmental behavior: A five-country survey. *Journal of Cross-Cultural Psychology, 29*(4), 540–558.

Shah, D., Kwak, N., & Holbert, R. L. (2001). "Connecting" and "disconnecting" with civic life: Patterns of Internet use and the production of social capital. *Political Communication, 18*, 141–162.

Shah, D., McLeod, J., & Yoon, S. (2001). Communication, context and community: An exploration of print, broadcast and Internet influences. *Communication Research, 28*(4), 464–506.

Shanahan, J. (1993). Television and the cultivation of environmental concern: 1988–1982. In A. Hansen (Ed.), *The mass media and environmental issues* (pp. 181–197). Leicester, England: University of Leicester Press.

Shanahan, J., & McComas, K. (1997). Television's portrayal of the environment: 1991–1995. *Journalism and Mass Communication Quarterly, 74*(1), 147–159.

Shanahan, J., & McComas, K. (1999). *Nature stories, depictions of the environment and their effects*. Cresskill, NJ: Hampton Press.

Shanahan, J., & Morgan, M. (1999). *Television and its viewers. Cultivation theory and research*. Cambridge, England: Cambridge University Press.

Shanahan, J., Morgan, M., & Stenbjerre, M. (1997). Green or brown? Television and the cultivation of environmental concern. *Journal of Broadcasting and Electronic Media, 41*(3), 305–323.

Shrum, L. J., Burroughs, J., & Rindfleisch, A. (2005, December). Television's cultivation of material values. *Journal of Consumer Research, 32*.

South, J. C. (2001). Online resources for news about toxicology and other environmental topics. *Toxicology, 122*, 153–164.

Teitler, J., Reichman, N., & Sprachman, S. (2003). Costs and benefits of improving response rates for a hard-to-reach population. *Public Opinion Quarterly, 67*, 126–138.

United Nations Environment Programme Annual Report. (2003). Retrieved January 12, 2005, from http://www.unep.org

Vining, J., & Ebreo, A. (1992). Predicting recycling behavior from global and specific environmental attitudes and changes in recycling opportunities. *Journal of Applied Social Psychology, 22,* 1580–1607.

Voinov, A., & Costanza, R. (1999). Watershed management and the Web. *Journal of Environmental Management, 56*(4), 231–245.

Weeks, P. (1999). Cyber-activism: World Wildlife Fund's campaign to save the tiger. *Culture and Agriculture, 21,* 19–30.

Wellman, B., Haase, A. Q., Witte, J., & Hampton, K. (2001). Does the Internet increase, decrease, or supplement social capital? Social networks, participation, and community commitment. *American Behavioral Scientist, 45*(3), 436–455.

Wild Wilderness. Policy objectives of the Wise Use movement. (2005). Retrieved April 24, 2005, from www.wildwilderness.org/wi/wiseuse.htm

Zelwietro, J. (1995). Progressive activism on the Internet. *Alternatives, 21*(30), 16–17.

Zelwietro, J. (1998). The politicization of environmental organizations through the Internet. *The Information Society, 14,* 45–56.

CHAPTER ELEVEN

Catalyzing Environmental Communication Through Evolving Internet Technology

Arno Scharl
Graz University of Technology
Graz, Austria

Information technology rapidly changes established political and economic processes (Wriston, 2004). Interactive media such as the World Wide Web and wireless devices in particular revolutionize the reach, speed, and efficiency of communication between individuals and organizations alike—from simple electronic mailing lists to activists broadcasting protests live via palmtop computers, to running an organization remotely while climbing K2, the second highest mountain in the world (Ward, 2003). Broadband connectivity and the ubiquity of mobile communication devices trigger social change and catalyze advanced economic systems (Castells, 1993; Wellman, 2001, 2002). Besides improving productivity and reducing transaction costs, new communication technologies enable people to actively participate in decision making (Doering et al., 2002) and increasingly align networked information systems with the visions of hypertext pioneers such as Vannevar Bush, Douglas Engelbart, and Ted Nelson (Bardini, 2000; Bush, 1945; Nelson, 1993). The proliferation of information networks powers the transition to a knowledge-based economy, connects the world's poor to entrepreneurial and educational opportunities, and creates opportunities for businesses, government, and civil society to scrutinize each other in collaborative, consensus-building processes (Doering et al., 2002).

ENVIRONMENTAL INFORMATION FLOWS

The importance of environmental communication through scientific exchange, educational programs, and the media has been recognized internationally at least since the United Nations 1972 Conference on the Human Environment in Stockholm (UNEP, 1972). Environmental communication benefits from rapid advances in gathering, transmitting, processing, accessing, and analyzing information. These advances are transforming the way society handles the explosive growth but diminished lifespan of information (Bell, 1973; Haddad & Draxler, 2002; Weingart, 2002), particularly in highly dynamic domains such as the environment (Pick, Menger, Jensen, & Lethen, 2000).

Climate monitoring, accessing remote sensing and satellite data, querying search engines, and aggregating news feeds from international media exemplify the fundamental role of information systems in gathering environmental information (understood as data with meaning to the recipient within an environmental context). The transmission of such information is influenced by emerging information brokerage platforms and peer-to-peer (P2P) networks, enabling two or more participants to collaborate spontaneously without the need for central coordination (Schoder & Fischbach, 2003). Such collaboration is facilitated by Tim Berners-Lee's vision of the Semantic Web (Berners-Lee, Hendler, & Lassili, 2001) and emerging standards that enable data interoperability over the Internet (Knobloch & Kopp, 2003).

Graphic user interfaces and interactive visualizations revolutionize accessing and interpreting information to explore dynamic environmental processes (Tochtermann & Maurer, 2000). Geographic information systems (GIS) in particular are an important application of environmental informatics, which project information from the Earth's three-dimensional curved surface onto two-dimensional media such as paper and computer screens. GIS provide intuitive mechanisms to access and manipulate georeferenced data on both the micro and the macro level. By offering public interfaces to maps, satellite data, or topographic information from previous space shuttle missions, projects such as *NASA World Wind* (worldwind.arc.nasa.gov) or *Google Maps* (maps.google.com) document the enormous potential and visual appeal of georeferenced projections.

Despite their limited screen resolution, mobile devices are also popular to access georeferenced environmental indicators such as water quality, air pollutant concentrations, or radiation levels (Westbomke, Haase, Ebel, & Lehne, 2004). Location-based services enable the customized reporting of such indicators—such services detect the user's present location by identifying the radio cell, or via the Global Positioning System (GPS).

THE EVOLUTION OF INFORMATION NETWORKS

The World Wide Web emerged in 1991 as an energetic, chaotic, quickly evolving, and largely anonymous environment. The unique combination of these qualities has attracted many of the Internet trailblazers who developed and propagated the first Web applications (Bucy, Lang, Potter, & Grabe, 1999; Psoinos & Smithson, 1999).

Web sites offering static hypertext documents embody the simplest type of Web communication. Static Web development activities are restricted to comparatively simple authoring tasks, although scripting languages offer a range of options for structuring content and user interface representations. In contrast to static systems, *dynamic* Web sites provide searchable content and interfaces to databases and archived information. *Adaptive* Web sites represent the next step in Web evolution by distinguishing between different user categories to provide better support and facilitate the execution of online transactions. Newer generations of adaptive systems address particular users individually and customize multiple aspects of their online experience such as content, layout, and navigational mechanisms (Scharl, 2000).

Early applications often transformed existing printed material into digital form without exploiting the World Wide Web's full potential. In the meantime, many developers have moved beyond traditional Web publishing by incorporating content management systems, semantic markup languages such as XML (eXtensible Markup Language), and advanced interactive components. The following sections outline these technologies and their environmental applications.

Managing Environmental Content

By separating content production from programming logic and layout decisions, *Content Management Systems* ensure rapid implementation, affordable maintenance, and intuitive interfaces for users and content providers (Lerner, 2003). User registration and login management handle the read, write, and execute permissions of different classes of users. Graphical editors allow creating, editing, previewing, and annotating documents via standard Web browsers. Authors can also specify when articles should appear and expire.

Content management systems are fundamental to building state-of-the-art environmental portals. The public availability of environmental indicators through such portals creates awareness and contributes to more informed decision making (Cash et al., 2003; Meadows, 1998). Online databases such as the World Resource Institute's *EarthTrends* (earthtrends.wri.org) and the United Nations Environment Programme's *GEO Data Portal* (geodata.grid.unep.ch) synthesize information gathered from a variety of institutions, geographical regions, and economic sectors (Petkova et al., 2002). At the same time, government reports on company emissions and pollution registers such as *Pollution Watch* (www.pollutionwatch .org) and the *Toxics Release Inventory* (www.epa.gov/tri) urge the corporate sector to disclose environmental data comprehensively and transparently.

Presentational Versus Semantic Markup Languages

Although presentational markup languages such as the *Hypertext Markup Language* (HTML) have been extended with frames, tables, and other formatting constructs, they remain optimized for the document rendering requirements of simple Web applications. HTML provides rich facilities for display, but lacks facilities to manage metadata effectively (metadata = data about data, e.g., information about the author, document version, language, copyright, or type of representation). HTML imposes a lowest common denominator for document rendering and inextricably mixes the content and its presentation—the wording of a heading, for example, and its font size and color. The move from HTML to semantic markup languages such as the *eXtensible Markup Language* (XML) addresses the rigidity of fixed HTML tags (Baeza-Yates & Ribeiro-Neto, 1999; Vasudevan & Palmer, 1999). XML complements content management systems by enforcing the separation of content and its presentation as described in the previous section. XML is extensible, validatable by external modules, and allows the definition of self-documenting tags; e.g. <title>Environmental Communication Yearbook</title>.

The widespread adoption of semantic markup languages paves the way for environmental Web services (W3C, 2003), remote environmental resources whose capabilities can be accessed via the Internet. One example of representing and transmitting environmentally relevant information via open Web standards is the *Exchange Network* (www.exchangenetwork.net), a recent initiative by the U.S. Environmental Protection Agency (EPA) to exchange environmental data across network nodes and demonstrate the reliability of XML for day-to-day operations (Dwyer, Clark, & Nobles, 2002; Stein, 2003).

The Emergence of Interactivity

Technology and its environmental impact do not exist in isolation, but are embedded in sociotechnical systems and connected to networks of actors (Kleef & de Moor, 2004; Xiang, Madey, Huang, & Cabaniss, 2004). Promoting sustainability and protecting natural ecosystems requires effective communication between these actors—either directly through electronic mail and instant messaging, or mediated through interactive Web sites, Web logs ("blogs"), and online discussion forums.

Although interactivity represents an inherent feature of these information networks, many environmental organizations are not taking full advantage of the latest interactive technologies. Interactivity represents a continuum, rather than a nominal variable. It measures responsiveness, the extent to which communication reflects back on itself (Newhagen & Rafaeli, 1996; Rafaeli & Sudweeks, 1997), and delineates the "media's potential ability to let the user exert an influence on the content and/or form of the mediated communication" (Jensen, 1998, p. 201). Arguably, interactivity may promote change and spur social action by means of interpersonal and group communication.

The emergence of interactivity within online environments can support democratic decision structures. Public participation adds value to decision making by pointing out the relevance of neglected issues, and by contributing to an honest accounting of the costs and benefits for different parts of the society (Petkova et al., 2002). Voting, lobbying, participating in public hearings, and joining environmental groups are typical ways individuals can influence environmental decisions. Challenging the status quo and benefiting from the synergistic effects of collective action, environmental movements have been formed in response to the perceived lack of accountability by existing power structures in dealing with specific issues that the citizenry deems important (Kutner, 2000). These movements envision a more enlightened public that is better able to connect its actions to environmental consequences, more likely to support policies that minimize environmental harm, and more able to hold governments and corporations accountable for their environmental performance (WRI, 2003).

When individuals interact within electronic environments, they leave a tangible trace. Specifically, server log files contain a detailed transcript of discourse based on their activities, which is an important source of feedback for the organization hosting the platform. The examination of traffic patterns across Web documents, for example, can identify "dead space" within a Web site, prioritize new content development, and improve the system's accessibility and navigability (Sullivan, 1997). Lacking the required technical expertise, many environmental organizations overlook this opportunity to optimize their Web site, despite an extensive array of available log file analysis tools.

CONCLUSION

This chapter summarized advances in networked information technology and their significance for environmental communication. These advances facilitate the transition from broadcast to interactive communication (Scharl, 2000), promote collaboration (Rheingold, 2002), and enable online communities to deal effectively with complexity, uncertainty, and risk (Oepen, 2000).

Although environmental online communication usually has an immediate positive effect on agenda setting within the target group, its long-term impacts are hard to evaluate. To a large extent, the long-term impacts depend on the quality, professional representation, and credibility of communicated content. Constant progress monitoring and participatory peer review among network members and external experts improve the quality of content and its representation (Adhikarya, 2000). Additionally, the credibility of information is supported by transparent processes respecting professional norms and procedural fairness (Keohane & Nye, 1998).

The emergence of interactivity may help increase environmental awareness. Interactive information systems encourage networking, public environmental discourse, and the social construction of meaning (Meppem & Bourke, 1999).

They improve the quality of decisions, build trust in institutions, and help resolve conflict among competing interest groups (WRI, 2003). Although the availability of the Internet among citizens in poor countries still lags behind Western societies, grassroots activist movements and hitherto marginalized or oppressed segments of the population benefit from an expansion of their ability to access, analyze, create, and disseminate information (Kutner, 2000). In the Internet, they find "a means of entering onto the world stage, of presenting their situations in their own words, of expressing their claims independently of governments and the channels laid down by the large media groups" (Marthoz, 1999, p. 73). These opportunities help explain the increasing popularity of "social software," such as electronic discussion forums, Web logs (*blogs*), and jointly authored document collections (*wikis*).

Addressing the importance of networked information technology for environmental communication as outlined in this chapter, the ECOresearch Network (www.ecoresearch.net) organizes an annual track on "Environmental Online Communication" at the *Hawaii International Conferences on Systems Sciences* (www.ecoresearch.net/hicss). The following chapter (chap. 12) is an extended and revised version of a paper submitted to this track. "Participatory Design as a Learning Process: Enhancing Community-Based Watershed Management Through Technology" describes building community capacity as one of the most beneficial outcomes of inclusive public involvement. The chapter presents a detailed case study of how consultants designed a participatory process as a learning experience. A nonprofit watershed management community group in Pennsylvania gained capacity in using information technology, particularly Web-site development. This made the watershed planning process transparent and inclusive, and provided the community with sustainable tools for productively handling future challenges.

ACKNOWLEDGMENT

This work represents a joint initiative of the Research Network on Environmental Online Communication (www.ecoresearch.net), Graz University of Technology, and the Know-Center (www.know-center.at). The latter is being funded by the Austrian Competence Centers Program K+ under the auspices of the Austrian Ministry of Transport, Innovation, and Technology.

REFERENCES

Adhikarya, R. (2000). Mainstreaming the environment through environmental education, training and communication. In M. Oepen & W. Hamacher (Eds.), *Communicating the environment* (pp. 82–92). Frankfurt: Peter Lang.

Baeza-Yates, R., & Ribeiro-Neto, B. (1999). *Modern information retrieval.* Harlow: ACM Press Books.

Bardini, T. (2000). *Bootstrapping: Douglas Engelbart, coevolution, and the origins of personal computing.* Stanford: Stanford University Press.

Bell, D. (1973). *The coming of post-industrial society: A venture in social forecasting.* New York: Basic Books.

Berners-Lee, T., Hendler, J., & Lassili, O. (2001). The semantic Web. *Scientific American, 284*(5), 28–37.

Bucy, E. P., Lang, A., Potter, R. F., & Grabe, M. E. (1999). Formal features of cyberspace: Relationships between Web page complexity and site traffic. *Journal of the American Society for Information Science, 50*(13), 1246–1256.

Bush, V. (1945). As we may think. *The Atlantic Monthly, 176*(1), 101–108.

Cash, D. W., Clark, W. C., Alcock, F., Dickson, N. M., Eckley, N., Guston, D. H., Jäger, J., & Mitchell, R. B. (2003). Knowledge systems for sustainable development. *Proceedings of the National Academy of Sciences of the United States of America, 100*(14), 8086–8091.

Castells, M. (1993). The informational economy and the new international division of labor. In M. Carnoy, M. Castells, S. S. Cohen, & F. Cardoso (Eds.), *The new global economy in the information age: Reflections on our changing world* (pp. 15–44). University Park: Pennsylvania State University Press.

Doering, D. S., Cassara, A., Layke, C., Ranganathan, J., Revenga, C., Tunstall, D., & Vanasselt, W. (2002). *Tomorrow's markets—Global trends and their implications for business.* Washington: World Resources Institute.

Dwyer, C., Clark, C., & Nobles, M. (2002, April). *Web services implementation: The beta phase of EPA network nodes.* Paper presented at the 11th International Emission Inventory Conference, Atlanta.

Haddad, W. D., & Draxler, A. (2002). The dynamics of technologies for education. In W. D. Haddad & A. Draxler (Eds.), *Technologies for education: Potential, parameters and prospects* (Vol. 1–17). Paris: United Nations Educational, Scientific and Cultural Organization (UNESCO).

Jensen, J. F. (1998). Interactivity: Tracing a new concept in media and communication studies. *Nordicom Review, 19,* 185–204.

Keohane, R. O., & Nye, J. S. (1998). Power and interdependence in the information age. *Foreign Affairs, 77*(5), 81–94.

Kleef, R., & de Moor, A. (2004). Communication process analysis in virtual communities on sustainable development. In A. Scharl (Ed.), *Environmental online communication* (pp. 209–220). London: Springer.

Knobloch, M., & Kopp, M. (2003). *Web design with XML.* Chichester: Wiley.

Kutner, L. A. (2000). Environmental activism and the Internet. *Electronic Green Journal, 12.* Retrieved on June 17, 2005, from http://egj.lib.uidaho.edu/egj12

Lerner, R. M. (2003). At the forge: Content management. *Linux Journal, 2003*(118). Available at http://portal.acm.org/dl.cfm

Marthoz, J. P. (1999). Freedom of the media. In M. Tawfik, G. Bartagnon, & Y. Courrier (Eds.), *World information and communication report 1999–2000* (pp. 72–82). Paris: United Nations Educational, Scientific and Cultural Organization (UNESCO).

Meadows, D. (1998). *Indicators and information systems for sustainable development: A report to the Balaton Group.* Hartland Four Corners, VT: Sustainability Institute.

Meppem, T., & Bourke, S. (1999). Different ways of knowing: A communicative turn toward sustainability. *Ecological Economics, 30*(3), 389–404.

Nelson, T. H. (1993). *Literary machines 93.1.* Sausalito: Mindful Press.

Newhagen, J. E., & Rafaeli, S. (1996). Why communication researchers should study the Internet: A dialogue. *Journal of Computer-Mediated Communication, 1*(4). Available at http://www.ascusc.org/jcmc/

Oepen, M. (2000). Environmental communication in a context. In M. Oepen & W. Hamacher (Eds.), *Communicating the environment* (pp. 41–61). Frankfurt: Peter Lang.

Petkova, E., Maurer, C., Henninger, N., Irwin, F., Coyle, J., & Hoff, G. (2002). *Closing the gap: Information, participation, and justice in decision-making for the environment.* Washington: World Resources Institute.

Pick, T., Menger, M., Jensen, S., & Lethen, J. (2000, October). *Access to environmental information: Towards a digital global knowledge marketplace.* Paper presented at the 14th Symposium Computer Science for Environmental Protection, Bonn, Germany.

Psoinos, A., & Smithson, S. (1999). *The 1999 World Wide Web 100 survey.* London: London School of Economics.

Rafaeli, S., & Sudweeks, F. (1997). Networked interactivity. *Journal of Computer-Mediated Communication, 2*(4). Available at http://www.ascusc.org/jcmc/

Rheingold, H. (2002). *Smart mobs: The next social revolution.* Cambridge, MA: Perseus.

Scharl, A. (2000). *Evolutionary Web Development.* London: Springer. Available at http://webdev.wu-wien.ac.at/

Schoder, D., & Fischbach, K. (2003). Peer-to-peer prospects. *Communications of the ACM, 46*(2), 27–29.

Stein, B. (2003). *Solutions for the environmental information exchange network.* Colorado Springs: XAware.

Sullivan, T. (1997, June). *Reading reader reaction: A proposal for inferential analysis of Web server log files.* Paper presented at the 3rd Conference on Human Factors and the Web, Denver, CO.

Tochtermann, K., & Maurer, H. (2000). *Umweltinformatik und Wissensmanagement—Ein Überblick* [Environmental informatics and knowledge management—An overview]. Paper presented at the 14th Symposium Computer Science for Environmental Protection, Bonn, Germany.

UNEP. (1972). *Report of the United Nations Conference on the Human Environment.* Nairobi: United Nations Environment Programme.

Vasudevan, V., & Palmer, M. (1999, January). *On Web annotations: Promises and pitfalls of current Web infrastructure.* Paper presented at the 32nd Hawaii International Conference on System Sciences (HICSS-32), Hawaii.

W3C. (2003). *Scalable vector graphics (SVG) full 1.2 specification: W3C working draft 13 April 2005.* Retrieved on June 17, 2005, from http://www.w3.org/TR/SVG12/

Ward, M. (2003). *Scot aims for high office.* Retrieved on June 17, 2005, from http://news.bbc.co.uk/1/hi/sci/tech/1387082.stm

Weingart, P. (2002). The moment of truth for science. *EMBO Reports, 3*(8), 703–706.

Wellman, B. (2001). Computer networks as social networks. *Science, 293*(5537), 2031–2034.

Wellman, B. (2002). Designing the Internet for a networked society. *Communications of the ACM, 45*(5), 91–96.

Westbomke, J., Haase, M., Ebel, R., & Lehne, D. (2004). Mobile access to environmental information. In A. Scharl (Ed.), *Environmental online communication* (pp. 11–20). London: Springer.

WRI. (2003). *World Resources 2002–2004: Decisions for the earth—Balance, voice, and power.* Washington: World Resources Institute.

Wriston, W. B. (2004). Freedom and democracy in the information age. *Technology in Society, 26*(2–3), 321–325.

Xiang, X., Madey, G., Huang, Y., & Cabaniss, S. (2004). Web portal and markup language for collaborative environmental research. In A. Scharl (Ed.), *Environmental online communication* (pp. 113–126). London: Springer.

CHAPTER TWELVE

Participatory Design as a Learning Process: Enhancing Community-Based Watershed Management Through Technology

Umer Farooq, Cecelia B. Merkel, Lu Xiao, Heather Nash,
Mary Beth Rosson, and John M. Carroll
The Pennsylvania State University

In general, the need to develop approaches to increase public participation in technological decision-making processes and technical policy decisions is well documented (Laird, 1993). In environmental communication, finding ways to increase the quality of technical expertise, while simultaneously increasing the inclusivity of decision processes, is the fundamental challenge for achieving social legitimacy and working through environmental conflict (Daniels & Walker, 2001). While information technology (e.g., Web sites, e-mail, interactive maps) offers great potential for environmental groups (Scharl, chap. 11, this volume) and community organizations in general (e.g., Gurstein, 2002; Schuler, 1994), a specific instance of the *fundamental paradox* (Daniels & Walker, 2001) applies to our research. Community organizations often want to use technology in effective ways to achieve their mission, but as situations become more complex, fewer organizational members have the technical competence needed to actively participate and engage in technology-related issues.

The broader problem we wish to explore is how to empower community organizations to achieve their mission by facilitating their development of skills and capabilities related to information technology. We are interested in finding ways to work with community organizations to help them envision new roles for technology in their organization and to take on the task of doing technology projects to achieve their goals. We are aiming for long-term shifts in practice within the organization by helping them focus on finding ways to encourage technology design and learning that can be sustained even after we fade as process collaborators.

In this chapter, we want to develop a process for working with community organizations that gives them greater control over their information technology and increase their capacity to manage technology projects. We report on our experience in working with a watershed management environmental group in collaboratively redesigning their Web site. We present an integrated approach, focusing on design and learning, as a way to work with this group. This approach emphasizes participatory design as a learning process in which we introduce hierarchical (cognitive apprenticeship) and lateral (collaboratively constructed Zones of Proximal Development; ZPD) aspects of learning. Using our approach, the technical competence of the environmental group increased. Moreover, the group gained control over their technology and actively participated in and guided the Web site redesign process.

This chapter is structured as follows. Section 1 describes related work on participatory design. Section 2 talks about our approach of participatory design as a learning process. Section 3 describes our overall project, gives background on the environmental group we worked with, and details the research methods. Section 4 provides description and analysis of our collaboration with the environmental group. Section 5 concludes our work and describes its contribution to environmental communication. Section 6 discusses future work to extend our case study toward reusable solution schemas called patterns.

RELATED WORK

Participatory Design (PD), originally emerging from sociotechnical systems theory (Mumford, 1983), is an evolving practice among design professionals that explores conditions for user participation in the design and introduction of computer-based systems in organizations (for detailed discussion, see Clement & Van den Besselaar, 1993; Greenbaum & Kyng, 1991; Kensing & Blomberg, 1998; Schuler & Namioka, 1993). In addition to design, we are also interested in community groups learning about technology. Thus, we explore new roles in participatory design that focus on hierarchical modes of learning in the form of cognitive apprenticeship (Collins, Brown, & Newman, 1989) and lateral modes of learning that take the form of collaboratively constructed Zones of Proximal Development (Sawchuk, 2003). Broadly speaking, our research is related to two distinct pieces of literature: participatory design with community groups and participative work in environmental communication.

Related work in participatory design has documented experiences in working with nonprofit community groups (for general discussion, see Benston, 1990). For example, Trigg (2000) related the experience of working on database redesign and development for a small nonprofit staffed by an empowered workforce. A similarity between this project and ours is the close work association between the researcher and one staff member. In our case, researchers work closely with the organizational stakeholders. One major difference is that Trigg, after 6 months into the project, became a staff member in the community group whereas we, as researchers and collaborators, continued in a capacity that is participatory rather than full membership. Moreover, our eventual goal is to fade, having established sustainable practices in the community organization and taking away key lessons that would guide research forward in this area. *CAVEAT* (McPhail, Costantino, Bruckmann, Barclay, & Clement, 1998) is another participatory design project where students worked with a volunteer organization to create a prototype, using database software, which could become a sustainable organizational information system. The difference lies in our focus on sustainability and empowerment. We are trying to get the end users in community groups to guide the technology process and not to build some tool or artifact for them. We also want to encourage reflection throughout the process, related to organizational shifts in practice, learning, and knowledge management.

Related work in environmental communication bears resemblance to our participatory design approach. Work by Bruner and Oelshlaeger (1998) and Cooper (1996), for instance, are key pieces related to environmental rhetoric. On the subject of power and expertise in public deliberation, Katz and Miller (1996) and Waddell (1996) specifically discussed environmental communication issues. Given the depth and breadth of the environmental communication domain, an elaborated literature review requires an extensive discussion that cannot be accommodated in this chapter. Instead, we refer to two specific pieces of work that deal with similar learning approaches in the area of environmental communication.

In their book, Daniels and Walker (2001) outlined a collaborative learning framework that addresses the fundamental complexity and controversy that define public policy decisions. The roots of collaborative learning are conflict management, learning theory, and systems thinking. Daniels and Walker's underlying premise is that social learning is a foundation for good decision making. Similarly, Laird (1993) proposed participatory analysis that requires a particular kind of learning process while people or groups are engaged in participation. Specifically, in this view, it is not enough that participants simply acquire new facts. They must begin, at some level, to be able to analyze the problem at hand, and eventually begin to learn how and when to challenge the validity of the asserted facts.

Although many parallels may exist among collaborative learning (Daniels & Walker, 2001), participatory analysis (Laird, 1993), and our approach, one difference is that participatory design as a learning process mandates that both design and learning are essential for community groups to build capacity in using and sustaining information technology. Participatory design may be distinguished from collaborative learning and participatory analysis in three ways: (a) it improves the knowledge base on which systems design is based, (b) it enables users to develop realistic expectations and thus reduce resistance to change, and (c) it increases workplace democracy by giving individuals the right to participate (Randall & Rouncefield, 2004). Our approach to participatory design as a learning process is elaborated in the next section.

PARTICIPATORY DESIGN AS A LEARNING PROCESS

As an old adage goes, "Tell me and I forget, show me and I remember, involve me and I understand." This is the guiding principle of our participatory design philosophy: to involve users in the design so that the process is sustainable and replicable. During this process, end users will learn about information technology, how to better use and sustain it, and how the process itself may become an integral constituent of their organizational practice.

Our project builds on previous work that takes a long-term participatory design approach in designing information systems to address local needs (Carroll, Chin, Rosson, & Neale, 2000). The assumption underlying this approach is that our community partners are active technology users, implementers, and learners. We also see users as the shapers and decision makers in their organization. We go beyond traditional participatory design models that seek to make users active participants in the design process. Our goal is for the community partners to take control of the design process itself by directing what should be done, by taking a central role in the doing, and by maintaining the technology infrastructure.

Learning is a prerequisite for our goal of helping the groups develop sustainable methods for promoting technology learning and planning in their organization. Therefore, we connect learning to the participatory process of design in two ways: in a hierarchical sense as a form of cognitive apprenticeship per Collins et al. (1989); and as a more lateral co-construction of areas of learning, or ZPD (Sawchuk, 2003). A ZPD is the area of learning bounded by what a learner can do with assistance from a more knowledgeable other and what a learner has mastered sufficiently to do alone. What follows is an elaboration on the two types of learning.

Cognitive Apprenticeship: Hierarchical Aspects of Participatory Design

Although cognitive apprenticeship was aimed at revamping pedagogical techniques in formal education, we believe that the idea can be adapted to account for hierarchical elements of the learning process in participatory design. *Cognitive apprenticeship* refers to the learning through guided experience on cognitive and metacognitive, rather than physical, skills and processes. Apprenticeship focuses closely on the specific methods for carrying out tasks in a domain. Apprentices learn these methods through a combination of what Lave and Wenger (1991) called observation, coaching, and practice, or what Collins et al. (1989) referred to as modeling, scaffolding, and fading. The idea is that learners start out on the periphery observing the activities and thought processes of experts and then slowly take on more responsibility for activities through the scaffolding (coaching) provided by experts. Eventually, learners take on full responsibility for carrying out activities with minimal help from experts.

In our work with community groups, we follow a more reciprocal form of cognitive apprenticeship, where participants have expertise in different domains. As researchers, we are design experts, whereas our community partners are experts in issues related to their organization. For example, in our work with the environmental group, we have expert knowledge of designing a Web site interface whereas members of that group have expert knowledge about the Web site content. In collaborating with this environmental group, we as members of a research group and they as members of a community group are both apprentices because we are coaching each other on different skills and knowledge domains.

Co-Construction of ZPD: Lateral Aspects of Participatory Design

We found that in our collaborations, there was a need to capture more lateral forms of learning in the participatory design process. There emerged areas for learning in which neither group was expert. As an example, during the design process, we needed to negotiate a functional collaborative space with the environmental group in which to work. The space had to allow, at the very least, for effective interpersonal contact, work on the technical aspects of the environmental group's Web site, learning about information technology by members of that group, and collection of data for the researchers' agenda. In this case, because we needed to work together to define the scope of our relationship and our work, the hierarchical relationships implied by the apprenticeship model was not the most appropriate model for the lateral learning that took place.

A more appropriate way to capture lateral learning is in the work of Peter Sawchuk (2003), who described a *co-constructed ZPD*. Sawchuk argues that a ZPD typically follows an expert–novice relationship in which a more knowledgeable other assists a learner in moving from emergent skill in an area to mastery of it. He suggests that a ZPD does not require this sort of relationship. Two novice learners intersubjectively construct a ZPD and then may work through it by calling on their combined skill sets, engaging in information-gathering activities, and calling on skilled others when necessary. He writes, "Each participant contributes to the formation of the conditions for the other's knowledge production process. It is analogous to seeing two people build a scaffold, communicating and working together to form a structure on which they both climb to new heights" (Sawchuk, 2003, p. 299).

Hierarchical and Lateral Relationships in Participatory Design

We advocate a more complicated view of participatory design that takes into account hierarchical and lateral relationships between the end users in community organizations and ourselves. At one end, we are coaching end users to develop sustainable methods for technology design and learning, and they are coaching us in terms of what is important for their organization (hierarchical shifts in learning). At the same time, much of the learning that is occurring involves mutual learning in which we and they work to define the scope of our work, the meaning of technology in this setting, and what it means to carry out a design process (lateral shifts in learning). For both types of learning models, a key element is the social context in which the learning takes place. In the hierarchical form of learning, apprentices are actually embedded in the subculture of target skills. In more lateral learning, the social context is the arena in which the ZPD is constructed and defines the rules by which it is done. In our participatory design process, we are working on authentic tasks that are defined and controlled by the community organization, which creates a socially conducive environment for leveraging the skills that different participating members bring to the table.

BACKGROUND AND RESEARCH METHODS

Background of Civic Nexus Project

As part of a 3-year research project known as Civic Nexus, we are working with community organizations to increase their ability to solve local community problems by leveraging and enhancing their capacity to use information technology (see Merkel et al., 2004, 2005). We work with about four community organizations each year. In negotiating our role with these groups, we have minimized our role as technology experts and have instead taken on the roles of facilitators and collaborators. For example, we will not create and deliver a Web site but will point the community organization to tools they might use to create and maintain a Web site (e.g., Microsoft Front Page) and help them think through some of the related technology decisions (e.g., how to choose a Web site host). After approximately a year's work, we gradually fade from the organizations, which means that we are not actively engaged in fieldwork per se, but continue to follow developments of the organizations.

Research access was initially negotiated in October 2003 when the Civic Nexus research group held an informational workshop, inviting community organizations that might be potential collaborators. One of the organizations that indicated interest in working with us was the Spring Creek Watershed Community.

Background of Group: Spring Creek Watershed Community (SCWC)

Spring Creek Watershed Community (SCWC hereafter; http://www. springcreek watershed.org) is a sustainable development community group located in Centre County, Pennsylvania. We have explored how technology can be used to promote SCWC's goal of sustaining watershed planning. Specifically, we report our experi-

TABLE 12.1
Key Players in SCWC

Name	Role	Background
Kathy	Lead coordinator	Limited technical background
Tim	Technical volunteer for Web design	Technically proficient in Web design and Web technologies
Dan	Technical volunteer for Web design	Consultant for Web design
Ned	Technical, unpaid intern for developing SCWC's online newsletter	Undergraduate student in a computer science related program, owns consulting company
Umer (first author)	Civic Nexus researcher	Technical computer science and qualitative research background

ence in working with SCWC to redesign their Web site. Their Web site is a vehicle to achieve their strategic goals of increasing public awareness of watershed issues through education and communication, and maximizing involvement and participation in Spring Creek Watershed Community actions.

SCWC is organized around a commitment to showing how regional environmental and economic planning by watershed is more effective than planning by municipality. The mission of the organization is to explain the basic terminology and information about watersheds and to demonstrate the impacts of watersheds on people's quality of life and local economy. Watershed planning is a challenge because the units of government charged with land use planning are different than the geographic units defining natural resources.

Similar to most nonprofits, SCWC has limited staffing and financial resources. Kathy (all names have been substituted for anonymity) works for Clearwater Conservancy, a stakeholder group of SCWC. She dedicates part of her time to SCWC in the capacity of a lead coordinator. Table 12.1 lists key players in SCWC.

Research Methods

The field research reported in this chapter was carried out during a period of 14 months, beginning in October 2003. Because we facilitated the Web site redesign process as participants of the Web site committee meetings, the primary method of data collection was participant observation. We attended eight scheduled Web site committee meetings, each lasting about an hour. During participant observation, we assumed active roles such as facilitators and consultants, consistent with our participatory design approach. We also made direct observations, during which we adopted more passive roles, only observing activities and their dynamics without taking part in them. Secondary sources of data collection included documentation (e.g., meeting agendas, meeting minutes, and newsletters), archival records (e.g., content of e-mails and Web sites), and physical artifacts (e.g., design mock-ups and scenarios).

We conducted two open-ended, focused interviews with Kathy that lasted approximately an hour each. We focused on Kathy because she was the primary stakeholder of SCWC and was a nonvolunteer member of the organization (paid staff member in charge of the decision-making process). The interviews focused on Kathy's perceptions of what happened and why in relation to SCWC's Web site, on how decisions and actions were influenced and made and conflicts resolved, and on our particular role. The interviews were tape recorded and subsequently transcribed. Additionally, many informal discussions, including both face-to-face interactions and phone conversations, were held with Kathy.

The analysis of the data collected was done using the general analytic strategy of developing a case description (Yin, 2003). Although the objective of the study was not a descriptive one, a descriptive approach has been followed to help identify the complex stages of redesigning a Web site and how the researchers scaffolded the process using the participatory design approach. Our approach of participatory design as a learning process was used to analyze the data, reflecting on important content, context, and process elements of redesigning a Web site. The multiple sources of data collection provide evidence of data triangulation. Investigator triangulation was achieved, as multiple researchers from Civic Nexus were part of the data collection process. The first author was the primary researcher. Also, the results of collaboration with SCWC have been compared and contrasted several times with other community organizations involved in Civic Nexus.

The reporting of the experience with SCWC in this chapter is from the researchers' perspective of the dynamics that occurred within the environmental group during their Web-site redesign process. Member checking was performed during the interviews, where the researchers presented their analysis for feedback from the environmental group.

A major practical issue was the tradeoff in our pursuit of encouraging technological sustainability and learning in the environmental group and the group's ability to absorb the induced methods and techniques. Forging ahead with technological solutions and unrealistic requirements for the group could have resulted in an unhealthy relationship that could undermine project goals. At a deeper level, this also raised questions about the research agenda itself and how much we were prepared to do in order to increase, maintain, or relinquish our control of the Web site redesign process. Reflexive practice within the Civic Nexus research group led the researchers to minimize biases by continuously rethinking, reframing, and reconsidering the approach toward participatory design while working with SCWC and other community organizations.

DESCRIPTION AND ANALYSIS

This section contains the description and analysis of our fieldwork, in which we applied our participatory design approach in working with SCWC. The next subsection provides an overview of the events during our collaboration with the environmental group.

Overview of Events

Prior to our involvement, SCWC's Web site was developed and maintained by a third-party commercial vendor. Whereas the goals of SCWC were local economic planning, influencing decision makers, and encouraging quality of life through watersheds, the Web site depicted SCWC as a stereotypical tree-hugger group. SCWC was dissatisfied with the Web site because it did not reflect their local cause and overall goals. Figure 12.1 is a screenshot of SCWC's Web site that was developed by the vendor. The logo (with four quadrants of a fish, bird, leaf, and water) and the background of the sidebar do not emanate any local feeling or reflect the group's mission. In one of Kathy's e-mails, she writes that their Web site "looked very 'environmental' and lost the other two primary components of the mission statement: quality of life and economy." Due to limited contract liability, the vendor did not provide any customer service, and thus, refused to make changes to the Web site as requested by SCWC.

The aforementioned situation highlights the paradox of SCWC: They wanted to use technology to promote their organizational mission, but they lacked the technology skills to do so and were dependent on a vendor. SCWC acknowledges that a causal factor for the situation was their lack of involvement in the Web-site design process. This realization was formative for the group. SCWC decided to take control of the Web site and redesign it. They established a Web site committee comprising the key players in Table 12.1 and a few other less regular volunteers. Committee meetings were held almost every 2 weeks. Kathy presided over these meetings. At least one researcher represented Civic Nexus in these meetings.

In the course of these committee meetings, SCWC commenced their Web-site redesign process by first focusing on site content and then site layout. They took down their old Web site (which was usable but not representative) and uploaded a placeholder home page that was not cosmetically pleasing, yet conveyed their group's core message (see Fig. 12.2). This was a key step toward redesigning their Web site because SCWC felt that they had corrected a misrepresentation implicit in the vendor-created Web site.

During the course of the Web-site redesign, we observed a shift in roles that SCWC adopted from being passive stakeholders to becoming active designers of their Web site. In our collaboration using participatory design as a learning process, we did not take control of SCWC's Web-site redesign process; rather, we provided scaffolding through methods and techniques to facilitate their goals.

The next three subsections describe and analyze the process of SCWC's building capacity in technology-related skills and activities through our conceptualization of participatory design. These subsections describe three facets of our involvement with SCWC.

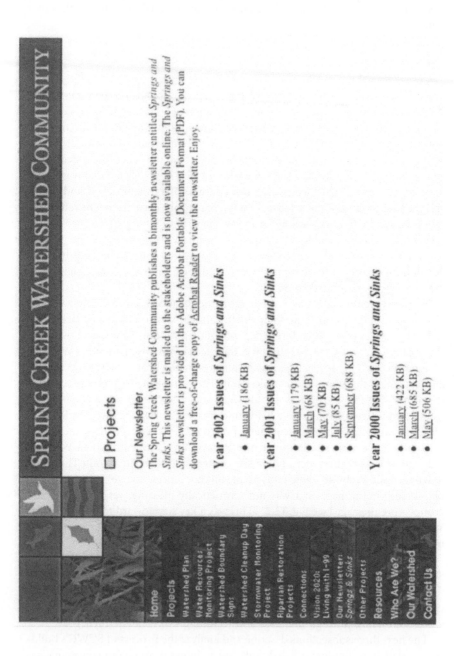

FIG. 12.1. Screenshot of SCWC's Web site that was developed by the vendor.

Mission

The Spring Creek Watershed Community promotes actions that protect and enhance the quality of life, environment, and economy throughout the watershed while maintaining and improving the high quality of Spring Creek and its tributaries.

Strategic Goals

- Increase public awareness of watershed issues through education and communication.
- Develop a vision for the future and implement it.
- Maximize involvement and participation in Spring Creek Watershed Community actions.
- Measure watershed quality and set goals for improvement.
- Increase intergovernmental and interorganizational cooperation.
- Promote business and economic development actions that support our mission.

SPRING CREEK
WATERSHED COMMUNITY

Phone: (814) 237-0400
Fax: (814) 237-4909
2555 North Atherton Street
State College, PA 16803

We're All in the Same Bathtub

FIG. 12.2. Screenshot of SCWC placeholder Web site. (Note that SCWC's mission statement was thrown out as the front page on their placeholder Web site, indicating their determination to get their message across this time.)

Scaffolding User Participation in Taking Control
of Technology

SCWC wanted to regain control of their Web site as they started to redesign it. This implied acquiring the Web site content from the vendor. At the same time, SCWC wanted to sever their connections with the vendor, because the rendered services were unsatisfactory. This created a tension for SCWC, as they anticipated that the vendor would not so easily relinquish control and content of their Web site. Legal efforts to acquire the Web site content were discussed; however, the cost of such a trajectory was an impediment for SCWC that lacked financial resources.

Kathy eventually decided to write an e-mail to the vendor. The e-mail was written in collaboration with Tim and Umer (first author), which showed evidence of lateral and hierarchical elements of learning. The following is the text of Kathy's e-mail addressing the vendor on December 13, 2003; this was blind carbon copied to Tim and Umer:

> Apparently I have misplaced the e-mail that you sent that explained what you had worked on before leaving for FL so I can't review what you did. Can you resend that e-mail or tell me again what you did? Also, a couple other things are going on. I looked into the billing question you had asked about. Our Executive Director asked me if we have a current copy of the Web site in house. I tried to open that CD that you gave me dated April 29 but I could not find everything that I was hoping/need to find. That coupled with the fact that I have an intern who is going to work on the Web site over break requires that I get some information from you this week.
>
> Please provide the following on a CD by Wednesday, December 17:
>
> • A copy of the current water quality database. In the previous CD you provided a large (12 MB) binary file before called "backup." The file did not have an extension and I was unable to read it. Please provide either tab delimited text files (1 per table) along with a clear diagram showing the relationships between the tables, or a Microsoft Access 2000 database with properly defined relationships,
>
> • A copy of the current bibliographic database (this should include any updates that my previous intern did this summer and what ever you did a week or so ago). Once again, please provide either tab-delimited text or MS Access 2000 format with all relationships defined, and
>
> • A backup of the entire Web site content and code.
>
> Please keep in mind that I do not have SQL server and cannot read SQL server files nor binary backups . . .

In this e-mail, Kathy referred to a nonexistent deadline that required SCWC to have all its Web site content available for their management as a matter of policy. Kathy's idea of using her boss's pressure (Executive Director) to request the content represents the overall motivation to redesign the SCWC Web site: Kathy wanted to regain control of the process and product. Following is evidence of lateral and hierarchical relationships during SCWC regaining control of their Web site.

Evidence of Lateral Relationships.   Kathy involved Tim and Umer in crafting the e-mail just displayed. She could have composed the e-mail by herself, or delegated the task to Tim and/or Umer. Instead, Kathy realized that writing the e-mail was a challenge because of her conundrum of wanting to leave the vendor while still regaining control of the Web site that the vendor had created. She anticipated that her skills as the primary stakeholder in SCWC's Web site, and Tim and Umer's skills as technologists, would not be sufficient alone to write an effective e-mail to the vendor. Therefore, Kathy elected to co-construct it with Tim and Umer.

The content of the e-mail message is evidence of an intersubjective combination of the two skill sets. Moreover, the process of writing the e-mail led to the emergence of a co-constructed ZPD. Neither Kathy, Tim, nor Umer were sure how to craft the e-mail because this was a new experience for everyone. Kathy had never dealt with a vendor in regaining control of information technology (in this case, SCWC's Web site). Tim and Umer had an even more vague understanding because they had never dealt with a Web site hosting vendor at all. Moreover, none of them had primary control of this construction process, which is a characteristic of constructing lateral relationships (Sawchuk, 2003). It was noted that although Kathy typed the e-mail, this did not imply primary control as Tim and Umer frequently asked Kathy to edit the content related to all parts of the e-mail. Suggestions from all three collaborators were negotiated and discussed; as a result, some suggestions were incorporated and some were not.

Evidence of Hierarchical Relationships.   In the joint effort of composing the e-mail directed to the vendor, evidence of hierarchical relationships was apparent. As in the cognitive apprenticeship model, Kathy displayed her expertise by using her boss's demand in pressuring the vendor to release the Web site content by a specific deadline (Tim and Umer were unaware of this aspect of the situation and would not have thought about composing the e-mail in this manner). Kathy's subtle way of extracting the Web-site content, while still working to sever ties with the vendor, was a new experience for Tim and Umer.

On the other hand, the situation flipped when Tim and Umer became experts in specifying the exact Web-site content requirements. In this case, Kathy automatically assumed the role of a novice. Kathy wanted to be technically precise and accurate in her request for the Web-site content, and therefore asked Tim and Umer to specify the details. The three bulleted points in the e-mail, requesting different content on a CD, were being typed by Kathy but uttered and explained in conjunction by Tim and Umer. In the e-mail to the vendor, technical jargon was used such as "binary file," "delimited text files," and "bibliographic database." At that time, Kathy admitted to not knowing these terms but included them in the e-mail. Realizing that she needs to be in more control of the Web site design process, Kathy had been inquisitive throughout our involvement. For example, in one instance during Web-site layout design, she asked what "WYSIWYG" ("What You See Is What You Get" type of interface) meant. Evidence of Kathy's hierarchical forms of learning was also noted in a subsequent interview. Referring to technologies and technical acronyms such as HTML (Hypertext Markup Language) and SQL (Structured Query Language), she said: "I think I'm marginally more aware of those things."

Listed below is an e-mail Kathy sent out to the Web site committee, updating the status of the Web site:

> I thought that I should update everyone since it has been a few weeks since we touched base . . . Umer, Tim, and I ended up meeting for a little while on December 13. What happened:
>
> 1. We outlined how I was going to leave our current host and then switch to our new host. Before the holiday, I worked with our new site host. They said it would take several business days to actually make the switch.
> 2. We finalized the draft mission statement.
> 3. I was assigned the task of compiling the first batch of content. . . .
>
> Next meeting: I will e-mail you in early January to let you know how I am making out with the content. It doesn't seem like it will be worth meeting until I have something to give you.

In this e-mail, Kathy's task of "compiling the first batch of content" establishes her as the expert in knowing the content of SCWC's Web site. She was assigned this task because of her expertise. Moreover, Kathy admits that the subsequent meeting with the Web-site committee will not be worthwhile until she has some draft of the Web-site content. This is because other members of the Web site committee are not experts in SCWC's Web-site content. Therefore, the Web-site redesign process can only move further once the expert (Kathy) provides a scaffold (draft of Web-site content) to the novices (Web-site committee less Kathy) so they may slowly take on more responsibilities in alignment with their respective skills.

Making Sense of and Collectively Participating in Web-Site Design

One of the more complex issues that community organizations face when trying to implement a technology project is making sense of the design process. As Collins et al. (1989) noted, part of the work that novices are doing is developing conceptual models that are needed to take on a task. This serves as an advance organizer for the user, helps them make sense of feedback, and provides a guide as they take on tasks alone.

We noticed that it took several meetings for the group to make sense of the Web site design process. The following subsections discuss the negotiation over the meaning of design and the subsequent collective effort in the design process respectively.

Negotiation on What Design Means. Group members had different perspectives on design that created tension between technical requirements and the need to organize information on the Web site effectively. One of the volunteers, Dan, was technically proficient in Web-site design and wanted to move directly into interface design of the Web site. Kathy was very excited and initially agreed to the idea. At this point, we as researchers intervened, suggesting that design is an iterative process. Further, we emphasized that the content of the Web site needs to be designed before the layout (interface). Tim agreed with us and talked about "arranging the Web site directory structure." He stressed the importance of "divorcing" the idea of how the Web site is going to be laid out from the current stage in the design process and focusing on the message of the organization. Kathy believed Tim's suggestion to be a good idea but did not know how to go on about it. She once remarked, "So where do we start?"

This tension between content and layout continued for a couple of meetings. In one meeting, Kathy suggested the idea of using video/audio in the Web site to make it more attractive. Tim advocated not dealing with the attractiveness of the Web site and attempted to direct all efforts toward "content design." Kathy seemed to struggle with the concept of separating the content from the layout and expressed her concern in a meeting: "Do we mean the same thing by the term 'design'?" This is evidence of the negotiation process between Kathy and the technically proficient volunteers—as we see it, an expert–novice ZPD was in play, helping to establish common ground and understanding. Kathy, being a novice, was bounded by her knowledge in content design. She could only expand her view on content design with the help of an expert (Tim). The difference between what Kathy knew herself about content design and what she could achieve with Tim's expert help created a ZPD. This resulted in Kathy appreciating and more concretely demarcating the boundary between content and layout design.

Similarly, we attempted to create an expert–novice ZPD to engage Kathy in the design process. As design experts, we encouraged the group to think about content with respect to different types of audiences that would probably visit the Web site. Our apprenticeshiplike scaffolding led to the identification of three types of Web-site visitors that Kathy elicited: new users, sympathetic members, and decision makers/stakeholders. This encouraged Kathy to think about the different content in which each of these three audiences would be interested. Again, Kathy transcended her individual ability to do design with the help of experts (us). In a later interview, Kathy acknowledged the value of undergoing the process of identifying types of audiences as a prerequisite for design: "Our target audience is so broad, how do you design so that each one of them wants to continue (being engaged in the Web site)?"

Eventually, the process was shaped by Tim's continuous push for designing content and "worrying" about layout in later meetings. To this end, he emphasized several times, "Presentation/layout can be done in next meeting, once the content is finalized."

Mutual Involvement Toward Collective Design Effort. As content started to emerge in an iterative fashion from contemplating SCWC's mission statement and thinking about the three Web-site audiences, Kathy asked in regard to design, "What things can we do as a group versus work by individuals and volunteers?" Tim suggested that the group should look at other environmental Web sites to structure the content. Kathy, trying her best to leverage everyone's skills, asked all the meeting participants to examine at least one environmental Web site and bring a hard copy of the site to the next meeting. Here is an excerpt from an e-mail message Kathy sent out on February 5, 2004, to the Web-site committee for the purpose of scheduling the next meeting:

> I have attached an agenda for this Saturday's meeting (February 7 from 10:30–12:00) . . . Since the focus of our meeting will be site design, we thought it would be a good idea if everyone could search for three Web sites that you feel are well designed and then bring a printout of each site's front page so we can discuss them at the meeting . . .

We think this e-mail provides evidence of three things. First, Kathy was attempting to keep everyone involved in the process and simultaneously leverage their individual skills. Second, she was learning how to manage the distribution of work among volunteers so as to offload some of her burden. Third, collecting Web sites represented an information-gathering exercise that can be translated as a co-construction of a ZPD between the differently skilled participants in the meeting. In the next meeting, Kathy printed several hard copies of environmental Web sites. Ned, the volunteer intern who owned his own technology consulting company, suggested, "We need to get people to where they want to go as quickly as possible," referring to the navigation scheme on the Web site. This interjection triggered Kathy's "content versus layout" dilemma. She was again confused about the process of design and said, "This seems like a chicken and egg problem of where to start."

Tim was once again persuasive, directing all efforts toward content architecture. He suggested that the group talk about "likes" and "dislikes" of the different environmental Web sites. Kathy drove this process because she was an expert in taking a quick look at Web sites and resolving their pros and cons. "It's easier to identify what you don't like versus what you do like," she said. In our view, this was an experiential reflection on how she went through the same disliking process for SCWC's original Web site. During the process of eliciting the pros and cons of Web sites, she did ask the group whether or not this particular approach to the problem was a good idea. Once again, Kathy was including everyone in the design process, and more importantly, she was adept at not being seen as the controller of the process but rather as a collaborator. Actions such as these are evidence that even if one is an expert (e.g., Kathy being an expert at pointing out pros and cons of Web sites), relinquishing control to other novices is essential for maintaining involvement and moving forward as a group. We hypothesize that this event was also driven by the fact that Kathy realized she is not the expert in everything, and by probing others now, their expertise will come into play later on issues where she is a novice. In this case, the experts at that point would return the favor to Kathy by asking her opinion. We refer to this interplay between experts and novices as the expert–novice cycle, where roles between these two states are dynamically switched during the design process with community organizations.

The result of this collective design exercise was that Kathy and other volunteers (who were initially pushing ahead with Web-site layout) had come to an understanding of what it means to do Web design. As a result of this meeting, Kathy sent out the following e-mail on February 21, 2004:

> Well, we had a good meeting this morning. We took our outline that we made at the last meeting and really focused in on content (see attachment called "desired Web site content"). We eliminated anything that showed up twice because we realized that we were confusing content with site structure . . .
>
> Agenda topics for our next meeting will include:
>
> 1. Review examples of other Web sites (this was homework for today's meeting but we didn't have time to review them);
> 2. Discuss basic design ideas (things we like, things we don't like);
> 3. Skill and time inventory . . .

Here is a sample of the content structure for their Web site taken from the attachment in the above e-mail:

<u>Main Content Areas for Web Site Design</u>
- Who We Are
  - Our roots (timeline)
  - Mission statement and strategic goals
  - Committees
  - Stakeholders
  - Project
- Why Should I Care about SCWC?
- Watershed Information

- Newsletter
- Plan by Watershed
- Contact Us

In a later interview, we asked Kathy about what she had learned so far through the content-design process. She referred to a product marketing metaphor, saying ". . . if you were a business, you would immediately understand the product we were trying to sell and I think that is really hard," acknowledging the complexity in design and the need to participate in the process for conveying the group's message.

Content design was neither a one-shot nor a unidirectional process. It was learning through guiding, as in cognitive apprenticeship, but was not limited to a mentor and his or her student. Acknowledging everyone's expertise in some skill and moving the process control around was key in the success of this exercise. We deliberately qualified our roles as facilitators and collaborators as the different situations demanded. This is not to say that our involvement was insignificant; our major contribution was to encourage the group in thinking about how the different types of audiences/users would interact with the Web site. We also learned the importance of ceding ownership to the full participants of the community organization. Only then may we fade from the process as the organization adopts the Web-site design strategies as everyday practice.

Providing Conceptual Tools

From the initiation of our collaboration with SCWC, we have observed the group transition from being dependent on a vendor to independent designers of their Web site. As we have already seen, Kathy has adopted multiple roles in the design process: meeting facilitator/coordinator, work distributor/manager, information/ Web-site content architect, novice and expert in various skills, and a learner (among others).

Next, we present the interesting case of Kathy adopting the role of a designer using a conceptual design tool and how she used it to convey her design concerns. The approach to design we adopted and inculcated in SCWC was scenario-based design (Carroll, 2000; Rosson & Carroll, 2001), partly because of our own experience and training in scenario-based design but primarily because of its comprehensible nature for nontechnical end users. In the subsections to come, we talk about how we introduced scenarios as a design practice in SCWC and how they subsequently adopted designer-like roles respectively. We then discuss the leverage of scenarios for community members.

Introduction of Design Into Organizational Practice. Once the content for the Web site was relatively stable, the process of content design segued into layout design. We introduced the concept of *designing* with scenarios to elucidate how the

Web-site front page should be laid out. The difficulty was to make the front page amenable to the three types of user audiences that Kathy elicited before: new users, sympathetic members, and decision makers/stakeholders. We suggested that a scenario for each audience might guide content customization for that user, which would inform the decision process of resolving how content should be displayed. To weave design into organizational practice, we had to take the first step as experts and enable novices, such as Kathy, to transcend their current level of expertise with our facilitation (example of an expert–novice ZPD). Specifically, one of the Civic Nexus researchers (Umer) sent this e-mail to Kathy on March 22, 2004:

> I have attached an example of a scenario for a new user who happened to arrive at the Spring Creek Web site by searching on Google. Please read it and perhaps refine it further.
>
> As we discussed last time, scenarios such as the one I have attached are evocative tools to design a Web site. As you will read through it, you will soon begin to realize what kind of things you need to put up on the front page of Spring Creek's Web site.
>
> We will discuss this further on Saturday. If you get a chance, take a shot at making some initial scenarios for decision makers and stakeholders . . .

The scenario for Web site used that we proposed was for new users, because we could easily imagine ourselves in this audience category. Here is the scenario that we wrote:

> Melissa just moved to Boalsburg and wants to get involved in the community by renewing her interest in preserving streams and watersheds. She searches for "Central Pennsylvania stream watershed preservation" on Google and clicks on the Clearwater Conservancy Web site link. As she reads their Web site, Melissa notices another link to related organizations and browses the Spring Creek front page. Being unaware of the environmental beauty that surrounds her new hometown, she is fascinated by the scenic pictures of Boalsburg Valley. Melissa clicks on the "More pictures" link underneath the picture and is directed to a collage of Happy Valley snapshots. As she browses the pictures, she reads the concise four to five word picture descriptions and begins to realize that Spring Creek is more than just preserving watersheds; they are somehow interested in town planning and economic development. Melissa goes back to the home page and finds out more about what Spring Creek does.

Community Members as Designers. Our scenario specifically targeted the local flavor of SCWC (e.g., local pictures of the valley) and their mission of town planning and economic development by watersheds. In our e-mail to Kathy about the scenario, we urged her to develop a similar scenario for the audience category of *decision makers/stakeholders.* We did this for two reasons. First, we wanted to encourage Kathy's participation in the design process. Second, she was most familiar with what this type of audience would look for on SCWC's Web site. We motivated the use of scenarios by alluding to benefits such as their evocative nature.

In a subsequent meeting on April 3, 2004, Kathy wrote a different user scenario based on the one we gave her:

> Joe is a recently elected Ferguson Township Supervisor. At a few of his first munici-
> pal meetings, he has heard people talk about three organizations: Spring Creek Wa-
> tershed Community, Spring Creek Watershed Commission, and the Clear Water
> Conservancy. He is unclear about the difference between the three organizations be-
> cause they seem to have similar purposes. He also has heard various people talk
> about multiple watershed issues and wants to find out what the hubbub is all about
> and if and how this stuff may affect the tax-paying residents of his township.
>
> One night after a meeting, he decided to go to the SCWC Web site to figure it out
> once and for all. As he skims their homepage, he can't help but notice attractive pic-
> tures of locally recognizable features that define our quality of life in the Penns,
> Happy, and Nittany Valleys. He quickly sees a link to the SCWC's mission state-
> ment, strategic goals, list of stakeholders, and an explanation of what a stakeholder
> is, and how to get your name or organization's name listed on the list. While viewing
> the list of stakeholders, he notices that Clear Water Conservancy is listed as one of
> the many participating stakeholders. His curiosity is peeked by the diverse listing of
> people and organizations so he decides to click on the "who are we" link. Once there,
> he sees a timeline of how the SCWC came to be and how it is related to other local,
> state, and federal initiatives. As he is navigating between these links, he sees the be-
> ginnings of an in-depth discussion of storm water, which is a hot topic in this urban-
> izing municipality because it is causing property damage. He has also heard that it
> might be harming the local streams so he decides to read on. Between the local pho-
> tos, the mission statement, the listing of stakeholders, and the explanation of "water-
> shed issues" [quotation marks deliberately used by Kathy for emphasis], he has got-
> ten the impression that SCWC is more than just preserving watersheds; they are
> somehow interested in smart planning and environmentally responsible develop-
> ment, both of which are good for the tax-paying residents of his township.

This scenario is evidence of Kathy's role as a designer. The information in the scenario is rich and evocative. Kathy, through the exercise of writing the scenario,
discovered that in addition to existing decision makers/stakeholders, *new decision
makers* such as Joe in the scenario, are another category of user audiences, that is,
decision makers who do not yet have a stake in the organization. During a subsequent interview, she remarked about her formulation of the scenario and how it
led her to identify a new user audience: "I think that was good [referring to scenario development], even though at first, I was feeling it's kind of an obvious
question, and as you do it, you just put somebody else's hat on for a minute."

In the aforementioned scenario, Kathy makes it clear that SCWC is a local organization by referring to the list of pictures and locally involved people. The scenario explicitly asserts the message of their new Web site to be more related to municipality planning in addition to the preservation of watersheds. The scenario
also delineates the navigation scheme through the Web site and the priority of various content.

"Designers are not just making things; they are making sense" (Carroll, 2000, p. 66). To this end, the rich information from the scenario affirms the expansion of Kathy's abilities toward design combined with our role as facilitators of this process.

Power of Community Members With Conceptual Tools. Kathy's scenario was the most striking example of not only demonstrating the effectiveness of its content, but also its impact on the design process. During a Web-site committee meeting (after Kathy wrote her scenario) when alternate Web-site layout designs were being formulated, Kathy asserted that there should be space on the front page for listing top watershed issues. Tim, who so far had been supportive of Kathy's suggestions, opposed the idea of having watershed issues listed on the front page. He suggested that watershed issues are not as important as the mission statement, and further, that the front page should be kept "simple" and "lean." Kathy insisted that watershed issues are core of the Web site by referring to them as "meat" of the front page.

At this point in time, Kathy took out the hard copy of her scenario (this scenario had been discussed only with Umer prior to this meeting), read it verbatim, and deliberately referenced the scenario as an example of a potential real-world situation, even though the scenario was completely hypothetical. In her explanation after reading the scenario, Kathy talked about the importance of having watershed issues as an answer to the "so what" question that a visitor might have when he or she visits the Web site. Moreover, Kathy said that she does not want to just mention watershed issues, but also tell the site visitor "why you should come" (i.e., why should the visitor explore the Web site). In her scenario, Kathy had emphasized watershed issues by putting the phrase in quotation marks and specifically wrote about an "in-depth discussion of storm water" as one of these issues.

As a consequence, Kathy's suggestion was accepted and incorporated into the layout design. We think the related event was an important instance of not only demonstrating the design power of scenarios, but also of how Kathy used the technique to incorporate her view into the design process. This is an indication of the viability of incorporating scenarios as part of a sustainable design process when working with community organizations (Farooq, 2005). As end users, community organizations are obviously more likely to imagine authentic scenarios that represent real-world situations because they are embedded in this context through their everyday work. We also believe that scenarios are an effective technique in decontextualizing knowledge so that it can be transferred and reused in other circumstances (Bransford, Brown, & Cocking, 1999).

CONCLUSION

We adopted the notion of participatory design as a learning process and applied it to a case study of an environmental group involved in the process of redesigning their Web site, which is central to achieving their organizational goals. Our participatory design approach offers a new way of looking at user participation and roles. This approach embodies both hierarchical modes of learning in the form of cognitive apprenticeship and lateral modes of learning that take the form of collaboratively constructing ZPD. We have also suggested the use of conceptual tools, such as scenarios, as evocative and powerful scaffolds to empower environmental groups in conveying design rationale, actively participating in the design process, and having a voice in design-related decisions.

We believe our chapter is of interest to social science and humanities scholars interested in the implications of information technology for environmental communication theory and practice. Scholars in environmental communication can also consider the role of participatory design in their own endeavors to make good environmental decisions and reflect on how our analysis and broader implications can be useful with respect to their own context (Daniels & Walker, 2001; Depoe, Delicath, & Aepli, 2004). We also believe that our chapter is valuable to the general audience of community practitioners and researchers interested in building community capacity using information technology.

FUTURE WORK: EXTENDING THE CASE STUDY TOWARD PATTERNS

In the Civic Nexus project, our collaboration with SCWC represents only one case study among others. The overall research design for Civic Nexus represents multiple case studies (Yin, 2003). Our rationale for using multiple case studies is to "predict similar results (a *literal replication*) or predict contrasting results but for predictable reasons (a *theoretical replication*)" (Yin, 2003, p. 47). A major criticism of case studies is that they provide little basis for scientific generalization. Yin (2003) said that an important step in replication procedures is the development of a rich theoretical framework. The framework needs to state the conditions under which a particular phenomenon is likely to be found (a literal replication) as well as the conditions when it is not likely to be found (a theoretical replication). The theoretical framework later becomes the vehicle for generalizing to new cases.

We argue that *patterns* are one way to develop such a theoretical framework for multiple case studies that are especially related to design (in our case, participatory design). Our conceptualization of patterns draws on discussions of design patterns in other disciplines, such as architecture (Alexander et al., 1977) and software (Gamma, Helm, Johnson, & Vlissides, 1995).

Patterns can be defined as general solutions to problems that recur repeatedly in many cases. Patterns generally include (a) a problem; (b) a description of the problem's context; (c) an analysis of relevant forces, that is, resources and trends that enable or constrain possible solutions to the problem; (d) a statement of our solution to the problem; (e) a discussion of how the resulting context, that is, how the problem context might be changed by adoption of our solution; and (f) examples of the solution, pointers to instantiations of the pattern in our ongoing work. Patterns provide a common language to be shared among domain experts for codifying and developing design knowledge.

From our case study, we can extract a pattern related to community-based learning of information technology. The problem this pattern addresses is that community organizations often experience a lack of control over their own technology. In our case study, SCWC did not have control of their Web site when the vendor was hosting it. The context is the rapid and pervasive growth of computing and the Internet that is a powerful information source for ordinary citizens. For SCWC, their Web site was the primary communication channel with their stakeholders. The forces are social capital and lack of resources. SCWC leveraged their social network of volunteers to redesign their Web site within constraints of their limited budget and time.

One pattern solution to address the problem is a self-sustained process of informal learning comprising three independent facets: (a) Reflection: realization that organization lacks participation in and control over information technology; (b) Analysis: identification and analysis of practices, needs, and issues related to information technology; and (c) Enactment: continuous engagement in meaningful activities. For SCWC, reflection was achieved after the vendor refused to update the Web site and its content. SCWC soon realized that that they needed to be in control of their Web site. Analysis started with the identification of the need to redesign their Web site. To this end, SCWC assembled a Web-site committee by identifying specific skills required to see the redesign process through. Enactment was apparent with SCWC being continuously engaged in all the Web-site committee meetings, and actively participating in and learning from the Web-site redesign process.

The resulting context of the pattern solution is the recasting of organizational practices related to information technology. Kathy, through reflection, analysis, and enactment, gained control of SCWC's Web site. As a consequence, SCWC's practice of developing their Web site has changed, from totally depending on a vendor to actively being involved in the process. Community organizations, such as environmental groups, need practical solutions to the problems they encounter everyday. In addition, because most technical interactions are participatory and about empowering people, the need for developing abstract solutions is clear for sharing reusable ideas and knowledge among practitioners and researchers. We believe that the development of patterns sets a research trajectory to develop such practical and abstract solutions by drawing on multiple case studies and other empirical work. Resources for community-oriented patterns already exist and are being developed (e.g., Carroll & Farooq, 2005; Schuler, 2002). We encourage the development of similar patterns in the area of environmental communication.

ACKNOWLEDGMENTS

An earlier version of this chapter appeared in the Proceedings of the 38th Hawaii International Conference on System Sciences (Waikoloa, HI, January 3–6, 2005; http://www.ecoresearch.net/hicss). The research was partially supported by the U.S. National Science Foundation IIS–0342547 to Penn State. Many thanks to members of our Civic Nexus research team, especially Craig Ganoe, Michael Race, Matt Dalius, Janet Montgomery, and Laurie Kemmerer. We thank the anonymous reviewers for their detailed comments and pointers to relevant literature that allowed us to refine the chapter. We would also like to extend our special thanks to Arno Scharl and Stephen Depoe for facilitating the publication of our chapter in *The Environmental Communication Yearbook*.

REFERENCES

Alexander, C., Ishikawa, S., Silverstein, M., Jacobson, M., Fiksdahl-King, I., & Angel, S. (1977). *A pattern language: Towns, buildings, construction.* New York: Oxford University Press.

Benston, M. (1990). Participatory design by non-profit groups. In *Proceedings of the Participatory Design Conference* (pp. 107–113). Seattle, WA, March 31–April 1, 1990, Palo Alto, CA: CPSR.

Bransford, J. D., Brown, A. L., & Cocking, R. R. (Eds.). (1999). *How people learn: Brain, mind, experience, and school.* Washington, DC: National Academy Press.

Bruner, M., & Oelshlaeger, M. (1998). Rhetoric, environmentalism and environmental ethics. In C. Waddell (Ed.), *Landmark essays on rhetoric and the environment* (pp. 209–225). Mahwah, NJ: Lawrence Erlbaum Associates.

Carroll, J. M. (2000). *Making use: Scenario-based design of human–computer interactions.* Cambridge, MA: The MIT Press.

Carroll, J. M., Chin, G., Rosson, M. B., & Neale, D. C. (2000). The development of cooperation: Five years of participatory design in the virtual school. In D. Boyarski & W. Kellogg (Eds.), *Designing interactive systems* (pp. 239–251). New York: Association for Computing Machinery.

Carroll, J. M., & Farooq, U. (2005). Community-based learning: Design patterns and frameworks. In H. Gillersen, K. Schmidt, M. Beaudouin-Lafon, & W. Mackay (Eds.), *Proceedings of the 9th European Conference on Computer-Supported Cooperative Work* (pp. 307–324), Paris, September 18–22, 2005. Dordrecht, The Netherlands: Springer.

Clement, A., & Van den Besselaar, P. (1993). A retrospective look at PD projects. *Communications of the ACM, 36*(6), 29–37.

Collins, A., Brown, J. S., & Newman, S. E. (1989). Cognitive apprenticeship: Teaching the craft of reading, writing, and mathematics. In L. B. Resnick (Ed.), *Knowing, learning and instruction: Essays in honor of Robert Glaser* (pp. 453–494). Hillsdale, NJ: Lawrence Erlbaum Associates.

Cooper, M. M. (1996). Environmental rhetoric in the age of hegemonic politics: Earth First! and the Nature Conservancy. In C. G. Herndl & S. C. Brown (Eds.), *Green culture: Environmental rhetoric in contemporary America* (pp. 236–260). Madison, WI: The University of Wisconsin Press.

Daniels, S. E., & Walker, G. B. (2001). *Working through environmental conflict: The collaborative learning approach.* Westport, CT: Praeger.

Depoe, S. P., Delicath, J. W., & Aepli, M. F. (2004). *Communication and public participation in environmental decision making.* Albany, NY: SUNY Press.

Farooq, U. (2005). Conceptual and technical scaffolds for end user development: Using scenarios and wikis in community computing. In *Proceedings of the IEEE Symposium on Visual Languages and*

Human-Centric Computing: Graduate Student Consortium on Toward Diversity in Information Access and Manipulation (pp. 329–330), Dallas, TX, September 20–24, 2005. Los Alamitos, CA: IEEE Computer Society.

Gamma, E., Helm, R., Johnson, R., & Vlissides, J. (1995). *Design patterns: Elements of reusable object-oriented software.* Reading, MA: Addison-Wesley Professional.

Greenbaum, J., & Kyng, M. (Eds.). (1991). *Design at work: Cooperative design of computer systems.* Hillsdale, NJ: Lawrence Erlbaum Associates.

Gurstein, M. (2002). Community informatics: Current status and future prospects—Some thoughts. *Community Technology Review* (Winter/Spring). Retrieved April 15, 2005, from http://www.comtechreview.org/article.php?article_id=56

Katz, S. B., & Miller, C. R. (1996). The low-level radioactive waste citing controversy in North Carolina: Toward a rhetorical model of risk communication. In C. G. Herndl & S. C. Brown (Eds.), *Green culture: Environmental rhetoric in contemporary America* (pp. 111–140). Madison, WI: The University of Wisconsin Press.

Kensing, F., & Blomberg, J. (1998). Participatory design: Issues and concerns. *Computer Supported Cooperative Work, 7,* 167–185.

Laird, F. N. (1993). Participatory analysis, democracy, and technological decision making. *Science, Technology, & Human Values, 18*(3), 341–361.

Lave, J., & Wenger, E. (1991). *Situated learning: Legitimate peripheral participation.* New York: Cambridge University Press.

McPhail, B., Costantino, T., Bruckmann, D., Barclay, R., & Clement, A. (1998). CAVEAT exemplar: Participatory design in a non-profit volunteer organization. *Computer Supported Cooperative Work, 7*(3), 223–241.

Merkel, C. B., Xiao, L., Farooq, U., Ganoe, C. H., Lee, R., Carroll, J. M., & Rosson, M. B. (2004). Participatory design in community computing contexts: Tales from the field. In *Proceedings of the Participatory Design Conference* (pp. 1–10), Toronto, Canada, July 27–31, 2004. Palo Alto, CA: CPSR.

Merkel, C., Clitherow, M., Farooq, U., Xiao, L., Ganoe, C. H., Carroll, J. M., & Rosson, M. B. (2005). Sustaining computer use and learning in community computing contexts: Making technology part of who they are and what they do. *The Journal of Community Informatics* (online), *1*(2). Retrieved April 15, 2005, from http://ci-journal.net/viewarticle.php?id=53&layout=html

Mumford, E. (1983). *Designing human systems.* Manchester, England: Manchester Business School.

Randall, D., & Rouncefield, M. (2004). Tutorial: The theory and practice of fieldwork for systems development. In *Proceedings of the Conference on Computer Supported Cooperative Work, Tutorial Notes* (pp.), Chicago, IL, November 6–10, 2004. New York: Association for Computing Machinery.

Rosson, M. B., & Carroll, J. M. (2001). *Usability engineering: Scenario-based development of human–computer interaction.* San Francisco: Morgan Kaufmann.

Sawchuk, P. H. (2003). Informal learning as a speech-exchange system: Implications for knowledge production, power and social transformation. *Discourse & Society, 14*(3), 291–307.

Schuler, D. (1994). Community networks: Building a new participatory medium. *Communications of the ACM (Association for Computing Machinery), 37*(1), 38–51.

Schuler, D. (2002). A pattern language for living communication. *Proceedings of the Participatory Design Conference* (pp. 434–436), Malmo, Sweden, June 23–25, 2002. Palo Alto, CA: CPSR.

Schuler, D., & Namioka, A. (Eds.). (1993). *Participatory design: Principles and practice.* Hillsdale, NJ: Lawrence Erlbaum Associates.

Trigg, R. H. (2000). From sandbox to "fundbox": Weaving participatory design into the fabric of a busy non-profit. In *Proceedings of the Participatory Design Conference* (pp. 174–183), Palo Alto, CA, November 28–December 1, 2000. Palo Alto, CA: CPSR.

Waddell, C. (1996). Saving the Great Lakes: Public participation in environmental policy. In C. G. Herndl & S. C. Brown (Eds.), *Green culture: Environmental rhetoric in contemporary America* (pp. 141–165). Madison: The University of Wisconsin Press.

Yin, R. K. (2003). *Case study research: Design and methods.* Thousand Oaks, CA: Sage.

Author Index

Subject Index

A

Advertising, 97, *see also* "Green" marketing, Social marketing campaigns and conservation psychology
as creative discourse, 100
as political discourse, 100
images and, 100, 111–112
Anthropocentric vision, 97, 110, *see also* Triad of centrisms
Anthropomorphism, 122–123, *see also* Orangutan protection campaign
Anti-toxic activism, 22
articulated with sexy, 22, 40
popular culture and, 22
Apocalyptic rhetoric, 117–118, 126
definition, 117
primary topoi, 117–118
use by environmentalists, 118
limitations of, 118, 131–132
tragic frame and, 120
Argumentation and speaker–audience agreement, 141
Aristotelian rhetoric, 140
enthymeme, 145
fallacies, 153
ethos, 140, 142

kairos, 140, 151–152
logos, 140, 143–146, 152–153
pathos, 140
Articulation theory, 24, 44
eclipse and, 28, 42
role in social movements, 24
AVEDA Corporation, 98, 102
commercial jeremiad advertising campaign, 105–109
connecting past and present, 106–107
environmental ethic of justice and, 110
holism and, 108–109
lamenting the decline of beauty, 105–106
promoting sustainability, 107–108

B

Burlesque frame of reference, 3, 4, 7, 14, 17, *see also* Environmental rhetoric, Poetic categories
comic corrective, 16
"formal admission of strictures" as alternative, 16
ironic perspective as alternative, 17
rhetorical strategy, 4–5, 13–14, 15

T - #0098 - 270225 - C0 - 229/152/16 - PB - 9780415652391 - Gloss Lamination